D1406547

Technology Interactions

Henry R. Harms
Industrial Technology Teacher
McBee High School
Chesterfield County, South Carolina

Neal R. Swernofsky
Technology Teacher
Lincoln Orens School
Island Park, New York

**Glencoe
McGraw-Hill**

New York, New York   Columbus, Ohio   Woodland Hills, California   Peoria, Illinois

Glencoe/McGraw-Hill

A Division of The **McGraw·Hill** *Companies*

Send all inquiries to:
Glencoe/McGraw-Hill
3008 W. Willow Knolls Drive
Peoria, IL 61614

ISBN: 0-02-838779-1 (Student Text)
ISBN: 0-02-677701-0 (Teacher's Resource Guide)

Printed in the United States of America.

3 4 5 6 7 8 9 10   027   03 02 01 00

Acknowledgments

The publisher gratefully acknowledges the assistance received from many persons during the development of *Technology Interactions*. Special recognition is given to the following.

Contributors

Shirley Blackburn
Metamora, Illinois

Charlene Crosby
Manistique, Michigan

Carolyn Gloeckner
Stuart, Florida

Jody A. James
Oviedo, Florida

Deborah Paul
Worthington, Ohio

Dr. Stuart Soman
West Hempstead Public Schools
West Hempstead, New York

Consultants and Reviewers

Dr. Michael R. Bachler
Technology Education Instructor
Russell Middle School
Winder, Georgia

Ronald G. Barker
Program Specialist
Technology Education
Georgia Department of Education
Atlanta, Georgia

Pamela J. Brown
Technology Education Instructor
Central Middle School
Newnan, Georgia

Michael F. Crull
Technology Education Instructor
West Jay Junior High
Dunkirk, Indiana

Gary Foveaux
Coordinator, Technology Education
Fairfax County Public Schools
Falls Church, Virginia

Raymond Haddix
Technology Education Teacher
Barren County Middle School
Glasgow, Kentucky

Robert C. Horan
Technology Education Department Head
Plant City High School
Plant City, Florida

Robert J. Kraushaar
Teacher
Pinckneyville Middle School
Gwinnett County, Georgia

Curtis Nelson
Industrial Technology Teacher
Washington Community High School
Washington, Illinois

Michael T. Peterson
Technology Education Instructor
New Smyrna Beach Middle School
New Smyrna Beach, Florida

Thomas N. Pitchford
Technology Education Teacher
Valdosta Middle School
Valdosta, Georgia

Mark Roberts
Technology Teacher
McLane Middle School
Brandon, Florida

Jacquelyn W. Rozman
Technology Teacher
Benito Middle School
Tampa, Florida

Jeffery Scott Salyer
Technology Education Teacher
Wolfe County Middle School
Campton, Kentucky

Nancy L. Smith
Technology Education Teacher
Booker Middle School
Sarasota, Florida

John R. Stover
Industrial Technology Teacher
Calhoun High School
Calhoun, Georgia

William A. Vallance
Teacher—Technology and Career Exploration
Russell Middle School
Russell, Kentucky

Steve Wash
Industrial Technology Teacher
Batesburg-Leesville High School
Batesburg, South Carolina

Raymond D. Wilson
Technology Education Instructor
McAlester Middle School
McAlester, Oklahoma

Contents in Brief

SECTION 1   **Introduction to Technology**   12
    Chapter 1   How Technology Works   14
    Chapter 2   Design and Problem Solving   30

SECTION 2   **Communication Technologies**   46
    Chapter 3   Computer-Aided Drafting (CAD)   48
    Chapter 4   Desktop Publishing   66
    Chapter 5   Computer Animation   84
    Chapter 6   Internet   98
    Chapter 7   Audio, Video, and Multimedia   114

SECTION 3   **Production Technologies**   136
    Chapter 8   Manufacturing   138
    Chapter 9   Structures   158

SECTION 4   **Power Technologies**   180
    Chapter 10 Flight   184
    Chapter 11 Land and Water Transportation   204
    Chapter 12 Fluid Power   228

SECTION 5   **Bio-Related Technologies**   246
    Chapter 13 Health Technologies   248
    Chapter 14 Environmental Technologies   264

SECTION 6   **Control Technologies**   286
    Chapter 15 Electricity and Electronics   288
    Chapter 16 Computer Control Systems   308
    Chapter 17 Robotics   326

SECTION 7   **Integrated Technologies**   348
    Chapter 18 Lasers and Fiber Optics   350
    Chapter 19 Engineering   370
    Chapter 20 Applied Physics   390

    Glossary   412
    Index   418

Table of Contents

SECTION 1   Introduction to Technology 12

    Chapter 1   How Technology Works 14

    Tools of Technology 15
    Resources of Technology 16
    Activities of Technology 18
    Activity: Design and Build a Lifting Device 22
    Systems in Technology 23
    Impacts..24
    The Future 25
    Activity: Design and Build a Profile Map of an Ocean Floor 26
    Careers in Technology 28
    Review 29

    Chapter 2   Design and Problem Solving 30

    The Design Process..31
    Activity: Design and Build a Floor Plan 32
    Activity: Design and Build a Time-Keeping Device 38
    Using the Design Process 41
    Activity: Design and Build a Spreadsheet 42
    Careers in Design and Problem Solving 44
    Review 45

SECTION 2   Communication Technologies 46

    Chapter 3   Computer-Aided Drafting (CAD)........................48

    CAD Advantages 49
    CAD System Components 50
    Cartesian Coordinate System 52
    Two-Dimensional and Three-Dimensional CAD Drawings 53
    Activity: Design and Build a Site Plan 56
    Using a CAD System 57
    Activity: Design and Build a Tic-Tac-Toe Board 58
    CAD/CAM...61
    Impacts...61
    The Future 61
    Activity: Design and Build Orthographic Projection 62
    Careers in Computer-Aided Drafting (CAD) 64
    Review 65

5

Chapter 4 Desktop Publishing 66

The Publishing Process 67
What Is Desktop Publishing? 67
Software and Hardware 68
Activity: Design and Build an Advertisement 70
Designing Documents 71
Activity: Design and Build a Document 74
Planning a Newsletter 78
Impacts 79
The Future 79
Activity: Design and Build a Newsletter 80
Careers in Desktop Publishing 82
Review 83

Chapter 5 Computer Animation 84

What Is Animation? 85
Types of Animation 86
Activity: Design and Build a Flip-Book 88
Activity: Design and Build a Thaumatrope 92
Activity: Design and Build a Zoetrope 94
Careers in Computer Animation 96
Review 97

Chapter 6 Internet 98

What Is the Internet? 99
Building the Internet 99
Getting on the Internet 100
Internet Safety and Appropriate Use 101
Activity: Design and Build an Internet Report 102
World Wide Web 103
Activity: Design and Build a Timeline 108
Impacts 109
The Future 109
Activity: Design and Build a Home Page 110
Careers in Internet Technology 112
Review 113

Chapter 7 Audio, Video, and Multimedia 114

Electronic Communication 115
Television 119
Activity: Design and Build a Storyboard 120
Video Recorders 125
Multimedia 126
Activity: Design and Build a Megaphone 128
Impacts 131
The Future 131
Activity: Design and Build a Tin Can Telephone............132
Careers in Audio, Video, and Multimedia 134
Review 135

SECTION 3   Production Technologies 136

Chapter 8 Manufacturing 138

Materials in Manufacturing...139
Processing Materials 140
The Manufacturing System..146
Activity: Design and Build a Mass-Produced Item.........146
Activity: Design and Build an Assembly System 150
Computers in Manufacturing 151
Impacts 152
The Future 153
Activity: Design and Build Product Packaging...............154
Careers in Manufacturing 156
Review 157

Chapter 9 Structures.......................................158

Building Is a System 159
What Is a Structure? 159
Forces on Structures 161
Structural Materials 162
Structural Members 163
Structural Shapes 166
Building a Residential Structure 166
Activity: Design and Build a Testing Station 172
Control Systems 173
Impacts 174
The Future 175
Activity: Design and Build a Small Structure 176
Careers in Structures..178
Review 179

SECTION 4   Power Technologies 180

Chapter 10 Flight 184

Forces 185
Aerodynamics 187
Airplanes 189
Activity: Design and Build an Airfoil 191
Activity: Design and Build a Rocket 192
Helicopters 194
Jet Planes 194
Activity: Design and Build a Propeller.........................200
Careers in Aviation 202
Review 203

Chapter 11 Land and Water Transportation 204

Modes of Transportation..205
Systems in Transportation 206
Land Transportation 207
Activity: Design and Build a Model Maglev System............208
Activity: Design and Build a Cam Operating System 214
Water Transportation 219
Impacts 222
The Future 223
Activity: Design and Build a Catamaran.......................224
Careers in Land and Water Transportation 226
Review...227

Chapter 12 Fluid Power...228

What Is Fluid Power? 229
Fluid Science...229
Fluid Power System Safety 230
Activity: Design and Build a Water Squirter 231
Types of Fluid Power Systems 232
Fluid Power System Diagrams...................................236
How Fluid Power Is Used 236
Activity: Design and Build an Air Cushion Vehicle 238
Impacts 240
The Future 241
Activity: Design and Build a Gameboard 242
Careers in Fluid Power...244
Review...245

SECTION 5   Bio-Related Technologies 246

Chapter 13 Health Technologies 248

Physical Enhancements 249
Human Technology Resources 249
Activity: Design and Build a Human Joint Replacement 251
Health Care Technology 252
Ergonomics 255
Activity: Design and Build an Assisted-Living Product..........256
Impacts 258
The Future 259
Activity: Design and Build a New Computer Keyboard..........260
Careers in Health Technologies 262
Review..263

Chapter 14 Environmental Technologies...........264

Agriculture and Environmental Technologies 265
Activity: Design and Build a Hydroponic Growing System.......268
Controlled Environment Agriculture (CEA) 273
Activity: Design and Build a Plant Watering Device..........275
Bioprocessing 279
Impacts 279
The Future...281
Activity: Design and Build a Casein Glue
     Manufacturing System 282
Careers in Environmental Technologies................284
Review..285

SECTION 6   Control Technologies 286

Chapter 15 Electricity and Electronics 288

What Is Electricity?..289
Activity: Design and Build an LED Warning System 292
Sources of Electricity.......................................294
Relationships among Voltage, Current, and Resistance 296
Electrical Circuits..297
What Is Electronics?..300
Electronic Systems 303
Impacts 303
The Future..303
Activity: Design and Build a Continuity Tester 304
Careers in Electronics 306
Review..307

Chapter 16 Computer Control Systems 308

Computers and Product Design 309
Computer Numerical Control 311
Activity: Design and Build a Product Concept........................312
Activity: Design and Build a Flowchart.................................316
Computer-Aided Manufacturing (CAM) Systems 319
Impacts 320
The Future 320
Activity: Design and Build Computer Numerical Control
      Coordinates 322
Careers in Computer Systems Control 324
Review..325

Chapter 17 Robotics 326

How Robotics Developed 327
Modern Robotic Systems 329
Power for Robotic Movements 332
Activity: Design and Build a Feedback Control Game 334
Controlling Robotic Systems 336
Impacts 343
The Future 343
Activity: Design and Build a Pneumatic Control Device 344
Careers in Robotics 346
Review..347

SECTION 7   **Integrated Technologies..................348**

Chapter 8   Lasers and Fiber Optics 350

The Nature of Light 351
Lasers 352
Activity: Design and Build a Prism System 354
Activity: Design and Build a Light-Carrying Device..............362
Impacts 365
The Future..365
Activity: Design and Build a Lighted Monogram 366
Careers in Lasers and Fiber Optics 368
Review..369

Chapter 19 Engineering 370

What Is Engineering? 371
Types of Engineering..372
Activity: Design and Build a Jack-in-the-Box 376
The Engineering Process 377
Activity: Design and Build a Shelf for a School Locker 380
Using the Engineering Process: A Case Study 381
Impacts 385
The Future..385
Activity: Design and Build a Toothpick Dispenser 386
Careers in Engineering 388
Review...389

Chapter 20 Applied Physics 390

Motion 391
Activity: Design and Build a Crane...........................392
Newton's Laws of Motion 395
Work, Power, and Machines....................................398
Sound Waves and Light Waves 402
Activity: Design and Build a Sound Wave Tester 404
Impacts 407
The Future..407
Activity: Design and Build a "Reaction" Rocket Racer.........408
Careers in Applied Physics...................................410
Review...411

Glossary 412

Index 418

Photo Credits..431

Dinamation Dinosaurs ©1997
Dinamation International Corp.

SECTION 1

Introduction to Technology

CHAPTER 1 *How Technology Works*

CHAPTER 2 *Design and Problem Solving*

Technology depends on design and problem solving. To complete a design or to solve a problem, you need resources. The chapters in this section will introduce you to the resources of technology. They will also show you how you can use these resources to create the tools you will need to solve problems. Design and problem solving are important in technology. They are also important in life.

Solving One Problem Can Create New Problems

Henry Ford applied the process of mass production to automobile assembly. He invented the automobile assembly line in 1913.

It now seems obvious that Ford's improvement to auto-making technology would greatly increase the number of cars on the road. This, of course, would increase the possibility of traffic accidents.

By the 1970s, Americans were dying in car accidents at a rate of 45,000 to 50,000 per year.

The Blowup Over Air Bags

Air bags were introduced in the early 1990s. It seemed that this new technology could help lower the traffic accident death rate.

Air bags, stowed in the steering column and dashboard, inflate within one-tenth of a second after an impact. They provide a cushion for people in the front seats. Holes in the air bags allow them to deflate barely one second later. The bags "give" and absorb the energy of a too-sudden stop.

By the mid-1990s accident reports showed that air bags were causing severe injuries. More than fifty deaths were linked to air bags, according to the National Highway Traffic Safety Administration.

In most instances, however, air bags did the job they were designed to do. By the mid-1990s, they had saved an estimated 1,500 lives.

Modifications to air bags will reduce their force of inflation. This will help reduce the risk of injury or death from too-rapid inflation. However, an air bag will still hit harder than the worst shot you ever took in a pillow fight.

Make a list of three inventions and the new problems they caused. What changes to the technology could solve the problems?

Linking to the COMMUNITY

Research the number of traffic accidents in your community within the past year. How many of the accidents resulted in death or injury? Did air bags prevent death or injury in any of the accidents? ▲

How Technology Works

OBJECTIVES

▶ identify the basic needs and wants.

▶ identify the seven resources upon which all technologies depend.

▶ describe and give examples of manufacturing, construction, transportation, communication, and bio-related technologies.

▶ explain the similarities shared by all technological systems.

▶ describe and give examples of technological impacts.

KEY TERMS

feedback

input

output

process

system

technology

A man sits by a fire. In one hand, he holds a large rock. In the other, he holds a small stone. He uses the small stone to strike chips from the large rock. After a short time, he holds the sharpened head of a stone axhead. He fastens the axhead to a short piece of wood. He now has a tool that he can use for many tasks.

In our own time, a woman sits in her home preparing a message that she will fax to her business partner 3,000 miles away. Rather than traveling to an office each day, the woman works at home. Her home has become an *electronic cottage*. It has become a place where she can work and still be close to her family. Here, she designs computer software that is used to control robots in a factory.

TOOLS OF TECHNOLOGY

Technology is using knowledge to develop products and systems that satisfy needs, solve problems, and increase our capabilities. All problem-solving tasks have much in common. This is true whether the task is making an ax head or writing a computer program. Technology is the link that ties various tasks together. Fig. 1-1.

A *tool* is an object used in carrying on work. Tools are the instruments of technology. A tool may be simple, such as an ax. It may be complex, such as a computer program. Tools help us satisfy our needs and wants. Fig. 1-2. All people have needs and wants. These include food, water, shelter, communication, protection, recreation, transportation, and health care.

The prehistoric man made his stone ax to meet various needs. Using the ax, he could hunt animals. In so doing, he could satisfy his need for food. He could also use the ax for protection and as a construction

tool. The man used a simple technology to help meet his basic needs.

Over time, people settled in villages. Their activities centered around farming and the raising of livestock. Agriculture changed how people lived and worked. New technologies were developed to satisfy new wants and needs. The hoe, plow, and sickle were the tools of this *Agricultural Age.*

▶ **Fig. 1-1** In technology, the effectiveness of any tool can be judged by how well it can focus energy. Note that the shape of these arrowheads will focus energy at the point.

▶ **Fig. 1-2** These electronic parts are tools. Like the arrowheads on the left, they allow energy to be focused. In this case, the energy is electronic. The parts will be used in an electronic circuit.

To be usable, most raw materials need to be processed. This led to the development of tools that could be used to process such materials. The loom is one example of such a tool. The loom was used to weave fabric from natural fibers such as wool.

As people learned more about their world, they undertook more complex tasks. Their needs and wants also became more complex. As trade developed, the need for transportation increased. To make transportation easier, transportation systems were developed and slowly improved.

Bridges and better roads made travel easier. They also allowed people to move products more quickly from one village to another. These early transportation systems were crude. However, they provided a base for new technologies that would make transportation more efficient.

New needs and wants continued to emerge as daily life became more complex. In the 1700s, new technologies and products exploded onto the scene during the *Industrial Age*. The steam engine enabled factories to satisfy consumer demand for new products. Transportation systems were improved to transport goods and people more rapidly.

Trade became more important. It became necessary to communicate more quickly and reliably. The technologies that led to the telegraph, telephone, and radio were developed in the Industrial Age.

Are we still in the Industrial Age? What makes a way of life change so much from one era to another? The answer lies in what people need. When the needs of people change, they develop new systems and new technologies. Today we live in the *Information Age*.

The Information Age began in the 1950s. It was brought about by the need to gather, organize, store, and share information. The Information Age was built upon the development of the transistor and the computer. The electronics of the Information Age improved all technologies. It made them faster and more reliable.

The technologies of the Information Age touch on all aspects of modern life. These technologies, for example, fuel the research that has advanced medicine. This research has helped us meet our need for health care. Doctors can now diagnose and treat illnesses that previously could not be cured.

Video games, portable stereos, CD players, videocassette recorders, and in-line skates are all products of technology. These items help us satisfy our need for recreation and relaxation.

People have always had the same basic needs and wants. What has changed through the ages is how we satisfy them. Technology is the process we use to create the products and services we use to meet our needs and wants. Fig. 1-3.

RESOURCES OF TECHNOLOGY

How would you make a stone ax? How would you write a computer program that controls industrial robots? You would rely on many resources to complete both tasks. *Resources* are all the things you may need to produce a product, provide a service, or solve a problem. Would you believe that it takes many of the same types of resources to produce a stone ax as it does to write a computer program?

NOURISHMENT

SHELTER

TRANSPORTATION

COMMUNICATION

RECREATION

HEALTH CARE

▶ **Fig. 1-3** Some of our needs in the Information Age: nourishment, shelter, transportation, communication, recreation, and health care.

All technologies rely on seven resources: people, information, materials, energy, tools and machines, capital (money), and time. When people use these resources to produce products or provide services that meet certain wants and needs, they are creating technology.

ACTIVITIES OF TECHNOLOGY

How are the seven resources of technology combined to produce hundreds of thousands of different products each day? This is done through complex activities. The *activities of technology* are used to produce these products. These activities include manufacturing, construction, transportation, communication, and bio-related technologies.

Every ninety seconds, a Ford Mustang automobile rolls off the assembly line. Let's examine each of the activities of technology as they relate to the manufacture of a Mustang.

Manufacturing technologies are the processes used in factories and shops to create products. Manufacturing activities involve the processes of forming, separating, combining, and conditioning materials. Fig. 1-4. Many of these processes are used in the manufacture of the Mustang.

Construction technologies are the processes used to build the structures in which manufacturing takes place. The activities of construction technology involve a series of tasks. These include preparing a site, building a foundation, building a frame, and enclosing the

▶ **Fig. 1-4** This manufacturing technology relies on robotic arms. The actions of robots can be precisely controlled. This allows their energies to be tightly focused in operations such as welding.

structure. Construction technologies were used to construct the factories in which the Mustang is built.

Transportation technologies are the processes used to move people or goods. Transportation outside the factory involves the use of trucks, railroads, and ships. Transportation technologies are responsible for bringing material resources to the Mustang factory. The material resources are processed there. Within the Mustang factory, cranes and conveyors transport the needed materials along the assembly line.

Communication technologies are the processes used to exchange information. Fig. 1-5. Communication technologies include the *computer numerical control* (*CNC*) of machines and *computer integrated manufacturing* (*CIM*). Both of these technologies control the manufacturing processes in the factory.

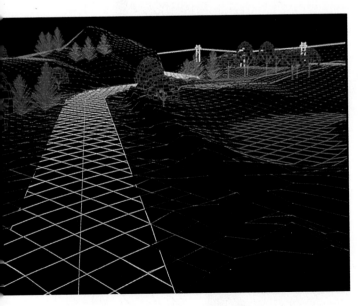

Fig. 1-5 Communication technologies have allowed the development of drawings such as this. This drawing was prepared by a landscape architect using a computer-aided drafting program.

Communication technologies ensure that the parts for the Mustang are available for assembly at the right time and place on the assembly line.

Bio-related technologies are the processes used to make or change products using living organisms. They are also used to engineer products and processes that have strong connections with organisms. Bio-related technologies are the result of combining biology and technology. Bio-related technologies are used in the manufacture of the Mustang. They are used to make sure that the design of the car meets human needs.

Some of the technologies that control industrial robots in factories have been adapted to human needs. For example, *myoelectric* (my-oh-e-LEK-tric) limbs use small motors and electronic circuits to control the humanlike movements of the artificial limb.

Technologies

The activities of technology channel resources to accomplish a goal. The goal may relate to manufacturing, communications, transportation, bio-related technologies, or construction. The successful completion of activities in technology is evident all around you.

MANUFACTURING

COMMUNICATIONS

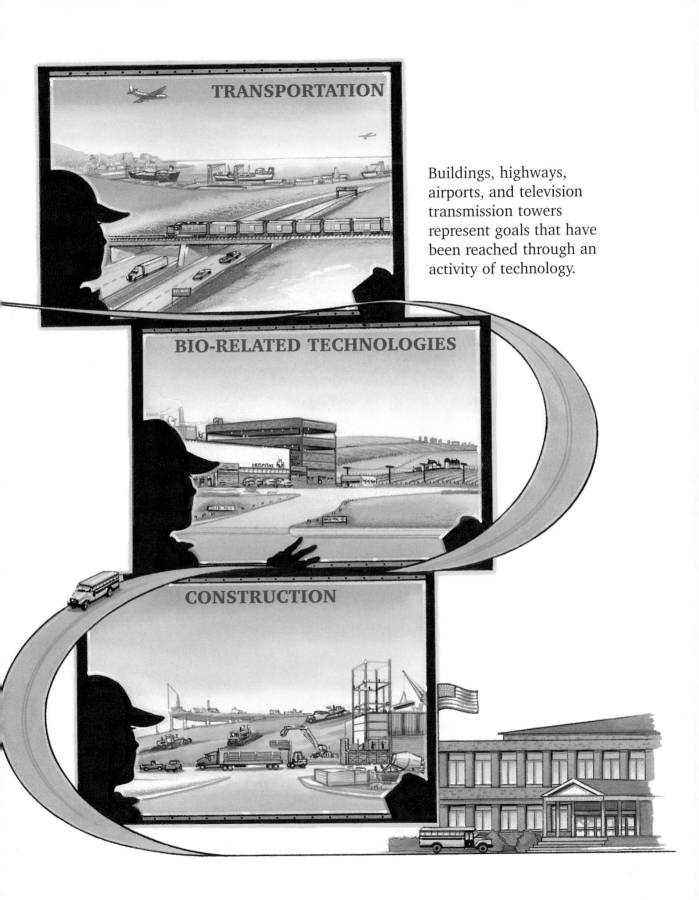

TRANSPORTATION

BIO-RELATED TECHNOLOGIES

CONSTRUCTION

Buildings, highways, airports, and television transmission towers represent goals that have been reached through an activity of technology.

Explore

Design and Build a Lifting Device

State the Problem

Design and build a lifting device that uses a lever. One possible design is shown in Fig. A. You should, however, prepare your own design.

Develop Alternative Solutions

The device must be strong enough to support the weight, regardless of how the weight is to be lifted.

The *fulcrum* (FULL-crum), or pivot, is the point on which the lever rests. What consideration must be made when determining the placement of the fulcrum?

The overall size and height of the device will be decided by your instructor. Its required strength will be in part determined by the weight that is to be lifted. Typically, the weight is two pounds or less.

Determine the placement of a lever needed to reduce lifting effort by more than half. For example, it should take less than one pound of effort to lift a two-pound weight.

Prepare several designs for the device.

Select the Best Solution

Select the design that you think will be the most effective.

Implement the Solution

1. Construct the device. Follow the design you have chosen.

2. Test the device for efficiency. Determine how much force is required to lift a given weight. Testing may be accomplished by pulling on the lever using a spring scale. This will indicate the amount of force being applied.

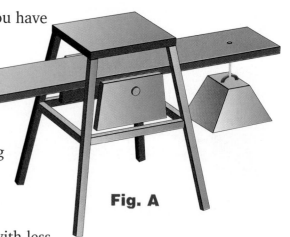

Fig. A

Evaluate the Solution

Using the device, can you lift the weight with less than one-half the force required to lift it directly?

SYSTEMS IN TECHNOLOGY

A **system** is an orderly way of achieving a goal. You probably use a system each morning to get ready for school. However, because you are so used to doing the same things every day, you may not know you are using a system. Systems are also used in technology. A system is used in manufacturing the Mustang. Such a technological system is designed to produce a desired result—one completed Mustang every ninety seconds.

All technological systems operate in the same way. Every system has an input, a process, an output, and feedback.

An **input** is something that is put into a system. In some systems, the input is a combination of the seven resources of technology. Think again about the resources required to manufacture an automobile. An input might also be a command that is given to the system. For example, you may command the heating system in your home to maintain the temperature inside your home at 67°F. You would input the command by setting the thermostat at 67°F.

The **process** is that part of the system during which something is done. The process is the "action" part of the system. A manufacturing system changes material input into products by using the energy resources of a machine. The manufacturing process can include combining, separating, forming, and conditioning techniques.

In a *home-heating system*, the process is a combination of the seven resources of technology. Fig. 1-6. This process might consist of burning fuel in the furnace to heat water. The hot water would then be pumped through pipes. The resources of

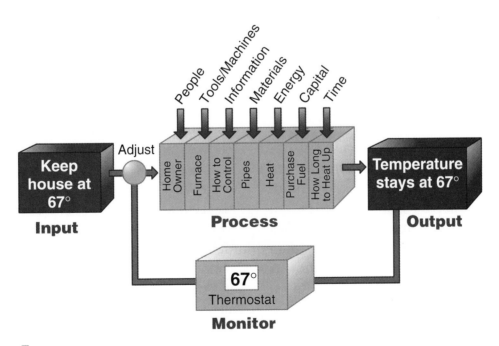

Fig. 1-6 The resources of technology.

this system would include the furnace system (machine), fuel (energy), and the money to pay for them (capital). What other resources can you identify?

The **output** of a system is the result of the system's process. If the system is working properly, the output will be the desired product or result. In Mustang manufacturing, the desired product would be a finished Mustang. In the home-heating system, the desired result would be a room temperature of 67°F.

Feedback is information about the output of a system. It is used to monitor how a system is working. When we know how a system is working, we can adjust the system to reach the desired goal.

Some systems are monitored and adjusted by people. Other systems, such as home-heating systems, control and adjust themselves. For example, the thermostat in a home-heating system monitors the temperature in the room where the thermostat is placed. It signals the furnace when to start and when to shut off. This type of system is known as a *feedback-control system* or a *closed-loop system*.

Systems and Subsystems

Complex systems are made up of smaller systems called *subsystems*. When subsystems are combined, a more powerful system results. Looking at the smaller parts of complex systems makes them easier to study. For example, a stereo system could be broken down into its subsystems. These might include a tuner/amplifier, speakers, a CD player, and a cassette player. Each of these subsystems could then be broken down into subsystems.

IMPACTS

An *impact* is the power of an event or idea. Since the first tools were made, technology has had both positive and negative impacts on society and the environment. Fortunately the positive impacts of technology have far outweighed the negative impacts. Predicting the desirable impacts of technology is usually easy. Predicting the negative impacts is a bit harder. Fig. 1-7.

As an example, in the twelfth century, eyeglasses were introduced in Europe. This technology had a positive impact on people with poor vision. Church scribes quickly saw the value of this new technology. It was the job of scribes to hand-copy the sacred books of the church. The use of eyeglasses extended the number of years a scribe could work. Failing eyesight was no longer a reason for a scribe to leave the job.

Unfortunately, those who wanted to work as scribes had to wait longer to obtain that work. This caused rioting between the scribes and those who wanted to work as scribes.

THE FUTURE

New developments in technology will be reflected in new products. The mass production of such products will bring

Linking to MATHEMATICS

Figuring Output An air-bag factory manufactures one air bag every sixty seconds. How many air bags will it manufacture in one year? Assume that the air-bag factory operates twenty-four hours a day, five days a week.

down their cost. Product advertising has been improved through the rapid growth of communication technologies. All of these developments will make it easier for businesses to sell their products worldwide. However, competition will force businesses to improve their products. Such improvements may introduce new technologies. This is only one pattern for future developments in technology.

▶ **Fig. 1-7** The excitement of using new technologies spurs new developments.

Apply What You've Learned

Design and Build a Profile Map of an Ocean Floor

State the Problem
Simulate mapping an ocean floor with sonar.

Develop Alternative Solutions
Does an ocean floor have mountains and valleys? Is it flat? Gather information on what an ocean floor might look like.

Prepare sketches of a model ocean floor you will make from clay. Be sure your model has many of the features actually found on the ocean floor.

Select the Best Solution
Select the sketch that is the most representative of a typical ocean floor.

Implement the Solution
1. Using the clay, model the ocean floor inside the shoe box. You do not need to cover the entire box bottom. You can use only a strip of clay 3" or 4" wide down the center of the box.
2. Place the lid on the box and tape it closed.
3. Cut and tape a piece of graph paper to fit the top of the lid.
4. Draw a line *lengthwise* down the center of the graph paper. Divide this line into 3/4" segments, or coordinates.
5. The second sheet of graph paper will become your profile map. On this sheet, draw a straight line lengthwise one inch from the bottom. Divide this line into 3/4" segments along its length.
6. Make a point on one end of the 10" dowel rod. Mark the rod at 1" intervals. Label the marks. Label the mark 1" from the pointed

Collect Materials and Equipment
Ocean Floor Model
shoe box with lid
clay
tape
Sonar
1/4" graph paper sheets (2)
1/4" dowel rod, 10" long

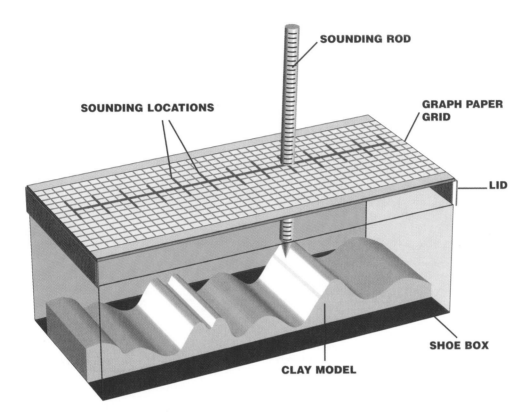

Fig. A. Conducting a sounding through the shoe box lid.

end of the dowel as 8". Label the next 1" mark as 7", and so on. Mark off each 1" interval in 1/8" increments.

7. Map the ocean floor model by making soundings. For each sounding, push the pointed dowel through the box lid at each of the coordinates you marked. Keep the dowel straight. Gently lower the dowel until you feel it come in contact with the clay. Fig. A.

8. Read the measurement on the dowel. Plot this number on the second sheet of graph paper. Make soundings on each of these coordinates. Plot these points on the second piece of graph paper.

9. Connect the points on the graph paper. You now have a profile map of the ocean floor model in the box.

Evaluate the Solution

1. Does the profile map look like the model you made?

2. How could you show even more detail of the ocean floor model?

CAREERS IN
Technology

CONSTRUCTION AND BUILDING MANAGER

General contractor seeks an experienced manager for a commercial construction project. Candidates must be experienced and have excellent contract administration skills. Should also possess strong communication and organizational skills. Fax resume with salary history and references to: Tri-C Construction Company, (330) 837-6983.

MULTIMEDIA PRODUCER

CD-ROM company needs multimedia producer to manage staff of graphic artists, writers, programmers, and testers. Will also have some programming responsibilities. Must be a take-charge individual with excellent people skills. Send resume to: Human Resources, Metas Tech Incorporated, 2878 Medina Road, Sacramento, CA 90822.

RESEARCH AND DEVELOPMENT ENGINEER

R & D engineer needed in product design and development for large consumer products company. Candidate will develop prototypes based on company recommendations and requirements. Working as a team, engineers will suggest ways to incorporate new features into basic design. Entry-level position for individual with bachelor's degree in engineering. Requires creativity and imagination. Submit resume to: J & J Plastics, 790 East Monroe Highway, Ada, MI 43002.

SOFTWARE ENGINEER

Opportunity to use project management, technical skills, and customer service skills in the planning, design, engineering, and support of bank's expanding voice and data network. We require three years of experience, professional communication skills, and a proven record of accomplishments. Comprehensive salary and benefits package offered. For consideration, mail or fax your resume to: Savings Bank, 1800 East 9th Street, Suite 200, Seattle, WA 99235, (812) 892-1559.

WEB MASTER

Large manufacturer and distributor of greeting products has an opening for a web master. This individual will work with project teams to develop applications for our web sites. Applicant must have strong knowledge of server scripting languages and database experience. A degree in computer science and one year of experience in Internet programming required. If you are interested in this exciting opportunity, please send your resume to: Greeting Corporation, 12 American Road, Knoxville, TN 22854.

Linking to the WORKPLACE

Changing technology has an impact on the job market every day. You may well retire from a job that doesn't even exist now. Your great-grandparents didn't have a word or concept for "software engineer." Can you imagine running a satellite recovery and repair service?

One very valuable skill is being able to spot trends. What job titles could you create if:

- The Moon and Mars were colonized?
- Robots were to perform all the routine and dangerous work?
- Developments in medical science increased the number of people who will live to be active and healthy at over 100 years of age?

Discuss the job titles with the class.

Chapter 1 Review

SUMMARY

▶ Technology is a way of making our lives better by meeting our needs and satisfying our everyday wants.

▶ Basic needs and wants include food, water, shelter, communication, protection, recreation, transportation, and health care.

▶ Resources are the things needed to produce a product, provide a service, or solve a problem.

▶ Technologies rely on seven resources: people, information, materials, energy, tools and machines, capital, and time.

▶ The activities of technology responsible for producing goods include manufacturing, construction, transportation, communication, and bio-related technologies.

▶ A system is an orderly way of achieving a goal. All systems have input, process, output, and feedback.

CHECK YOUR FACTS

1. List the basic needs and wants common to all people.

2. When people use resources to produce products or provide services that meet certain wants and needs, what are they creating?

3. What are the seven resources on which all technologies depend?

4. List and describe each of the activities of technology.

5. List and describe the things that all technological systems have in common.

CRITICAL THINKING

1. At one time, the stone ax represented the peak of technological development. Today, computers are thought to be the height of technological development. Explain how both products can be considered examples of the "height of technological development" when they are so different.

2. List two examples of each of the seven resources of technology. Describe how each resource would be used to manufacture engraved keychains.

3. What would be the desired output of a home-security system?

4. Describe three positive impacts and three negative impacts of the automobile on today's society and the environment.

CHAPTER 2

Design and Problem Solving

OBJECTIVES

▶ define design.

▶ explain how problem solving is part of designing.

▶ explain the forces that are involved in motion.

▶ explain the steps of the design process.

▶ describe a variety of modeling techniques.

▶ use the design process to solve real problems.

KEY TERMS

brainstorming
design
design brief
design process
ergonomics
innovation
invention
prototype

Do you sometimes imagine a world without problems? Do design and problem solving sound like familiar activities? You have been practicing these activities for years. In everyday life, you must solve problems as you go along. You may not realize it, but you use problem solving whenever you make a decision.

Some decisions are easier to make than others. Because of this, you may not always be aware that you are using the problem-solving process. Problem solving is part of the design process.

THE DESIGN PROCESS

A **design** is a plan for making something. Design is used to create technology. The **design process** is a process that uses problem solving to arrive at the best solution, or design. It is also used to improve products and services that are already in use. Fig. 2-1.

The process of designing new products is called **invention**. Thomas Edison was a very successful inventor. He conducted more than 800 experiments as he worked to develop a practical lightbulb.

Improving an existing technology creates innovation. **Innovation** (in-o-VAY-shun) occurs when something new is introduced. Today we are still trying to improve on Edison's invention. We are designing new bulbs that last longer and use less energy.

The process of designing may seem complicated. This is because so much information may have to be considered. In technology education you will do design-and-build activities. These experiences will help you become a successful designer and problem solver.

▶ **Fig. 2-1** The design process frequently involves teamwork.

Being able to recognize and solve problems is a valuable skill. You will be able to use this skill in other classes and outside of school. A good way to solve problems is to use the design process. In this chapter the steps of the design process are explained in step-by-step order.

The steps in the design process are:

1. Identify the need.
2. Gather information.
3. Develop alternative solutions.
4. Select the best solution.
5. Implement the solution.
6. Evaluate the solution.

Explore

Design and Build a Floor Plan

State the Problem

Design a bedroom in a house. The initial required drawing is a floor plan. This is a view of a room looking down from the ceiling. The scale of the drawing should be 1" = 1'-0". The bedroom will have the following:

- Floor area of 168 square feet
- One door (standard 30-inch width)
- Two outside walls
- Two windows, 12 square feet each
- One desk
- One desk chair
- One bed
- One dresser
- One bookcase
- One night stand

Develop Alternative Solutions

Ask yourself a few questions. Are all rooms in a house the same size? Do all bedrooms have the same layout? How

Collect Materials and Equipment
CAD program or graph paper T-square standard triangles pencil common templates (circles, furniture, etc.) eraser

---

The activities in this book will give you experience in design and problem solving. As you become an experienced problem solver, you may want to use the steps in the design process in slightly different ways.

Identify the Need

Identifying the need is a good place to start. Some needs are easy to understand and lead to simple solutions. For example, you may want to keep together several sheets of paper. A paper clip or staple may be all that is needed.

Other needs are more complex. For example, many communities have problems with traffic congestion. This may show the need for more roads or a new mass transit system. Reducing traffic congestion may require changes that take years to put into practice. Difficult

important is the placement of windows and doors? Draw several floor plans for the bedroom. Remember that the room has 168 square feet. The floor plans will vary, depending on the length of the walls and where you place the door and windows.

Select the Best Solution
Select a floor plan that you think will allow you to place your furniture as you want it. Determine standard sizes for the furniture you will have in your room.

Implement the Solution
1. Cut out templates representing each piece of furniture. Label all furniture templates.
2. Place the templates on the floor plan. Move the templates until you have a pleasing arrangement. Seek comments from your classmates.
3. Sketch or draw the final floor plan.
4. Present your completed drawing to the class.

Evaluate the Solution
1. Is the floor plan neatly drawn?
2. Are the furniture templates labeled?

problems such as this affect many people. They are usually solved by people working together in teams. Each team member is expected to contribute to the solution to the problem.

Some needs are unique to one person. For example, suppose your family is moving to a new home. You are to help plan your new bedroom. What would you need to include?

As you know, some needs affect many people. Consider the needs of persons who are elderly. As we age we may need help with tasks like climbing stairs, turning faucets, and opening jars. Today people are living longer. This means that the population of elderly people is increasing. This is creating opportunities to design products and services to meet the needs of older citizens.

| Identify the need. | Gather information. | Develop alternative solutions. |

Fig. 2-2 The design process.

The design process is shown in Fig. 2-2. Once a need is established, the designer should clearly state the problem that is to be solved. Fig. 2-3. A statement of the problem that is to be solved is called a **design brief**. The design brief should include all the information that the designer needs to understand the problem. It should include details about the materials to be used, how much can be spent, and when the solution is needed. Fig. 2-4.

Gather Information

After you have a good understanding of the problem, look for information that will help you design a solution. Talk to people who have knowledge and experience that will help. Friends, parents, and teachers may offer good ideas. Engineers and other workers in local industry may be helpful. They often welcome the chance to help with a challenging technology activity. Ideas can also come from the library,

Fig. 2-3 This engineer is studying the design of an offshore oil platform. This will help him suggest improvements.

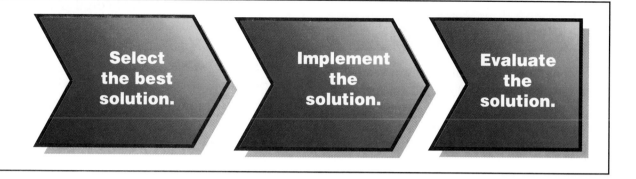

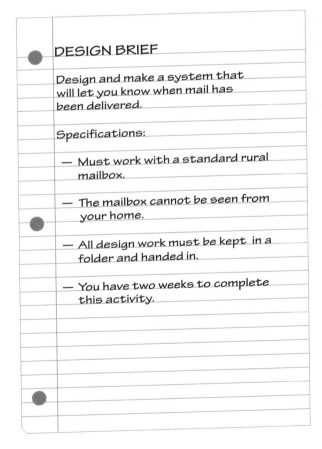

stores, museums, and the Internet. (Chapter 6 discusses the Internet as a research tool.)

Develop Alternative Solutions

When you are working alone, sketching is a good way to explore possible solutions to a design problem. As you refine your ideas, add details to those sketches that represent ideas you think will work.

Most problems have more than one possible solution. One way to generate alternative ideas is to use brainstorming.

Linking to COMMUNICATION

The Effect of Design. Interview an older person. Ask that person to tell you about his or her buying habits. Ask what effect design had on his or her buying of various items. You might mention, for example, refrigerators, cars, and televisions. You might ask about books and magazines. Plan to record your interview on cassette or videotape. Include an introduction that states your purpose and identifies the person you interview. Conclude with a summary of the interview. ▲

Brainstorming is a process in which group members suggest ideas as they think of them. It is a group problem-solving method. One member records the

> DESIGN BRIEF
>
> Design and make a system that will let you know when mail has been delivered.
>
> Specifications:
>
> — Must work with a standard rural mailbox.
>
> — The mailbox cannot be seen from your home.
>
> — All design work must be kept in a folder and handed in.
>
> — You have two weeks to complete this activity.

▶ **Fig. 2-4** A design brief should help you understand what is needed. The specifications give additional information that must be considered during the design process.

ideas. Usually a chalkboard or large pad is used so everyone can see the list. Each idea is then discussed. The team selects the ideas that show the greatest promise. Fig. 2-5.

Select the Best Solution

Look at all possible solutions. Review the advantages and disadvantages of each idea. Then narrow the list of possible solutions. Models can help you choose the best solution.

Modeling Techniques

Model building is a good way to develop and record ideas during the design process. Two-dimensional (2D) and three-dimensional (3D) models are useful. For many projects, both are needed.

Two-dimensional models include sketching, drawing, and rendering. Many good ideas start with quick pencil sketches. They help the designer change ideas as they are imagined. Later, more careful drawings can be prepared. As ideas become more specific, *renderings* (drawings) can be made. Colored pencils and markers are used. Renderings get the designer ready for 3D models. Fig. 2-6.

Appearance models, *scale models*, and *prototypes* are three-dimensional. Appearance models resemble finished products, but they do not work. Scale models are small, accurate representations of a finished product. Architects use scale models. These models help them communicate their design ideas to clients.

Good materials for scale models include clay, cardboard, foamboard, and foam block. These materials are easy to work with. They can be shaped using craft knives, abrasive paper, and files. Paint and markers can be used to add color and details.

A **prototype** (PRO-tow-type) is a working model. It looks and functions just like the finished product. Other useful modeling techniques include computer-aided design (CAD), computer simulations, mathematical models, and construction kits.

CAD systems enable designers to electronically draw anything that can be drawn on paper. Changes can be made quickly and easily. Drawings that used to take days may now take only hours.

Newly designed products can be tested using computers. For example, simulated wind-tunnel testing of a new airplane design can be done. Such testing may suggest changes in the wing shape.

Mathematical models include charts, graphs, and spreadsheets. Construction kits are useful for building working models of machines and manufacturing systems. The knowledge gained can be applied to building a full-size system.

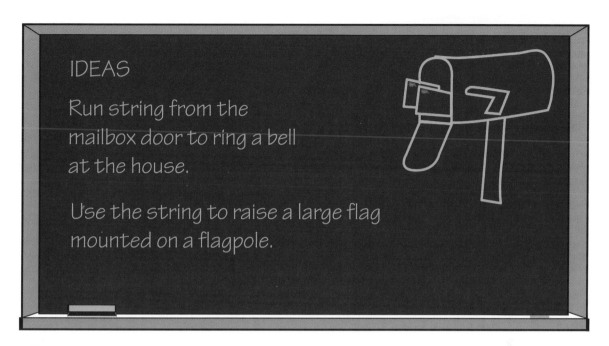

IDEAS

Run string from the
mailbox door to ring a bell
at the house.

Use the string to raise a large flag
mounted on a flagpole.

▶ **Fig. 2-5** In brainstorming, accept all ideas. Later, you may want to combine some ideas and reject others.

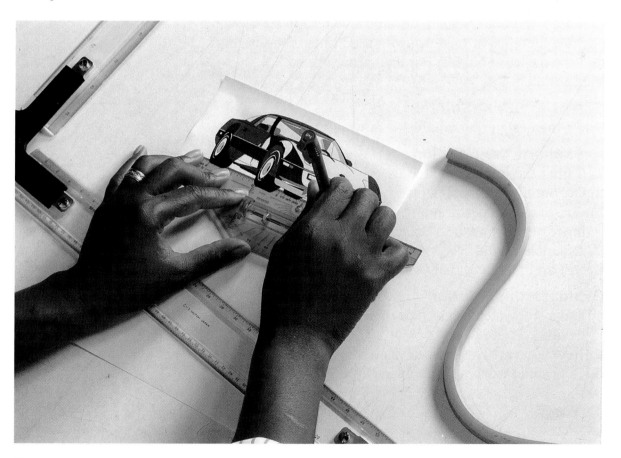

▶ **Fig. 2-6** Two-dimensional modeling includes rendering, or making a realistic drawing.

Explore

Design and Build a Time-Keeping Device

State the Problem

Technology has been used for thousands of years to keep track of time. From early sundials and sand clocks to modern atomic clocks, people have engineered time-keeping devices to help measure their time.

Design and model a time-keeping device that will accurately keep time for a one-hour period.

Develop Alternative Solutions

Research early time-keeping devices. Devices developed before gear and spring-drive clocks used a variety of materials and methods to measure time.

Based on your research, develop some sketches showing possible methods of measuring elapsed time. Keep in mind that the device should be durable, accurate, and transportable.

Select the Best Solution

Select the design that you feel is most appropriate.

Implement the Solution

1. Gather the building materials.
2. Lay out any measurements or patterns.
3. Cut, drill, and perform any separation or forming processes that need to be done.
4. Assemble the device.
5. Test and modify the device.

Evaluate the Solution

1. Use a stopwatch or clock to time the accuracy of the device.
2. How accurate was your device?
3. How could it be made more accurate?

Implement the Solution

As models of possible solutions are considered, important questions need to be answered. Fig. 2-7. Some of these questions are:

- Aesthetics—Does it have a pleasing appearance?
- Function—Does it do what it is supposed to?
- Durability—Will it last as long as it needs to?
- Cost—Is the cost within acceptable limits?
- Ergonomics—Is it comfortable to use?

Linking to MATHEMATICS

Data Display. Use a computer and appropriate software to show how data you collect can be displayed in a chart, graph, or spreadsheet.
▲

Ergonomics is the matching of design to human needs. Today ergonomics influences the design of many products. One example of an application of ergonomics is developing "user-friendly" computer software. Other examples include conveniently locating controls on an automobile dashboard and making desk chairs that are comfortable. Fig. 2-8.

▶ **Fig. 2-7** Wind tunnels can be used to test the design concepts represented by three-dimensional models.

▶ **Fig. 2-8** Virtual reality can be used in design. Here, it is being used to design an automobile dashboard. The position of the driver's hand is indicated at center by the electronic glove. The driver can operate the dashboard controls and make recommendations for design changes.

After the best solution is selected, the design may still need further testing. Government regulations require testing of products such as the pollution-control and safety systems of new automobile models. Firms such as Underwriters Laboratories test products to make sure they are safe to use. New computer programs are tested by persons who want to be among the first to use a piece of software. Their feedback can help improve the program.

Once testing is complete, the new product is ready for production.

Evaluate the Solution

Even after a product or process is in use, it needs to be checked to see if it is working properly. Sometimes products are randomly selected from a production line for testing. Another way to evaluate products is to survey consumers. A consumer is a person who buys products or services. Consumers may be contacted to get their opinion about a product. Complaints and warranty claims also indicate that changes may need to be made in the product.

Evaluating the solution is the final step of the design process. This step can reveal information that will make the production more efficient. Fig. 2-9.

USING THE DESIGN PROCESS

You can use the design process to develop solutions for problems of interest. Read the problem situations described below. Select one problem or get teacher approval to work on a different problem, either by yourself or as part of a team. Your solution should include sketches and an appearance model. Present your model to the class.

▶ **Fig. 2-9** Evaluation requires a good knowledge of materials and processes.

1. Does your community have an intersection where a large number of crashes have occurred? If so, design and build a model that shows how the intersection can be improved. Consider the use of signs, signals, and turn lanes.

2. Help improve school spirit. Design and implement a system to keep students informed about upcoming events. You might want to use a newsletter, local television or radio program, a message board, or a combination of these.

3. Many people want to pick up a few items on their way home from work. Often they are in a rush. A local company wants to open a convenience store that offers only drive-through service. Select a name, design a logo, and build a scale model.

4. Imagine you are a snack food manufacturer. Recently you surveyed consumers and learned that they would switch to a new brand of potato chips if all chips in the package were large and unbroken. Design and build a new type of package that will protect the chips.

Linking to SCIENCE

The Scientific Method. If you have not already learned the scientific method, do research to identify its main steps. Then compare the steps in the design process with the steps in the scientific method. ▲

Apply What You've Learned

Design and Build a Spreadsheet

State the Problem

A *spreadsheet* is an accounting tool. It allows you to perform automatic calculations. Spreadsheets have columns, rows, and cells. A spreadsheet can help you efficiently estimate the cost of construction and furnishings for a residence. Design a spreadsheet that could be used to track the cost of building a small house.

Develop Alternative Solutions

Determine what services and items you will need to price. Place these services and items in the first column. There will be one item per cell. They might include:

- New construction
- Renovation work
- Kitchen appliances
- Bedroom furniture
- Living room furniture
- Bathroom
- Miscellaneous furniture

How will these services or items be sold—by the unit or the foot? At the top of each column add the appropriate categories. The spreadsheet should include column categories such as:

- Number needed
- Cost per unit/square foot
- Total cost

The spreadsheet should include formulas in the appropriate cells. After indicating the number of units for any service item, the total cost should appear in the desired cell. A grand total should also appear. This total should be clearly labeled. You may have several possible designs.

		A	B	C
		Number Needed	Cost per Unit/ Square Foot	Total Cost
1	New Construction	2000 sq ft	140 sq ft	280,000
2	Renovation Work	400 sq ft	50 sq ft	20,000
3	Refrigerator	1	500	500
4	Gas Range	1	450	450
5	Freezer	1	600	600
6	Bed	3	350	1050
7	Dresser			
8	Desk			
9	Desk Chair			
10	Couch			
11	Chair			
12	TV Stand			
13	Coffee Table			
14	Bookcase			
15	GRAND TOTAL			

Select the Best Solution
Select the spreadsheet design that you think will be most effective.

Implement the Solution
1. Set up the column and row headings.
2. Insert the individual row entries.
3. Insert formulas in the correct cells.
4. Insert the costs associated with each individual service or item. (Your instructor will offer guidance.)
5. Complete the entries for a typical single-family residence.

Evaluate the Solution
1. Does your spreadsheet contain sufficient categories to estimate the cost of a residence?
2. Are the formulas correctly written and placed within the correct cells of the spreadsheet?
3. Is the grand total clearly labeled?

CAREERS IN
Design and Problem Solving

SAFETY INSPECTOR

Manufacturing company seeks safety inspector with degree in industrial or mechanical engineering or business. Must be computer-literate with formal training in OSHA regulations, insurance regulations, and state EPA regulations. Prefer candidate with experience in health and safety programs in a manufacturing setting. Forward resume to: Rontech Manufacturing, Human Resources Manager, 8975 High Street, Dallas, TX 87120

ECONOMIST

Private research company has opening for an experienced economist with ability to present findings clearly, both orally and in writing. Must have ability to make decisions and forecast trends based on data. Will assess economic trends for international trade. Computer experience helpful. Please send resume to: Lovett Research, Inc., 802 Oak Street, Bellevue, WA 92022.

SYSTEMS ANALYST

Growing consulting company has created a new position in the development department. Use your design, development and testing skills to create quality systems. A computer science degree and 2+ years of programming experience required. We offer competitive salary and benefits with extensive growth opportunities. Please submit your resume to: Ava Technology, 4045 Embassy Parkway, Baltimore, MD 27786.

HUMAN RESOURCES SPECIALIST

Health insurance company seeks HR specialist with expertise in personnel policies, wage, salary/benefit administration, and legal compliance. Must be able to recruit, interview and select staff. Strong communication and interpersonal skills required. Computer knowledge essential. We offer competitive salary and benefits. Please submit resume to: Corporate Health Insurance, Human Resources Department, 4848 Riverside Drive, Portland, OR 76555.

MARKET AND RESEARCH ANALYST

High-tech company seeks market researcher for internal staff. Develop market reports by analyzing companies and forecasting future trends. Must be comfortable asking questions and making presentations. Requires four-year degree with strong math aptitude. Competitive salary and benefits. Submit resume to: Semiconductors, Inc., 3200 Phoenix Circle, Scottsdale, AZ 80022

Linking to the WORKPLACE

Congratulations! You have just been promoted to Human Resources Specialist. You need to hire a new Systems Analyst. Make a list of ten words that describe the person you would hire for the job. After a class discussion about positive work habits, go back and identify those words on your list that you think describe your work habits. Are there any work habits you would like to improve? What are your best work habits?

Chapter 2 Review

SUMMARY

▷ Problem solving is a part of the design process.

▷ A design brief is a statement of the problem that is to be solved.

▷ Brainstorming is a process by which group members suggest ideas as they think of them.

▷ A prototype is a working model.

▷ Model building is a good way to develop and record ideas during the design process.

▷ Ergonomics is the matching of a design to human needs.

CHECK YOUR FACTS

1. Define *design*.

2. How is problem solving used in designing?

3. Explain the difference between *invention* and *innovation*.

4. List the steps in the design process. Choose one step and explain what is done during that step of the process.

5. Identify several important items that should be included in a design brief.

6. Name three sources of information that may help solve a problem.

7. Name and describe 2D modeling techniques.

8. Explain the difference between an appearance model and a prototype.

9. What is *ergonomics*?

CRITICAL THINKING

1. A group of students wants to design a new logo for their school. Name and describe a process that should help them get started.

2. Select a common product that you think needs to be redesigned. Express your ideas for improving it in writing and by using sketches.

3. Investigate the work of a famous inventor such as Alexander Graham Bell. If the inventor were alive today, what do you think he or she would be working on?

4. Identify a need that exists in your home, school, or community. Write an appropriate design brief. Use the design process to propose a good solution.

SECTION 2

Communication Technologies

CHAPTER 3 *Computer-Aided Drafting (CAD)*

CHAPTER 4 *Desktop Publishing*

CHAPTER 5 *Computer Animation*

CHAPTER 6 *Internet*

CHAPTER 7 *Audio, Video, and Multimedia*

Few technologies have changed our world as much as communication technologies. They have streamlined publishing techniques and allowed us to produce more accurate drawings. They have expanded our entertainment choices. They have enabled us to contact people thousands of miles away. The chapters in this section will discuss the uses of communication technologies.

Technology and Society

The Paperless Office?

Many predicted that computer usage would reduce our need for paper. Well, virtually every desk in the industrialized world has a computer on it. Sales of plain white paper continue to climb.

One reason for these paper sale increases is the appeal of laser and inkjet printers driven by desktop computers. Meanwhile, improved photocopy machines make it easier to run off copies. According to a Gallup poll, the paper-eating/paper-spewing fax machine is the favorite means of communication in corporations. It is preferred over voice mail, e-mail, and overnight courier.

Not surprisingly, sales of file cabinets rose 30 percent during the first half of the 1990s.

Paper Doesn't Crash

One notion about electronic documents has become commonplace. We take it for granted that computers can suddenly fail. These four words are now a universally accepted excuse for delay: "The system is down."

People despise "paperwork." However, they know it won't disappear because of a hiccup in the power grid.

Recycling Fever

Even though people are using more paper, they feel they should be using less. Two-thirds of large businesses collect office paper for recycling. Unfortunately, much of this collected paper ends up in the landfill. The toner used in copiers, fax machines, and printers is difficult to scrub out. Researchers are pursuing a process to break down toner. Such a process would help us recycle more paper, even if we can't begin using less paper.

Linking to the COMMUNITY

Observe users of a photocopy machine in a busy office, store, or library. Keep a count of how many good copies are made versus rejected copies. Compare your results with the observations of others. ▲

Computer-Aided Drafting (CAD)

OBJECTIVES

▶ explain the advantages of CAD.

▶ identify the components of a CAD system.

▶ discuss how CAD is used in industry.

▶ describe the basic CAD system commands.

Do you enjoy drawing? Even if you don't, you probably realize that drawing is an important form of communication. Drawings have always been used to record ideas.

Architects use drawings to show how to construct buildings and what materials to use. Civil engineers draw the plans for new roads and other projects like bridges and water treatment plants. Electronic engineers use drawings to show the circuits of new computers and stereo systems. Many successful products began as sketches on a scrap of paper.

KEY TERMS

CAD/CAM

Cartesian coordinate system

command

computer-aided drafting (CAD)

coordinate pair

drafting

random-access memory (RAM)

read-only memory (ROM)

CAD ADVANTAGES

Drafting is the process of representing three-dimensional objects in two dimensions. Drafting is called the language of industry. Traditional drafting is done directly on paper. The tools used are drafting machines, pencils, rulers, triangles, erasers, and compasses. Today, traditional drafting is being replaced by computer-aided drafting (CAD).

Computer-aided drafting (CAD) is the process of using a computer to create drafted documents. In this chapter, you will learn about the CAD system components, the different types of CAD, and the basic CAD commands.

CAD has a number of advantages over traditional drafting. Using up-to-date hardware and CAD software, a drafter can prepare drawings much more quickly than by manual drafting. Here's an example: Suppose an architect is drawing plans for a home that has ten identical windows. Using traditional drafting techniques, the same window would have to be drawn ten times. When a CAD system is used, the desired window can be retrieved from a *library*. It can then be placed in the ten locations in a matter of seconds.

CAD drawings are neater and more accurate than manually prepared drawings. Fig. 3-1. For example, CAD drawings can be accurate to more than 1/10,000 of an inch. Even highly skilled drafters cannot work to that degree of accuracy.

Corrections to manually prepared drawings are made with an eraser. CAD drawings never need erasing. Changes are made on-screen.

Manually preparing additional original drawings is a time-consuming process.

▶ **Fig. 3-1** Using CAD software, drawings can be prepared with great precision. Details can be changed more quickly and easily than on drawings prepared by traditional drafting methods.

Because CAD drawings are electronically stored and retrieved, additional copies can be produced quickly.

Despite CAD's many advantages, it is important to remember that drawings created using CAD reflect the talent and skill of the designers and technicians who use the systems. CAD systems help designers and drafters be more productive. However, these systems can't change a poor design into a good one.

CAD SYSTEM COMPONENTS

CAD systems need two components—hardware and software. The hardware includes the computer itself. It also includes the devices used to enter (input) information into the computer, and devices used to display (output) the drawings. CAD software is the computer program that enables a computer to be used as a drawing tool.

Hardware

The *central processing unit (CPU)* is the "heart" of the computer. It determines the speed of the computer. This affects how quickly the computer can process information. CAD work requires a fast, powerful computer.

Computers have two types of memory. **Read-only memory (ROM)** is permanent memory. It usually includes the computer's operating system. **Random-access memory (RAM)** is temporary memory. Many CAD programs require a computer with at least sixteen megabytes of RAM.

Information is entered into the CAD program through an input device. The keyboard and mouse are examples of input devices. In CAD, the *digitizing* (DIJ-uh-TIE-zing) *tablet* and its *puck* are important input devices. The puck is moved on the digitizing tablet. The puck resembles a mouse, but is much more accurate. The puck has crosshairs that are used to mark points accurately. Fig. 3-2. The digitizing tablet is also called a *digitizer*. A digitizer can also be used to enter an existing drawing into a CAD system.

A monitor displays the drawing while the CAD operator works on it. Large monitors are preferred for CAD work.

Computer data is stored in a number of different ways. Hard disk drives store the computer's operating system, software programs such as CAD, and information in the form of files. Drawing files require a lot of room on the hard drive. Many programs are now provided on CD-ROM (compact disc read-only memory) rather than on floppy disks. Floppy disks provide convenient and inexpensive storage for data.

One advantage of a CD-ROM is that it can store a large amount of data. Today, most personal computers cannot save data to CD. However, this is expected to change. Tape drives and removable hard drives are used when large amounts of data must be saved.

Printers and plotters transfer the drawing to paper. These drawings are called "hard copy." Standard laser printers are usually not used for printing CAD drawings. They can print only relatively small drawings. Plotters that use ink pens, felt-tip markers, or ink jets are used to produce most CAD drawings. Fig 3-3.

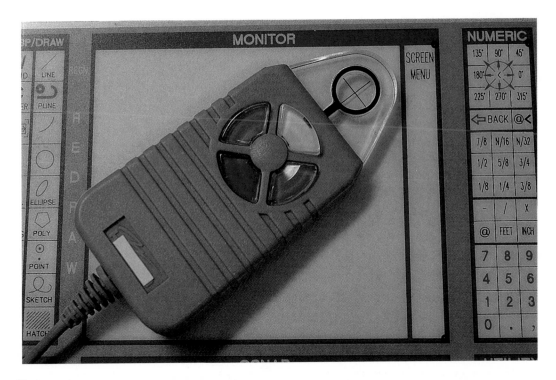

▶ **Fig. 3-2** The puck, shown here on the digitizing tablet, is moved over the tablet to place points on the drawing. The movements of the puck on the digitizing tablet are tracked on the computer screen.

▶ **Fig. 3-3** An electronic pen on a plotter. The pen can operate at 24 inches per second. Optional sensors set the pressure, height, and speed of the pen.

Software

Dozens of different CAD programs are available. Some of the programs can be used by beginners. Others require that users have extensive training. CAD software developers regularly release new versions of their programs. They also offer training at technical colleges throughout the United States.

Before purchasing CAD software, make sure that it includes all the needed features. Check to be certain that the software will work with the hardware you plan to use. Also, find out if the software company offers technical support and regular upgrades.

CARTESIAN COORDINATE SYSTEM

CAD systems produce accurate drawings because the user can select exact points, such as the ends of lines and the centers of circles. The system used is called *Cartesian* (kahr-TEE-shun) *geometry.*

The **Cartesian coordinate system** is a system that allows you to plot points on a drawing. The system is based on an imaginary grid. For two-dimensional drawings the grid has two axes, the X axis and the Y axis. (*Axes* is the plural of *axis*.) The X axis is used to plot width. The Y axis is used to plot height. A third axis (Z), showing depth, can be added for three-dimensional drawings.

Figure 3-4 shows the Cartesian coordinate system. The X axis and Y axis are at right angles to each other. They

meet at a point called the *origin*. Note that each axis has a positive (+) side and a negative (−) side. Note also that the axes create four imaginary *quadrants* (KWAD-runt). A quadrant is one-quarter of a circle.

To locate any point, you need to specify two numbers. Assume that the numbers are 4 and 2. These numbers (4,2) are called a coordinate pair. A **coordinate pair** is a set of two numbers that will locate a point on a grid. The first number (4) is the X coordinate. The second number (2) is the Y coordinate. If (2,4) is used as the coordinate pair, will a different point be selected?

Linking to MATHEMATICS

Plotting Coordinates. In the Cartesian coordinate system, the X value is always the first number. The Y value is the second number in the pair (X, Y).

In the example (3,5), 3 would be the X coordinate; 5 would be the Y coordinate.

If the third-dimensional Z axis is used, the number to be plotted on that axis is the third number. In the example (X, Y, Z), Z would represent the Z axis.

Use an 8" x 11" sheet of 1/4" grid paper. Place the X axis horizontally in the center of the page. Place the Y axis vertically in the center of the page. Label the X axis and Y axis. Write the positive and negative numbers from the origin on both axes. Plot the following points. Then connect them in order of plotting.

1. (0,4)   5. (0,-4)
2. (3,2)   6. (-3,-2)
3. (4,0)   7. (-3,2)
4. (3,-2) 8. (0,4)

▲

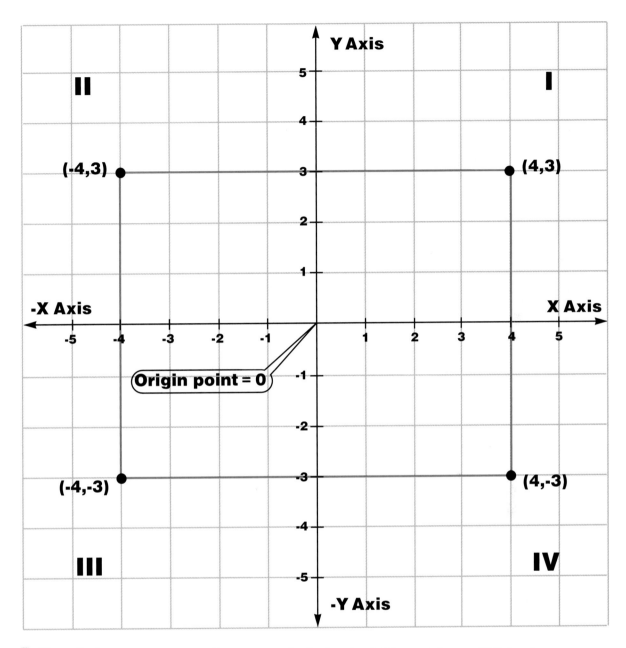

▶ **Fig. 3-4** The Cartesian coordinate system is the key to plotting points on a CAD drawing.

TWO-DIMENSIONAL AND THREE-DIMENSIONAL CAD DRAWINGS

A two-dimensional drawing shows width and length or width and height or length and height.

Most CAD programs can quickly create a three-dimensional drawing from one that is two-dimensional. They do this by adding depth. Depth is the Z axis in the Cartesian coordinate system. The drawing in Fig. 3-5 is two dimensional. Which dimensions does it show?

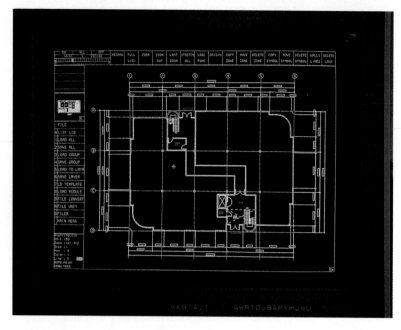

Fig. 3-5 A two-dimensional CAD drawing.

Figure 3-6 shows the Z axis. It also shows how the Cartesian coordinate system can be used to draw a rectangle. Three important kinds of three-dimensional CAD drawings are wireframes, surface models, and solid models.

Wireframe drawings look like wire sculptures. They are see-through stick drawings that show the length, width, and height of an object. Fig. 3-7.

Surface models are more advanced than wireframe models. They are easier to understand and show solids instead of just lines. Fig. 3-8.

Solid models have several advantages over wireframe and surface model drawings. They can be so realistic that they resemble a photograph or the work of a skilled artist. Also, the CAD operator can specify that the object be made of a particular material. The model of an object, such as a beam, can then be tested on the computer to determine important qualities such as weight and strength. Fig. 3-9.

Drawings of newly designed products are used to make *prototypes* (PRO-tow-types), or models. Detailed drawings describe the final product and the assembly line that will be used to manufacture it in quantity.

Linking to SCIENCE

Beam Strength. A beam is a structural support. For example, a beam is used in a simple form of the bridge. You can study the forces on a beam bridge using a wooden yardstick. Place two chairs back to back about a foot apart. Bridge them with the yardstick. Gently press the center of the yardstick until the wood bends slightly. Now move the chairs another foot apart to make a longer model bridge. Is less or more force required to bend the yardstick? Now turn the yardstick on its edge and apply force. How does this change the strength of the beam? Can you see how a CAD program can be used to rework a poor design?

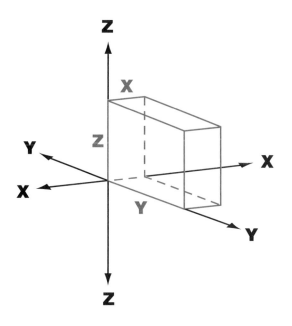

Fig. 3-6 The Z axis.

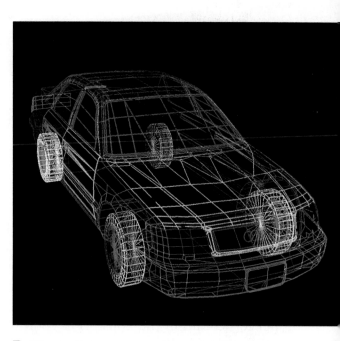

Fig. 3-7 A wireframe drawing presents the surface of the object in a series of interlocking shapes. Design changes can be made quickly on such a drawing.

Fig. 3-8 A surface model.

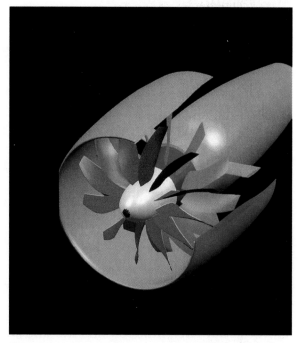

Fig. 3-9 A solid model of a propfan power unit used in aviation.

Explore

Design and Build a Site Plan

State the Problem
Use CAD to create a site plan.

Develop Alternative Solutions
The house is to be built on a 30' x 35' lot. Use the LINE command to create the outline of the house on the site plan. The house must sit on the lot exactly as shown. Using a calculator and/or graph paper, find the coordinate pairs you will need to create the outline of the house. (Hint: The upper left corner of the house is at coordinates (5,25).) Experiment with different starting points and order of coordinate entry.

Select the Best Solution
Decide where you will start (with which coordinate pair). In what order will you enter the coordinate pairs?

Implement the Solution
Note: The first three steps make the drawing area large enough to show the entire site plan.
1. Enter the LIMITS command to set the drawing area for this drawing.
2. Press the ENTER key to accept a lower left corner of 0,0.
3. Enter 50,40 to set the upper right corner.
4. Enter the LINE command.
5. Enter the first coordinate pair to start the floor plan.
6. Enter the rest of the coordinates in the order you planned.

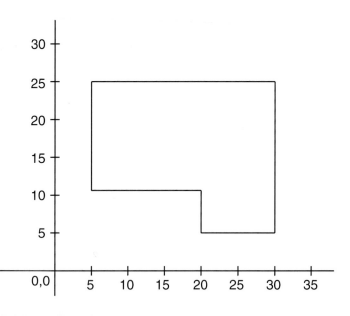

Evaluate the Solution
1. Does your site plan look like the one shown? If not, how does it differ?
2. If you had to create another site plan, what might you do differently? Why?

Collect Materials and Equipment
calculator
graph paper
computer workstation
   with AutoCAD Release
   12 or later installed

USING A CAD SYSTEM

CAD drafting is done through the use of commands. A **command** directs the software program to perform specific drawing tasks. Although they are not exactly the same, the basic commands used by popular CAD programs are similar.

The most frequently used commands are in the following three categories: drawing commands, editing commands, and utility commands. The commands are entered by using an on-screen menu, a digitizing tablet, a mouse, or a keyboard. Most CAD programs include tutorials that help new users learn and practice the commands.

Drawing Commands

Drawing commands are used to create lines, circles, and other geometric shapes.

Each line or shape is called an *object* or *entity*. The commands described below make it possible to draw almost any object and modify the drawing.

The LINE command is used to make lines by connecting points. The points are selected by using the mouse, puck, or keyboard.

The CIRCLE commands create circles. Circles can be drawn by specifying the radius or diameter of the desired circle. Another way to draw a circle is to select several points on the edge of the circle you want. Fig. 3-10.

The ARC command creates arcs. The length and radius can be specified in several different ways.

TEXT commands are used to add labels and notes to drawings.

DIMENSIONING commands calculate the dimensions of objects.

SELECT CENTER AND
ENTER RADIUS OR
DIAMETER

SELECT THREE POINTS
ALONG THE EDGE OF THE
CIRCLE

SELECT AND CONNECT
THREE POINTS

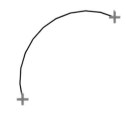

SELECT THE ENDPOINTS
AND THE RADIUS

▶ **Fig. 3-10** The CIRCLE command.

Explore

Design and Build a Tic-Tac-Toe Board

State the Problem
Explore CAD commands to create a tic-tac-toe board made up of 2-inch squares. The lower left corner of the board must be placed at coordinates (1,2).

Develop Alternative Solutions
Most CAD programs have more than one way to create objects such as squares and circles. In AutoCAD, you can create a square using the LINE, POLYGON, or RECTANG command. Find out more about each of these commands. (Hint: Refer to the Help pull-down menu in AutoCAD.) What must you know to use each command?

Using a calculator or graph paper, sketch the tic-tac-toe board. Find the coordinate pairs for each of the nine squares.

Select the Best Solution
Select the command that seems most efficient in this situation.

Implement the Solution
The following steps allow you to try each of the three commands. Each set of steps places the first square in its correct location. Follow the steps for all three commands. Decide which command is most efficient. After you have created the first square, use the command you prefer to finish the tic-tac-toe board.

LINE Command
1. Enter the LINE command.
2. Enter these coordinate pairs (in order): (1,2), (3,2), (3,4), (1,4), (1,2). When you enter the coordinates in AutoCAD, do not include the parentheses. Do not type a space after the comma.

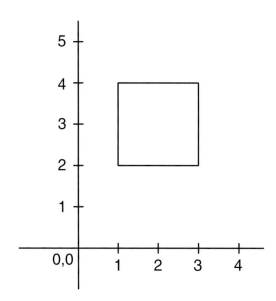

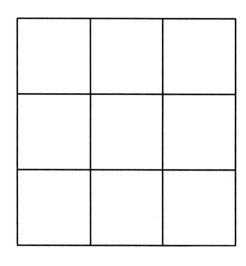

POLYGON Command

1. Enter the POLYGON command and specify 4 sides.

2. Enter E to specify the edge of the polygon.

3. Enter coordinate pair (1,2) for the first endpoint and (3,2) for the second endpoint.

RECTANG Command

1. Enter the RECTANG command.

2. Enter the coordinate pair (1,2) for the first corner and (3,4) for the second corner.

After you finish drawing the tic-tac-toe board, play a game of tic-tac-toe with your partner. How will you claim squares? (Hint: You do not have to use Xs and Os. You can use any symbol, such as a diagonal line.)

Evaluate the Solution

1. Which method is the most efficient for drawing the tic-tac-toe board? Is this the method you thought would be best?

2. If you thought another method would work better, explain why. Why did you change your mind?

Editing Commands

Editing commands are used to make changes on drawings.

The ERASE command removes objects from the drawing. It does what an ordinary eraser does, but it does it electronically.

The MOVE command is used to change the location of objects on the drawing.

The COPY command is used to make another copy of an object on the drawing. The copy can be placed in one or more additional locations. The original object remains in place.

The ROTATE command is used to turn an object around a specific point. Fig. 3-11.

The MIRROR command will produce a mirror image of an object. For example, suppose you have drawn the right wing of an airplane. Using the MIRROR command, you can then quickly create the left wing. Fig. 3-11.

Utility Commands

Three frequently used utility commands are ZOOM, PAN, and PLOT.

The ZOOM command works like the lens of a camera. It allows the drafter to reduce or magnify a drawing or part of a drawing as it appears on the monitor. The actual drawing is not changed, but details of it are easier to see.

The PAN command is used with the ZOOM command when the drafter wants to view other parts of the drawing at the same magnification.

The PLOT command is selected to make a hard copy of the drawing.

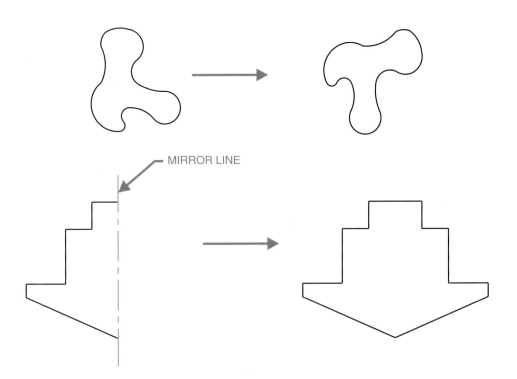

MIRROR LINE

▶ **Fig. 3-11** The ROTATE and MIRROR commands.

THE FUTURE

Advanced CAD systems will use *virtual reality*. For example, you will be able to view every room of a new home during the design process. Like some computer games and theme park rides, this imaginary trip will seem real. People working with an architect to design a new home will be able to do a "walk through." They will be able to suggest changes before the plans are complete. Virtual reality also will be used to show the inner workings of complex machines before they are built. Fig. 3-12.

CAD/CAM

CAD/CAM is a process that combines computer-aided drafting and computer-aided manufacturing. In CAD/CAM, output from the computer used to design a product is used to operate the machines that manufacture it. Relatively simple drill presses and complex *computer numerical control* (*CNC*) machines can be controlled in this way.

IMPACTS

CAD has replaced traditional drawing in many industries. Traditional drafters have needed to learn new skills. Fortunately, many have been able to transfer their skills from the drafting table to the computer. Many companies offer the training needed to make this transition.

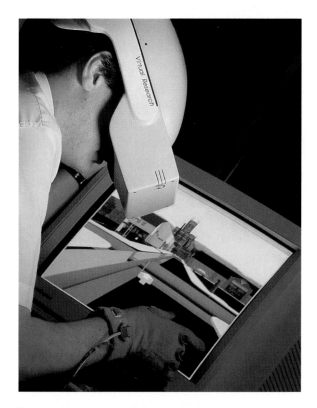

▶ **Fig. 3-12** This urban planner is using a virtual reality system. It allows him to assess traffic flow in a redesigned intersection.

Apply What You've Learned

Design and Build Orthographic Projection

State the Problem
Create a three-view drawing of the step block.

Develop Alternative Solutions
An *orthographic projection* is a drawing that contains several views of a three-dimensional (3D) object. It contains enough two-dimensional (2D) views to describe the object completely.

Study the step block shown in Fig. A. Use a calculator to find the missing dimensions. On graph paper, sketch the three views you need to describe the block completely. Place the top view at the top of the paper. Place the front view below the top view so that their edges align. Then place the side view on the right side of the front view so that their edges align.

Plan ways to create the drawing using CAD. What command or commands will you need? Assume that the lower left corner of the front view is at the origin (0,0). What coordinate pairs should you enter to create the three views? Remember to leave space between the views.

Select the Best Solution
Decide where you will start (with which coordinate pair). In what order will you create the views?

Implement the Solution
Note: The first three steps make the drawing area large enough to show the entire orthographic projection.

1. Enter the LIMITS command to set the drawing area for this drawing.
2. Press the ENTER key to accept a lower left corner of 0,0.

Collect Materials and Equipment

calculator
graph paper
computer workstation
with AutoCAD Release
12 or later installed

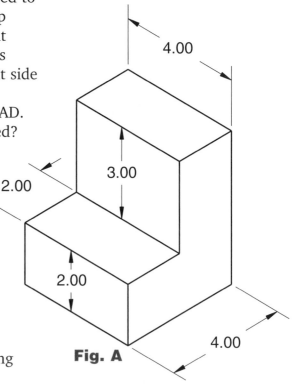

Fig. A

4.00

3.00

2.00

2.00

4.00

3. Enter 50,40 to set the upper right corner.
4. Enter the LINE command.
5. Enter the first coordinate pair to start the first view you plan to draw.
6. Enter the rest of the coordinates to finish the first view.
7. Using the coordinates you determined on graph paper, create the other two views.
8. Enter DIM at the Command line to enter the Dimensioning mode.
9. Enter HOR (for Horizontal), press ENTER to select a line, and pick the bottom line of the top view.

TOP

FRONT   **SIDE**

Fig. B

10. Move the cursor away from the line and pick a point to place the dimension. Press the ENTER key to finish the command.
11. Reenter HOR and create the horizontal dimension on the side view.
12. Enter VER (for Vertical), press ENTER to select a line, and pick the right side of the top view.
13. Move the cursor away from the line and pick a point to place the dimension. Press the ENTER key to finish the command.
14. Reenter VER to create the remaining dimensions, as shown in Fig. B.

Evaluate the Solution

1. Is your orthographic projection similar to the one shown?
2. Did the coordinates you chose for the top and side views leave enough space between views?
3. Can you think of an easier way to create the three-view drawing? Explain.

CAREERS IN
Computer-Aided Drafting (CAD)

DRAFTER

Growing manufacturer needs drafter with CAD skills. Duties include making new drawings and revising both CAD and paper drawings that range from simple machine components to detailed assemblies and technical illustrations. Will use standard drafting techniques and devices as well as computer-assisted design/drafting equipment. Send resume to: Dreisson International, 5610 Barberton Avenue, Detroit, MI 68901.

ARCHITECTURAL DRAFTER

Retail design firm seeking drafter with training in CAD systems to draw architectural and structural features of buildings. Architectural experience and knowledge of construction a plus. Entry-level position offering excellent growth opportunities. Salary plus comprehensive benefits. Submit resume to Chase and Brass Architects, Inc., 190 North Union Street, Littleton, CO 76658.

CIVIL DRAFTER

Construction company seeks drafter to prepare and update drawings, site plans and maps used in highway engineering projects. Some field work required for revisions of plans. Experience with CAD required. Excellent salary and benefits. Forward resume to: Allen Construction Company, 3490 Branch Highway, Gainesville, FL 22385.

DESIGN TECHNICIAN

Steel fabricator seeking a disciplined self-starter with experience in a diverse, demanding manufacturing environment. The successful candidate must have drafting skills, CAD experience, knowledge of fabricating and welding. Strong communication skills and computer literacy a must. CNC knowledge a plus. We offer an excellent salary and benefits package. Send resume to: Twinsburg Steel, 1902 Huron Avenue, Toledo, OH.

TECHNICAL ILLUSTRATOR

Electronics manufacturing company has an immediate opening for a technical illustrator. Will lay out and draw illustrations for use in technical manuals dealing with assembly, installation, and operation of equipment. Responsible for preparing drawings from blueprints, designs, mockups and photoprints using drafting and optical equipment. Please submit resume to: Electro Industries, 1800 Ross Street, Allen, PA 56628.

Linking to the WORKPLACE

When selecting a career you should think about your personal interests, your skills, the educational requirements, and the work setting for the job—just to name a few. However, there are other considerations. One area that people focus on is money. How much will the job pay? Choose one of the careers listed above. Pretend that it is your job. Now develop a budget based on your salary. How much will you be able to save? How much will you spend each month on rent, car, food, clothes, and entertainment? Do you think the job you selected will pay enough? Will you need to cut back on spending?

Chapter 3 Review

SUMMARY

▶ Computer-aided drafting (CAD) is replacing traditional drafting. Compared with traditional drafting, CAD is faster, more accurate, and neater.

▶ CAD systems require hardware and software. The hardware includes a computer, an input device, and an output device. CAD software allows a drafter to use the computer as a drafting tool.

▶ The Cartesian coordinate system is based on an imaginary grid with X, Y, and Z axes.

▶ CAD can be used for two-dimensional and three-dimensional drawings.

▶ Important kinds of three-dimensional drawings are wireframe drawings, surface models, and solid models.

▶ Commands direct the software program to perform specific drawing tasks. Frequently used commands include drawing commands, editing commands, and utility commands.

CHECK YOUR FACTS

1. Name three advantages of CAD when compared with traditional drafting.

2. Identify and describe the hardware and software that CAD systems require.

3. What is the difference between 2D and 3D drawings?

4. Identify some uses of CAD in industry.

5. What are the three main categories of CAD commands?

CRITICAL THINKING

1. Describe how CAD can be used in manufacturing and construction.

2. Make a labeled sketch to show the hardware components of a CAD system.

3. Describe the Cartesian coordinate system.

4. Explain two advantages of solid model drawing over other kinds of three-dimensional CAD.

5. Draw a chart that shows the basic drawing commands.

Desktop Publishing

OBJECTIVES

▶ define desktop publishing.

▶ identify the software and hardware needed for desktop publishing.

▶ identify the three main categories of typefaces.

▶ describe some of the important factors that must be considered when designing a document.

KEY TERMS

desktop publishing

digital camera

leading

page layout programs

publishing

resolution

scanners

typeface

Have you ever considered the impact of printing in your life? Can you imagine a world in which there were no books, magazines, or newspapers? The development of movable type by Gutenberg in the 1500s was a major event in the history of technology. It gave birth to the printing industry.

In our own century, the development of the computer has changed the printing industry. Laser printers and desktop publishing programs now allow easy and affordable printing and publishing.

THE PUBLISHING PROCESS

Publishing is the process of presenting material in printed form. Fig. 4-1. Until recently, most publishing was done by large companies. For years, the process they used to get books and other documents ready for printing included these steps:

- The writer wrote a manuscript.
- A publisher accepted the manuscript for publication.
- In the publisher's office, an editor checked spelling and grammar. The editor also made sure that the writer's ideas were clearly expressed.
- A typesetter set the type of the edited manuscript. The typeset text was set in sheets called galleys.
- The typeset copy was cut apart to fit the page. Space was left for drawings and photographs. The typeset copy was pasted in place.
- The "camera-ready" copy was sent to a printer.
- The text and artwork were photographed to make a negative. This was used to make a printing plate.

Does getting a document ready for printing sound complicated? Computers

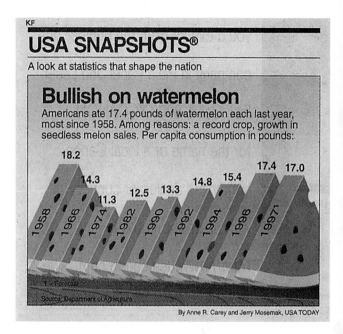

> **Fig. 4-1** This illustration was reproduced from a newspaper. It combines text and artwork. Notice that the choice of artwork reinforces the topic.

have dramatically changed and simplified publishing. Publishers now use desktop publishing to get documents ready for printing. Not only do publishers use desktop publishing—you can, too.

In this chapter you will learn the basics of desktop publishing. You will learn its uses and what is needed to get started. You will also learn some of the important things that need to be considered when designing something you want to publish.

WHAT IS DESKTOP PUBLISHING?

Desktop publishing is the use of a computer and special software to produce documents. Desktop publishing makes it possible to combine text and graphics (photographs and drawings) on the same page. The documents can be small, such as a business card or a single-page newsletter. Desktop publishing can also be used to produce long documents, such as catalogs and textbooks.

In commercial publishing, desktop publishing can produce high-quality documents that are ready for printing in quantity. For office and school work, it can be used to produce advertising fliers, announcements, and newsletters that can be reproduced on a copy machine.

SOFTWARE AND HARDWARE

Desktop publishing requires computer software and hardware. Some desktop publishing programs are inexpensive and easy to use. Others are expensive. These are designed for production of high-quality documents, including books and magazines. Personal computers similar to those you have probably used can produce attractive publications. More powerful hardware is needed to prepare documents that include many pages and features such as color photographs. Fig. 4-2.

Software

Desktop publishing uses page layout programs. **Page layout programs** are software programs that combine text and graphics in a document. Using such software, you can place text in columns and add headlines. You can also add photographs, drawings, charts, and graphs.

Desktop publishing projects usually begin with text prepared on a personal computer (PC) using a word processing

▶ **Fig. 4-2** Desktop publishing software allows you to include photographs in your document. You can also size the photographs to fit a certain space.

program. You have probably used word processing in school or at home to create school reports or to write letters. Word processing is an important part of desktop publishing. Word processing puts *text* (words) into printed form. Some page layout programs can be used for word processing, but most work best with text created with a word processor. The word-processed material is saved to a disk and *imported* (brought into) the page layout program.

Several other types of programs are used with page layout programs. *Drawing programs* are used to create illustrations. In addition, artists use them to create logos, modify typestyles, and add special effects such as shadows.

Image editing programs modify drawings and photographs. They can change the size, shape, color, and brightness of images. They can also *crop* (remove unwanted portions of) photographs. Fig. 4-3.

Clip-art software contains photographs and drawings that can be imported into a document. Hundreds of clip-art programs are available. Some contain thousands of images. The Internet can also be used as a source of clip art.

When people buy a license to use these programs, the price usually includes permission to use the artwork for certain personal purposes. However, some clip art designed for commercial use can be expensive. Other collections are available as samples. To use the photographs, publishers have to get permission from the artist or photographer. They also usually need to pay a fee.

▶ **Fig. 4-3** You can use image-editing software to remove unwanted parts of a photograph. You can also use it to select just one part of a photograph for reproduction. Note that the woman in the photograph on the opposite page (second from right) is shown here in a single photo.

Hardware

The hardware needed for desktop publishing depends on the types of documents to be produced and the software package being used. For school or home use, most people buy basic, easy-to-use software. With this software, a powerful computer is enough to get you started in desktop publishing.

Professional desktop publishing programs require computers with a lot of memory and hard disk space. It is important to choose hardware and software that are compatible. Because large documents require so much space, professional desktop publishers often use computers with removable hard disks. Removable hard disks are also good for making backup (spare) copies of the computer data.

Computers used for desktop publishing should include a CD-ROM drive, because most software is now available in this format. CD-ROMs are fast. They are also usually the most convenient way to install software onto the computer.

Design and Build an Advertisement

State the Problem

Create a one-page advertisement on an 8 1/2" x 11" sheet of paper using desktop publishing. The advertisement should demonstrate the wrapping of text around art. In text wrapping, the text "flows" around the graphic. Consider using several fonts and importing clip art or other artwork into the document.

Develop Alternative Solutions

Consider what you want to say in your advertisement. Decide on the ad's overall appearance. The ad should be informative and entertaining. Sketch several rough layouts. Remember that several fonts can be used. Remember also that the ad should be attractive. Consider importing art into the document.

Select the Best Solution

Select the layout that you think will provide the most effective advertisement.

Implement the Solution

Keyboard your ad copy. Import any graphics. Tag the text. Create your ad. Make sure that it includes text wrapping.

1. Compose your ad on the computer. Your instructor may provide guidelines regarding the number of fonts and use of clip art.
2. Print your ad.
3. Display your ad and ask the class for comments.

Evaluate the Solution

1. Was the advertisement generally effective? Did your classmates respond favorably?
2. Did you use the desktop publishing program to its full advantage?

Collect Materials and Equipment
Computer system with desktop publishing software. If such a system is not available, a word processor may be used. Cut-and-paste techniques can be used to insert artwork. If computers are not available, the project may be completed by hand.

Photographs for desktop publishing can be handled in several different ways. To place an existing photograph into a document, a scanner is used. **Scanners** are devices that change images such as photographs into an electronic form that computers can use. After a photograph is scanned, it can be modified using an image editing program.

Three types of scanners are commonly used for desktop publishing. *Hand-held* scanners are the least expensive, but they need to be moved manually across the material being scanned. *Flat-bed* scanners work well for larger images. The artwork is placed face-down on the scanner. The scanner head moves across the image. *Sheet-fed* scanners accept one page at a time.

A **digital** (DIJ-uh-tuhl) **camera** is a camera that can produce electronic images that can go directly into a computer. After a digital photograph is taken, it is stored in the camera's memory or on a disk. Today, digital cameras are relatively expensive. However, they are expected to drop sharply in price. Fig. 4-4.

Large-screen monitors allow the user to view an entire page. These are ideal for publishing work. Some professionals use monitors that allow them to see two facing pages at the same time.

The quality of documents prepared by desktop publishing depends on resolution. **Resolution** refers to the number of dots per inch (dpi) of ink on printed images. As the number of dots per inch increases, images become clearer. Such clearer images can show more detail. Home and business office laser printers typically print at 300 or 600 dpi. Commercial publishers use printers that have a much higher resolution.

Color printers are useful for previewing color pages and for printing small quantities of documents. They are not usually used to print many copies because they are slow. The ink they require is expensive.

DESIGNING DOCUMENTS

A well-designed publication will attract the reader's attention. This allows the publication to communicate the desired message. Some important things to consider when designing a document include:

- the style and size of type that will be used.
- the kinds of graphics that will be used.
- whether color will be used.
- features that will encourage people to read the document.

Type

The printing and publishing industries use a measurement system based on points and picas. Twelve *points* make a

▶ **Fig. 4-4** The images captured by a digital camera can be input directly into a computer. The images can be copied onto the conputer's hard drive.

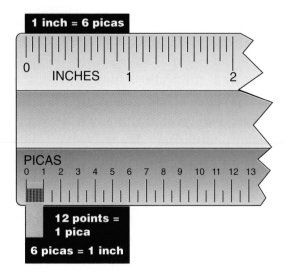

Fig. 4-5 In publishing, measurements are in points and picas. The pica (PIE-kuh) is used to measure the length of a line of type. The point is used to measure typefaces.

several different publications to see how line length affects design.

Typefaces

A **typeface** is a set of letters, numbers, and symbols that look the same. Many computer programs allow the user to change the typeface by selecting from a menu. The three main categories of typefaces are serif, sans serif, and decorative.

The difference between a serif and a sans serif typeface is shown in Fig. 4-7. Decorative typefaces are designed to capture attention. Dozens of decorative typefaces are available. Some are difficult to read when they are used for more than a few lines.

Typefaces are often available in type styles such as **bold** and *italic*. Sometimes **bold** and *italic* styles are combined to give ***bold italic***. In many publications, including this book, these type styles are used to call attention to new terms and important ideas.

Graphics

The graphics used in publications fall into two categories: drawings and

pica. There are 6 picas per inch. As you work with desktop publishing software, you will come across these terms many times. Fig. 4-5.

Printed letters are measured by their height, and type size is specified in points. Perhaps some of the computer programs you have used allowed you to choose type size by specifying "points." The text used in books and newspapers is usually between 8 and 12 points. Large type is used where there is a need to capture attention. Such type is used in headlines, headings, and in advertising. Fig. 4-6.

In addition to type size, desktop publishers must select the length of each line and the space between lines. The spacing between lines is called **leading** (LED-ing). Choosing the right leading is important. Lines of type that are set too close or too far apart are difficult to read.

Line length must also be considered. Long lines of small type and short lines of large type are difficult to read. Look at

This type is 6 points.

This type is 12 points.

This type is 18 points.

This type is 24 points.

This type is 36 points.

▶ **Fig. 4-6** The smaller the point size, the smaller the type size.

photographs. Drawings are sometimes referred to as *line art.* Maps, charts, and cartoons are examples of drawings frequently used in desktop publishing. Photographs can be in black and white or in color.

The use of illustrations should be planned carefully so that they add interest to the publication. A good illustration can present a great deal of information in a small space. For example, the newspaper *USA Today* includes a feature called "USA Snapshots" on the first page of each section. These illustrations are designed to capture attention and provide information about a timely topic.

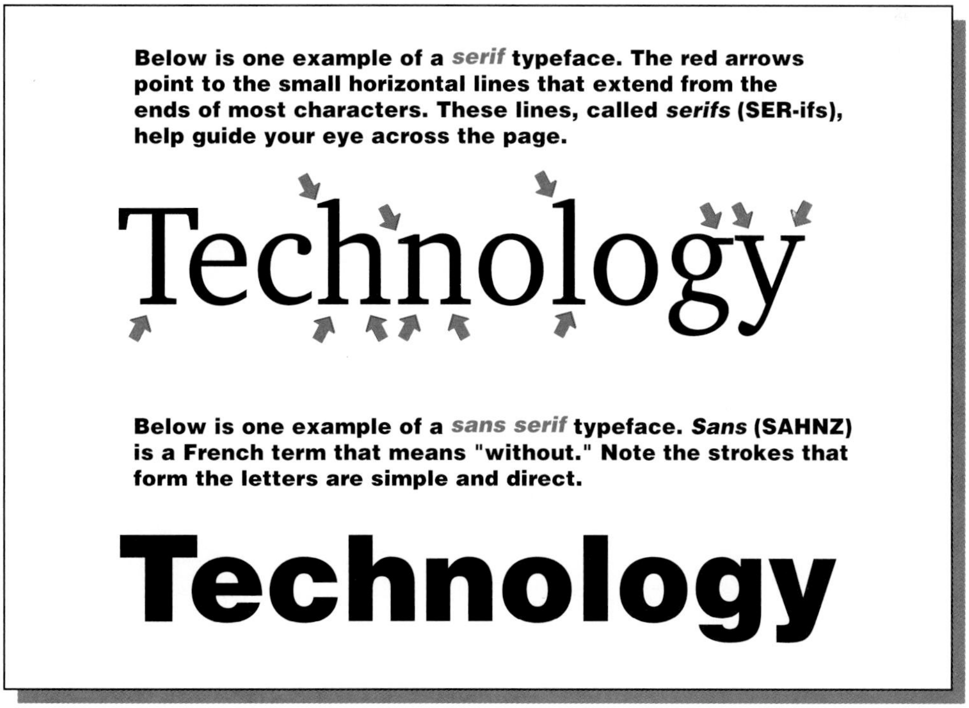

Below is one example of a *serif* typeface. The red arrows point to the small horizontal lines that extend from the ends of most characters. These lines, called *serifs* (SER-ifs), help guide your eye across the page.

Technology

Below is one example of a *sans serif* typeface. *Sans* (SAHNZ) is a French term that means "without." Note the strokes that form the letters are simple and direct.

Technology

▶ **Fig. 4-7** A serif typeface compared with a sans serif typeface.

Design and Build
a Document

State the Problem
Use desktop publishing to compose a single-page document that incorporates the following items:
- Music, film, book, product review
- Headline(s), by-line, and body text
- Picture or other graphic surrounded by body text that conforms to the shape of the graphic (often described as "text wrap")

The document should be printed on an 8 1/2" x 11" sheet of paper.

Develop Alternative Solutions
Sketch several possible designs for your document.

Select the Best Solution
Select the design that best combines all of the items.

Implement the Solution
1. Create your document using desktop publishing software or an advanced word processing program.
2. Print the document and review it.
3. Seek comments from classmates regarding the appearance of your document.
4. Revise the document. Reprint it. Submit the revised document to your instructor.

Evaluate the Solution
1. Did the text wrap the way you planned?
2. Did you use the desktop publishing program to its full advantage?

Collect Materials and Equipment
Computer system with desktop publishing software. An advanced word processing program can also be used for this project.

REVIEWS
By Suzie

The best thing about last week's concert

~ Had a good time!

There are several different ways to add graphics to publications. One way is to use *clip art*. Clip art is predrawn art. Some page layout programs include some clip art. You can also purchase clip-art programs. These are often on CD-ROMs that contain thousands of illustrations. CD-ROMs of photographs are also available. Fig. 4-8.

Drawing programs can also be used to create illustrations. They can be used to draw any shape or provide any background that can be imagined. Some drawing programs are designed for specific purposes, such as making charts. CAD (computer-aided drafting) illustrations can also be imported into desktop publishing documents.

Today, the most common way of importing photographs into a computer and publication is to scan an actual photograph. An image editing program is then used to modify it as needed. As the quality of digital cameras improves, they will be used more frequently to import a photo directly from the camera.

Color

Color encourages people to read a document. For example, color makes pie charts and graphs more effective. It must be used carefully, however. Color increases the cost of production. It can also make reading difficult if it is not used properly. Two methods of printing in color are spot color and process color.

Spot color refers to the use of one or two additional colors to emphasize certain parts of a document. Spot color is useful for holiday-related artwork and for backgrounds.

Fig. 4-8 Clip art can be decorative and informative. Note the many different elements shown here. Clip-art collections offer an inexpensive way to add graphs, borders, emphasis, and design interest.

Newsletters

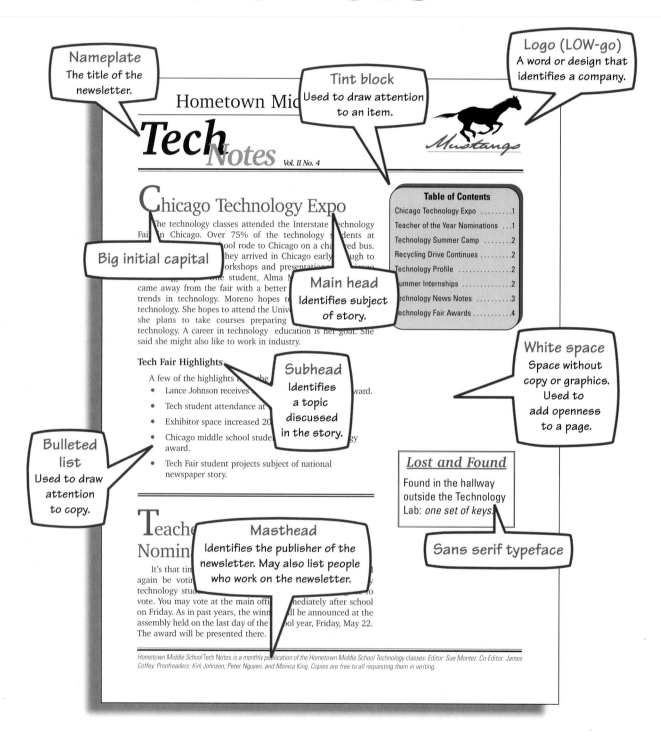

Nameplate — The title of the newsletter.

Tint block — Used to draw attention to an item.

Logo (LOW-go) — A word or design that identifies a company.

Hometown Mid...

Tech Notes
Vol. II No. 4

Mustangs

Big initial capital

Main head — Identifies subject of story.

Chicago Technology Expo

The technology classes attended the Interstate Technology Fair in Chicago. Over 75% of the technology students at ...ool rode to Chicago on a cha...red bus. ...hey arrived in Chicago early ...ugh to ...orkshops and presentation... ...one student, Alma M... came away from the fair with a better ... trends in technology. Moreno hopes to ... technology. She hopes to attend the Univ... she plans to take courses preparing ... technology. A career in technology education is her goal. She said she might also like to work in industry.

Subhead — Identifies a topic discussed in the story.

Tech Fair Highlights

A few of the highlights ... the...

- Lance Johnson receives ... ...ward.
- Tech student attendance at ...
- Exhibitor space increased 20...
- Chicago middle school stude... ...gy award.
- Tech Fair student projects subject of national newspaper story.

Bulleted list — Used to draw attention to copy.

White space — Space without copy or graphics. Used to add openness to a page.

Table of Contents

Chicago Technology Expo 1

Teacher of the Year Nominations ...1

Technology Summer Camp 2

Recycling Drive Continues 2

Technology Profile 2

...ummer Internships 2

...echnology News Notes 3

...echnology Fair Awards 4

Lost and Found

Found in the hallway outside the Technology Lab: *one set of keys.*

Sans serif typeface

Masthead — Identifies the publisher of the newsletter. May also list people who work on the newsletter.

Teacher... Nomin...

It's that tim... ...again be votin... ...technology stud... ...vote. You may vote at the main offi... ...mediately after school on Friday. As in past years, the winn... ...ll be announced at the assembly held on the last day of the ... ...ool year, Friday, May 22. The award will be presented there.

Hometown Middle School Tech Notes is a monthly publication of the Hometown Middle School Technology classes. Editor: Sue Montez. Co-Editor: James Coffey. Proofreaders: Kirk Johnson, Peter Nguyen, and Monica King. Copies are free to all requesting them in writing.

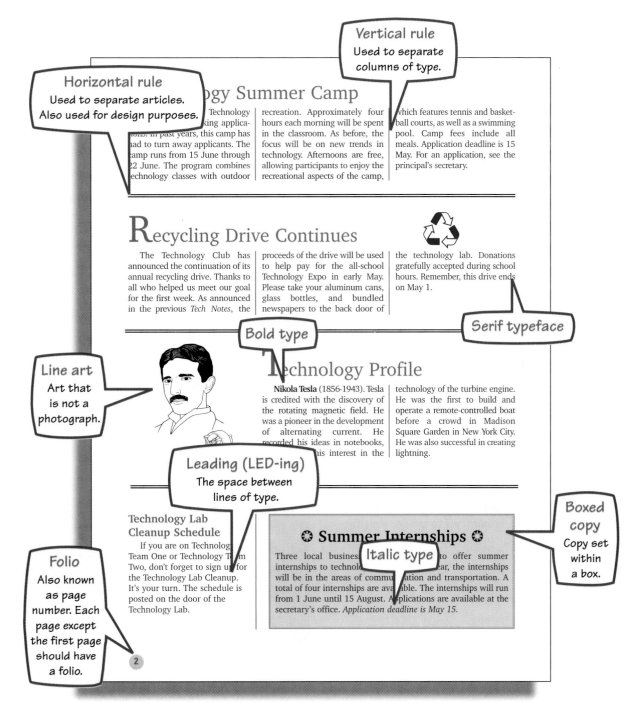

Process color is used to reproduce color photographs and art in high-quality publications. Process color was used in printing this book. The use of process color requires a page layout program that is more advanced than those used in most technology classes. Fig. 4-9.

Don't print blocks of text in color. Such text can be difficult to read.

PLANNING A NEWSLETTER

Newsletters are among the most common products of desktop publishing. Companies use newsletters to provide information to employees and to promote their services and products. You might use a newsletter to inform the school community about activities in your technology class.

Some important decisions should be made before starting a newsletter. If the newsletter is a class project, brainstorming and the problem-solving process can be used to answer some important questions. Consider the following:

• Who is the intended audience? Will the newsletter be read by students, teachers, and/or the community?

• What kinds of articles will interest them?

• What would be a good name for the newsletter?

• How often will it be published?

• What will the page size be?

• How many pages will it have?

• How will it be reproduced?

Thousands of newsletters are produced by individuals and organizations each week. Although they may focus on different topics, most newsletters have a number of things in common.

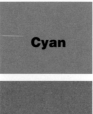

Cyan

Magenta

Yellow

Black

▶ **Fig. 4-9** The four blocks of color on the left show the four process colors. Nearly any color can be made by combining cyan (process blue), magenta (process red), and yellow. Black is added to sharpen detail. In color printing, the process colors are combined to produce all the colors in this photo.

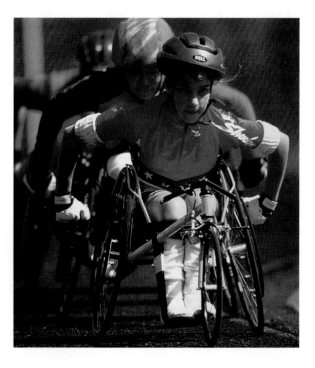

Design Guidelines

Gather a collection of newsletters. Identify the elements they have in common. Try to identify why some newsletters are more attractive than others. What is responsible for this? Below are a few things to consider.

- Use only two typefaces. Select a serif typeface for the body text. For occasional emphasis, use the same typeface in bold. For headings and headlines, use a sans serif typeface.
- Include white space on every page to avoid a cluttered look.
- Print with black ink. If you want to add color, consider using black ink on paper of an interesting color. Color paper is less expensive than color ink.
- Select drawings and photographs carefully. Many beginners use too many graphics. Place them in several locations before making a final decision.
- Experiment with margin size and column width. On 8 1/2" x 11" paper, most readers prefer two columns.
- Maintain a consistent look throughout the document. This means that the design of page 2 and the design of page 6 should be similar.
- Seek feedback from readers. This will allow you to accept credit for successes and to plan improvements.
- Review past issues to plan future newsletters.

IMPACTS

Typesetting is inexpensive in desktop publishing. This has had a great impact on the typesetting and publishing industries. Desktop publishing does not require the purchase of expensive equipment. If the publisher is also the typesetter, changes and corrections can be made easily at no charge. Publications can be printed in very small quantities. This makes it possible to publish items that might not otherwise have been printed.

Desktop publishing has also given the small publisher more control over expenses.

THE FUTURE

Printing has been the key method of transmitting information. It is just as important in the Information Age. Desktop publishing is a very effective printing tool. It can be quickly learned. It provides an efficient way to provide information. Desktop publishing will prompt further growth in the communications industry. It will continue to provide a means for people to express their ideas in an inexpensive and attractive format.

Linking to SCIENCE

Color Vision. Color pictures and graphics are made by using four ink colors: black, yellow, cyan (sigh-ANN), which has a blue-green color, and magenta (ma-GENT-uh), which has a purple-red color. Tiny dots of these inks form areas of color on the printed page. A pale green, for example, is created by cyan and yellow dots, with some uninked (white) spaces. The human eye cannot resolve such tiny dots of different colors. It sees them instead as large areas of a single color.

Using a magnifying glass, observe the colors on the color comics page of a newspaper. What colors of ink are used to create red, brown, orange, and blue?

Apply What You've Learned

Design and Build a Newsletter

State the Problem

A newsletter is a publication containing items of interest to a particular group of people. Your newsletter should include the following:

- 4 pages of text and graphics
- At least two font styles
- Bold, italic, and regular typefaces
- One photograph (from a digital camera, or output from a scanner)
- One or more electronic clip-art images
- One or more original art pieces (drawn using the computer, such as with a paint program)
- One or more images surrounded by text wrap formatting
- Two or three columns of text on a page
- Table of contents on the first page
- Attractive nameplate (title)
- Use of color (if possible)

The organization of your publishing teams will be indicated by your instructor.

Collect Materials and Equipment

computer system with desktop publishing software
digital camera
scanner

Develop Alternative Solutions

Identify the information you plan to present in your newsletter. Determine how long you would like each article to be. Identify possible illustrations (including photographs, clip art, and electronic art). Prepare rough sketches showing the placement of text and art on each page.

Select the Best Solution

Compare your sketches. Select the one that you think is the most effective layout for your newsletter.

Implement the Solution

1. If you are working in a team, make sure that responsibility has been assigned for each part of the newsletter. Each person must know exactly what he or she is expected to accomplish.
2. Use the desktop publishing and drawing programs to compose or illustrate the part of the newsletter for which you are responsible.
3. Print the newsletter.
4. Review the copy.
5. Make any needed changes. Print the newsletter again.
6. Assemble all pages of the newsletter.
7. Submit the newsletter for evaluation.

Evaluate the Solution

1. Does the newsletter meet all specifications?
2. Do the four pages seem to blend in well with each other?
3. Is the newsletter easy to read?
4. Is the newsletter pleasing to look at?

CAREERS IN
Desktop Publishing

⁂ GRAPHIC DESIGNER ⁂

Immediate opening for graphic designer at leading manufacturer of plumbing products. Applicants must have a bachelor's degree in Graphic Design and three years of experience. Responsibilities include high degree of technical illustration, design and layout of installation manuals and parts catalogs. Send resume to: Miller Incorporated, 25300 Miller Parkway, Evanston, IL 60023.

DESKTOP PUBLISHER

Commercial printer is expanding its state-of-the-art desktop publishing department. Must have thorough computer applications knowledge. Understanding of offset printing process and electronic prepress experience helpful. Growth-oriented company offers full benefits and salary. Call us at (702) 982-8798 or send resume to: Image Graphics, P.O. Box 800, Louisville, KY 82302.

ELECTRONIC PAGINATION SYSTEM OPERATOR

Busy printer in Southville needs capable individual with knowledge of printing industry and electronic page layout system experience. Will set up and transmit pages for production directly to plates. Send resume to: Superior Printing, 33001 Station Street, Southville, RI 33208.

BOOK DESIGN ASSISTANT

Major publishing company has opening for an entry-level assistant skilled in the technology of book design. Knowledge of printer capabilities and design, as well as computer programs and equipment needed. A bachelor's degree is required. Excellent opportunity with salary and benefits. Send resume to: Literary Publishers, 2323 Grand Street, New York, NY 10020.

WORD PROCESSING SPECIALIST

Large publishing company seeks full-time word processing specialist for evening shift. Understanding of variety of word processing programs and typing speed of 70+ words per minute required. Must be able to handle multiple tasks with attention to detail and accuracy. Professional work environment with excellent benefits. Please send resume to: Human Resources, P.O. Box 144041, Syracuse, NY 14411.

ELECTRONIC PREPRESS SYSTEM OPERATOR

Highly proficient in computer use with working knowledge of line screens, screen angles, high resolution image setter output and process color separations. System troubleshooting required. Salary and benefits for the right candidate. Send resume to: The Graphic Shop, Box BA978, New Orleans, LA 44565.

Linking to the WORKPLACE

Select one of the desktop publishing careers listed above. Pretend that you have a job interview for that position. Identify and write down questions that you would ask the employer about the job.

Make a list of questions that you think the employer would ask you. How would you answer? How would you dress for your interview?

Chapter 4 Review

SUMMARY

▶ Desktop publishing is the use of a computer and special software to produce publishable documents.

▶ Page layout software combines text and graphics on a page. Other programs used in desktop publishing include drawing, image editing, and clip-art software.

▶ Scanners and digital cameras can be used to place photographs into desktop-published documents.

▶ Important things to consider when designing documents include type size, type style, line length, and the use of graphics.

▶ Graphics fall into two categories: drawings and photographs.

▶ Some of the decisions that must be made when planning a publication include who will read it, what kinds of articles will interest the readers, how long it should be, and how often it will be published.

CHECK YOUR FACTS

1. Define desktop publishing. Name three different kinds of documents that it can be used to produce.

2. Identify the hardware needed in desktop publishing.

3. What is a typeface?

4. How is type measured?

5. Identify the three main categories of typefaces.

6. What kinds of graphics can be used in desktop publishing? Explain two different ways of obtaining them.

7. Name three types of scanners. How are they used?

8. What factors must be considered when designing a document?

CRITICAL THINKING

1. Explain the difference between word processing and desktop publishing.

2. Describe three different kinds of software that can be used with a page layout program.

3. Identify five things that should be considered when planning a newsletter. Why is it important to think about these things before beginning the newsletter?

4. Make a sketch to show how you would like the first page of a class newsletter to look.

CHAPTER 5

Computer Animation

OBJECTIVES

▶ discuss the uses of animation.

▶ identify the information included on a storyboard.

▶ compare the three types of animation.

▶ list the six steps in three-dimensional computer animation.

KEY TERMS

animation

key frame

model animation

persistence of vision

primitives

storyboard

What have you done for entertainment in the past week? Did you watch an animated video? Did you play a computer game? Did you read the comic section in your local newspaper?

Many of the things you do for entertainment did not exist when your grandparents were your age. Not everything has changed, though. Some things, such as reading the comic section, have been popular for generations.

Did you know that many early cartoons were created by animating popular newspaper comic strips? In the 1940s and 1950s cartoons were usually shown before the main feature in most movie theaters.

Cartoons, video games, and many popular television commercials are created using animation. In this chapter, you will learn about the animation process. You will also learn about different kinds of animation.

WHAT IS ANIMATION?

Animation is the creation of the illusion of movement in a series of still images. For many years, animation has been used to produce cartoons and films. Today, it is also used to produce special effects in commercials, live-action films, and video games.

Animation was first used in the 1800s in toys. One toy, the *zoetrope* (ZO-uh-trope), used a cylinder and a long strip of paper with a series of images. When the cylinder was rotated, the images could be viewed through slits in the cylinder. The images appeared to move. The zoetrope and similar toys were important. They led to the development of motion pictures.

Another simple animation device is a *flip-book*. A flip-book is easy to make. It consists of a sequence of drawings placed on top of each other and fastened together along one edge. Each drawing is slightly different from the drawing before. When the pages are flipped rapidly, the image appears to move. Fig. 5-1.

▶ **Fig. 5-1** Flip-books provide a simple way to create motion from a series of pictures.

Linking to SCIENCE

Experiencing Persistence of Vision. You can recognize separate images if they are viewed at twelve (or fewer) per second. If you see more images per second, they begin to merge into a single image. You can demonstrate this effect, called *persistence of vision*.

Place a coin on an index card. Draw a circle around the coin. Cut out the circle. Draw a bird on one side and a cage on the other. Tape the circle to the eraser of a pencil so the circle sticks up above the pencil. Twirl the pencil slowly between your palms, so you can see first the bird, then the cage. Twirl the pencil faster. The two images will become one. You will see the bird inside the cage.

▲

The image appears to move because the eye sends signals to the brain faster than the brain can process them. Before the brain has finished processing one image, it receives another. This blending of individual images into one image that seems to move is called **persistence of vision**. Television programs and movies create the illusion of movement by showing about 30 still pictures per second.

FASCINATING FACTS

The Disney cartoon character Mickey Mouse was originally named Mortimer Mouse. Walt Disney named the character after his pet mouse. Mickey Mouse starred in *Steamboat Willie*. Released in 1928, this was the first animated cartoon with matching sound.

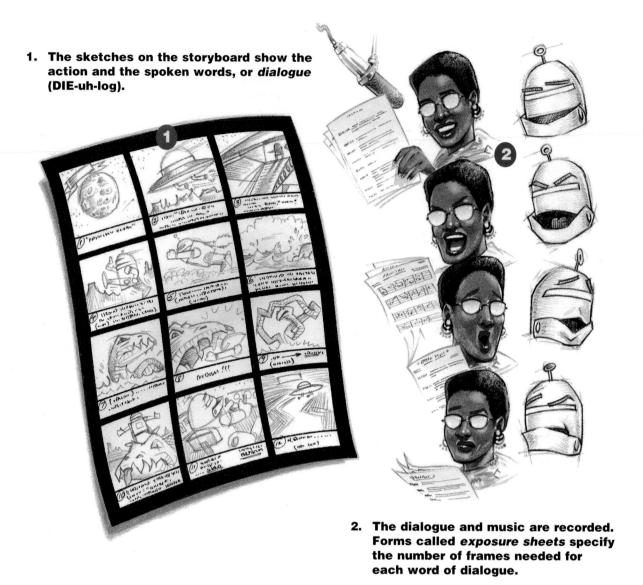

1. The sketches on the storyboard show the action and the spoken words, or *dialogue* (DIE-uh-log).

2. The dialogue and music are recorded. Forms called *exposure sheets* specify the number of frames needed for each word of dialogue.

▶ **Fig. 5-2** A storyboard is a series of sketches that can be used as a guide for making a film. This shows how a storyboard is used.

TYPES OF ANIMATION

Modern animation techniques can be divided into three basic types:
- hand-drawn animation.
- model animation.
- computer animation.

Hand-Drawn Animation

In hand-drawn animation, a series of drawings are photographed. Each drawing makes up one frame of the film. Fig. 5-2. The position of the character or object changes very slightly from frame to frame. This technique is called *cel animation*

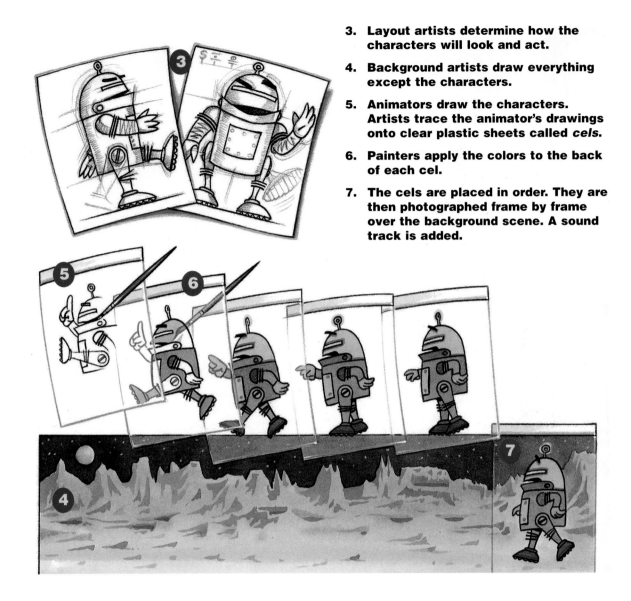

3. **Layout artists determine how the characters will look and act.**

4. **Background artists draw everything except the characters.**

5. **Animators draw the characters. Artists trace the animator's drawings onto clear plastic sheets called *cels*.**

6. **Painters apply the colors to the back of each cel.**

7. **The cels are placed in order. They are then photographed frame by frame over the background scene. A sound track is added.**

because the first animated characters were drawn on thin plastic sheets called *celluloid*. Cel animation has been used to create many well-known cartoon and feature films.

In 1937, Walt Disney released *Snow White and the Seven Dwarfs*. This was the first full-length animated film produced using cel animation. Over the years, the cel animation process has been improved. Many of the techniques used in cel animation are now used in model and computer animation.

Explore

Design and Build a Flip-Book

State the Problem

Design and build a flip-book. This is a book composed of pages that have a series of pictures. Each picture is slightly different from the picture on the page before and the page after. When the pages are flipped the image appears to be moving, as if it were animated.

Develop Alternative Solutions

Prepare designs for the flip-book. You will need to decide on page size. You will also need to decide on an image.

Select the Best Solution

Select the page size and image that you think will provide the most effective solution.

Implement the Solution

1. Prepare the pictures. Remember that each picture must be slightly different from the one preceding it. For example, to show an analog clock indicating the passage of time, you would draw several pictures showing the minute hand as having moved from one number to the next in each picture. (Don't forget that the hour hand also moves slightly.)
2. Staple or glue the pages together. You may want to put the book on a stiff backing, such as cardboard.

Evaluate the Solution

1. Do the images appear to move smoothly when the pages are flipped?
2. Are the pages of the flip-book fastened tightly enough to allow repeated use?

Collect Materials and Equipment

Flip-book pictures can be drawn by hand or by computer. It is relatively easy to draw a starting image on a computer. You can save the image as a separate slide and then modify it slightly, save it again, and so on. The computer software allows you to alter each slide without redrawing the entire image.

The pages of the flip-book may be stapled together along one edge. If the book is too thick to be stapled, padding compound may be brushed along one edge. (Padding compound is a substance used to make pads.)

Model Animation

Model animation is animation that involves the use of three-dimensional figures called *puppets*. This animation technique has been used to create many short films and commercials. It has also been used in major films, including *Star Wars*. When the models are made of clay, the process is called *clay animation*.

During model animation, the puppets are photographed on a set one frame at a time. After each photograph, the models are adjusted. Fig. 5-3. Then another photograph is taken. When the frames are played back rapidly, the models appear to move.

Computer Animation

Computers have many possible uses in animation. For example, they can speed up the traditional cel animation process.

Computer animation uses animation software to create an animated scene, a cartoon, or even a full-length film.

Another kind of computer animation can be used to make three-dimensional graphics and interactive games. In interactive cartoons and games, the user has control over some of the action taking place.

Computers can greatly reduce the time needed to produce an animated feature. Several skilled computer artists can replace the dozens of animators needed to draw and color the individual frames. Some artists prefer to begin by drawing the characters on paper. The drawings are then scanned into the computer. Other animators use the computer both for the initial drawings and for the entire animation process.

Skills in layout, design, and timing are important in traditional animation. These

▶ **Fig. 5-3A** In clay animation, lip-sync and movement depend on a series of small adjustments.

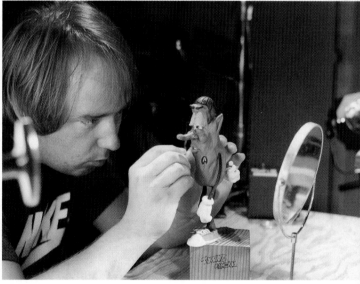

▶ **Fig. 5-3B** An animator makes an adjustment to a character. The wooden block that holds the character is fixed to an L-shaped bracket bolted to the set. This allows the character to be removed, adjusted, and then replaced in exactly the same position on the set.

same skills are also important in computer animation. Animators with traditional experience can learn to use the computer as an animation tool.

Computer animation programs vary in price and quality. Several good programs are available for use with personal computers. These will allow you to learn the basic points of animation. Professional quality animation requires expensive software and fast computers with a lot of memory.

Two-Dimensional (2D) Animation

Two-dimensional (2D) animation programs eliminate the need for tracing and coloring. The original character drawings can be made by hand or directly on the computer. Computers can store the drawings electronically. This makes it easy to change the order of their use. The computer can also be used to add the sound track.

Software programs called paint programs are used to add color to the drawings. Even inexpensive paint programs can provide hundreds of different colors. After choosing the color, the artist moves the cursor to the desired area on the object or character. The artist adds shading and texture in a similar way.

Two-dimensional computer animation programs have other advantages. One major advantage is their ability to create "in-between" movements. For example, assume that a character is to be shown getting up from a chair. The character can be drawn in the sitting and standing positions. Using the computer software, the artist can then create the frames in between to show the movement from sitting to standing. However, all the artist has to do is specify how many intermediate (in-between) frames should be created. The software then creates these frames. Computers also make it easy to change the time required to perform a particular action.

Three-Dimensional (3D) Animation

Three-dimensional (3D) animation adds realism and excitement. It is used in films, commercials, games, and architectural designs. The same program, in the hands of someone with artistic and computer skills, can be used to create scary alien creatures. It can be used to create full-length feature films. It can also be used to take a customer on a tour of a building that does not yet exist.

Linking to MATHEMATICS

Comparing Costs. Assume that a state-of-the-art computer system costs $30,000. One computer system is needed for each skilled computer artist. A company employs 14 skilled computer artists (each with an annual salary of $40,000). Assume that these computer artists are employed in place of 112 animators (each with an annual salary of $30,000). What cost savings would result?

To answer this question, you will need to consider the following:
1. What is the annual salary expense for the 14 computer artists?
2. What is the total expense for a state-of-the-art computer for each computer artist?
3. What is the annual salary expense for 112 animators?
4. Is there a cost savings? If so, how much?

Feature Film Production

Producing a feature film using three-dimensional computer animation involves six steps:

- storyboards.
- modeling.
- animation.
- shading.
- lighting.
- rendering.

Storyboards

A **storyboard** is a series of sketches that can be used as a guide for making a film. This was shown on pages 86 and 87. A typical full-length animated feature requires more than 4,000 storyboard drawings to describe the action and dialogue of the film. Storyboards can be fairly simple or very detailed. Fig. 5-4.

Modeling

Modeling refers to the use of computer software to create 3D computer models of characters, props, and sets. The process usually begins with computer-generated solid objects such as balls, cubes, cylinders, and cones. These basic geometric shapes are called **primitives**.

New shapes are created by changing the sizes of the primitives and combining them to form the desired objects. Drawing tools similar to those used in computer-aided drafting (CAD) programs are also used to draw free-form shapes.

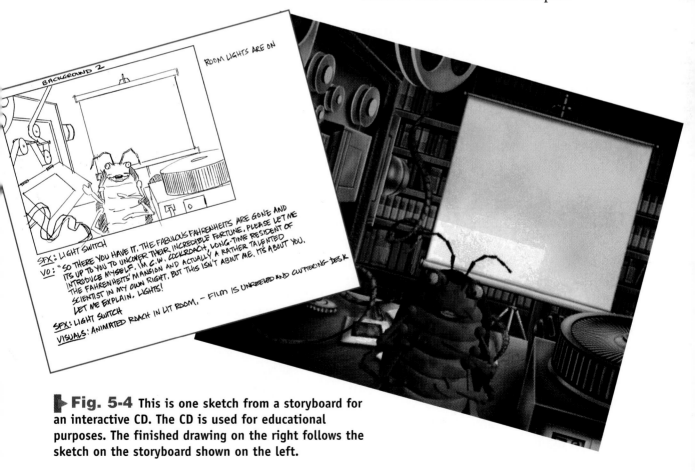

▶ **Fig. 5-4** This is one sketch from a storyboard for an interactive CD. The CD is used for educational purposes. The finished drawing on the right follows the sketch on the storyboard shown on the left.

Explore

Design and Build a Thaumatrope

State the Problem
Design and build a *thaumatrope* (THOM-uh-trope), also known as a "Wonder Turner." This device, popular as a toy in the early 1800s, demonstrates persistence of vision.

Develop Alternative Solutions
Prepare several designs for the thaumatrope.

Select the Best Solution
Select the design that you think will be most effective.

Implement the Solution
1. Create a set of two images that are a "matched set." Examples might be a bird and a birdcage or a dog walking into a doghouse. Fig. A. Other examples are pictures of a plant and a vase or a pencil and some handwritten text. The pictures may be generated by computer, drawn by hand, or clipped from magazines. When "optically merged," the two pictures will appear to blend into one. It is important that the pictures relate to one another.
2. Attach the two pictures back to back. The methods for doing this will vary. Fig. B.
3. Spin the pictures and watch the two images merge. You might wish to experiment with the rotation speed to identify the speed at which the images appear to blend.

Evaluate the Solution
1. Does the device spin easily?
2. Are the pictures aligned properly, so that the resulting image is a blend of the two different images?
3. Is the mechanism made well enough to be used again and again?

Collect Materials and Equipment

Materials for this project can be selected by the designer. The backing for the pictures may be paper, cardboard, plastic, or any other suitable material. The pictures will be attached to string or thin dowels. Your instructor will indicate what materials are available.

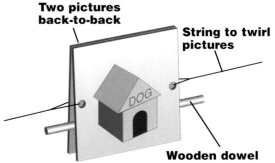

Two pictures back-to-back

String to twirl pictures

Wooden dowel

Fig. A

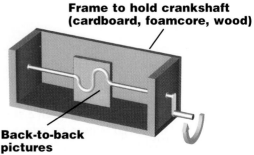

Frame to hold crankshaft (cardboard, foamcore, wood)

Back-to-back pictures

Fig. B

Animation

A **key frame** is a frame that shows a beginning or ending point in an action sequence. The 3D models are animated by drawing key frames. The computer then generates the "in-between" frames needed to simulate motion. The process is similar to that used for 2D frames, except that it involves 3D images.

Imagine a scene that shows two characters playing catch with a ball. The key frames might show each character either throwing or catching the ball. The frames in between determine how fast the ball is thrown, how many times the ball goes back and forth, and how long the game continues. The action shown in these frames can be controlled by the

▶ **Fig. 5-5** The popularity of animated films has spurred the development of new animation techniques.

computer software. Some animation programs can even show how the ball changes shape when it bounces. The computer has been a powerful tool for animators. Fig. 5-5.

Shading, Lighting, and Rendering

Shading programs add colors and textures to objects. Unlimited colors are possible. A variety of materials can be simulated, including glass, wood, and metal.

Lighting is used to "light up" the scene in each frame. Such lighting is similar to stage lighting.

Rendering software combines computer information from the modeling, animation, shading, and lighting steps to create the final images. The final images are then transferred to videotape, film, or CD-ROM.

Linking to COMMUNICATION

Writing Dialogue. *Dialogue* (DIE-uh-log) is conversation. In comic strips, dialogue is printed in a "balloon." In writing, dialogue is indicated by paragraphing, quotation marks, capitalization, commas, and end punctuation. Within your study group, design and draw a short comic strip that demonstrates an event in your technology class. Draw stick figures for your characters and "balloons" to contain their dialogue. Prepare a written copy of the dialogue in standard composition form, using paragraphing and quotation marks.

A *synonym* (SIN-oh-nim) is a word that has a meaning that is the same as or similar to the meaning of another word. Try using synonyms for "said." Possible synonyms include words such as asked, questioned, murmured, whispered, warned, and yelled. Can you think of others?

▲

Apply What You've Learned

Design and Build a Zoetrope

Collect Materials and Equipment

paper
markers
pencils
cardboard
wood
tape
scissors
glue

State the Problem

Design and build a zoetrope, one of the earliest devices to suggest moving images by using still pictures. Your teacher may divide the class into teams.

Develop Alternative Solutions

Decide on the size of the zoetrope. Prepare several design sketches. Your designs should also specify the images that you will be using.

Select the Best Solution

Select the design that you think will be the most effective.

Implement the Solution

1. Establish the length of the strip of pictures. The length will be determined by the circumference of the zoetrope wheel.

2. Draw the pictures using markers or pencils. You might also use a computer software paint program. The images on the picture strip should show at least one complete cycle of motion. In the example of a jumping jack, the picture strip would show the figure in several stages of jumping. The number of images should match the number of slots in the wheel.

3. Prepare the zoetrope wheel, using a strip of cardboard, oak tag, or a similar material. The wheel should have a height of 2" to 6", with vertical slots cut along its length. The slots should be about 1/4" in width. Each slot should be about two-thirds the height of the wheel. The space between slots should be approximately 2". The number of slots should match the number of images on the picture strip.

4. Attach the strip to a base that can be spun on a sturdy hand-held shaft.

5. Attach the picture strip to the inside of the slotted ring with the pictures facing inside. Spin the wheel. The images, when viewed through any one slot, should appear to move.

Evaluate the Solution

1. When the wheel is spun, the images on the zoetrope should present the illusion of motion. If they do not, what can be done to make the zoetrope work correctly?

2. What are the effects of using too few pictures on the picture strip?

3. Why must the wheel spin within a certain range of speed to function properly?

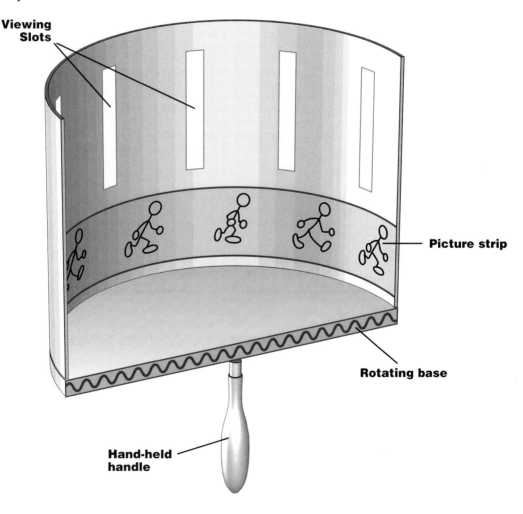

Viewing Slots

Picture strip

Rotating base

Hand-held handle

CAREERS IN
Computer Animation

ANIMATOR

Imagination combined with strong artistic and technical skills required for animator position at computer game development company. Formal art training required. Finalists will be required to take drawing tests. Must be able to work well under tight deadlines. Apply to: National Computer Games, 100 East Samson Avenue, Suite 34, Austin, TX 70033.

COMPUTER PROGRAMMER

Opportunities for computer programmers who enjoy state-of-the-art technology and new application development. Two years of programming experience preferred. Bachelor's degree in computer science or related field and good communication skills required. Competitive salary and benefits package. Send resume to: Software Solutions, Inc., 6090 Front Street, San Francisco, CA 90022.

SOFTWARE WRITER

Multimedia software publishing company has an opening for a software writer to write code in COBOL or C language. Work as part of a team to create applications. Bachelor's degree in computer science or management information systems preferred. Must be logical thinker and be able to set goals. Competitive salary. Send resume to: Media Publishers, 6604 Strickland Street, Minneapolis, MN 40502.

SOFTWARE INTEGRATION ENGINEER

Software consulting firm seeks integration engineer to work with clients in the field. Will be required to write code that will act as interface between software programs. Must have experience in writing code and be familiar with a variety of software languages and programs, including animation software. Excellent working environment. Apply to: Fisher Software Consultants, 55 West Eagle Drive, Atlanta, GA 30032.

CD-ROM PRODUCER

Major publishing house needs a CD-ROM producer. Will be involved in all facets of production. Must create concepts and work with editors, authors, designers, and programmers. Experience as a production assistant helpful. Degree and experience in animation, programming, or graphic arts required. Submit resume to: Lane Street Publishers, Inc., 344 East 95th Street, New York, NY 02205.

Linking to the WORKPLACE

There are many routes to career success, many paths from school to work. Taking the right courses in school can start you in the direction of your future career. Think about each of the jobs listed above. Write down three classes that you think would be helpful and/or required in preparing for each of the jobs presented here.

Chapter 5 Review

SUMMARY

▶ Animation is the creation of illusion by assembling a sequence of still images.

▶ Modern animation includes hand-drawn animation, model animation, and computer animation.

▶ Storyboards are used as "blueprints" to guide the action and dialogue in each scene of animated films.

▶ Model animation involves the use of puppets that are photographed on a set, one frame at a time.

▶ Computer animation programs can reduce the time needed to produce an animated film. They can be used for every step of the animation process.

▶ The steps in producing a feature film using three-dimensional (3D) computer animation are: storyboards, modeling, animation, shading, lighting, and rendering.

CHECK YOUR FACTS

1. Identify several uses for animation.
2. Explain persistence of vision.
3. What information do storyboards include?
4. Compare the three different types of animation techniques.
5. What is claymation?
6. List the six steps in three-dimensional computer animation.
7. What is a cel? How is a completed cel used?
8. What is a primitive?
9. What is a key frame?
10. What is the purpose of shading in animation?

CRITICAL THINKING

1. Make a series of drawings that show action in a sport. Use the drawings to create an animated flip-book.
2. Why is a storyboard important in preparing all types of animation?
3. Describe how an artist can use a computer animation program to assist in traditional animation.
4. Describe a technology class activity that you think could be clearly explained through computer animation.

Internet

OBJECTIVES

▶ identify what is needed to access the Internet.

▶ explain Internet safety and etiquette.

▶ explain the purpose of a hyperlink.

▶ explain the parts of an e-mail address.

Imagine sitting at your desk at home and being able to buy a pair of shoes, get help with your homework, or view paintings in a museum thousands of miles away.

Today, the Internet can be used for these activities and for many more. People are finding new uses for the Internet. Many schools now have home pages. In this chapter, you will learn how the Internet works, how it can help you, and how it affects your life.

KEY TERMS

communication

computer network

domain

e-mail

hyperlinks

hypertext

modem

Uniform Resource Locator (URL)

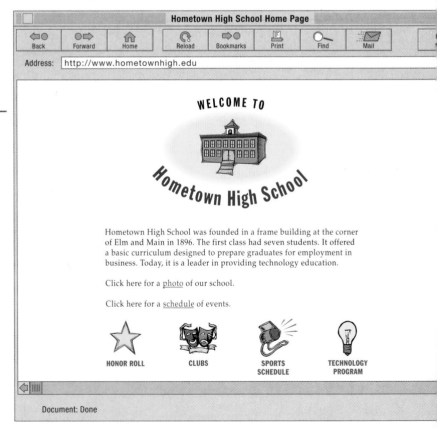

WHAT IS THE INTERNET?

Communication is the process of giving or exchanging information. The telephone, the television, and the United States Postal Service are all familiar communication systems. The Internet is a fairly new communication system. To some, it is a community—a place to visit every day. To others, it is a tool for obtaining information without going to the library.

The Internet is the world's largest computer network. A **computer network** is a communication system created by connecting many computers. Networks, including the Internet, allow two-way communication between any two computers on the network. The Internet is made up of networks established by private companies, universities, and government agencies. In reality, the Internet is a supernetwork that connects many individual networks into a huge system.

BUILDING THE INTERNET

In the late 1960s, the United States Department of Defense set up an experimental computer network to connect government agencies with universities and private companies that were doing military research. This network grew rapidly. It included an important feature that rerouted messages if one part of the network failed. Fig. 6-1.

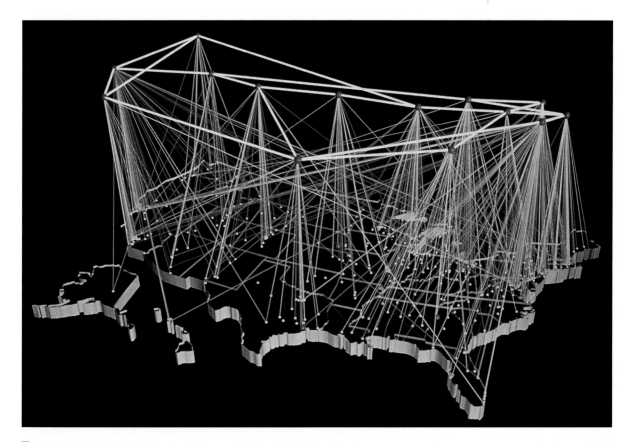

▶ **Fig. 6-1** The volume of traffic over the computer network of one agency of the federal government. The various colors represent varying amounts of computer traffic.

By the 1980s, universities throughout the United States were connecting their computers and sharing information over networks. When the National Science Foundation set up a larger network, electronic communication among universities increased. Soon commercial firms connected their networks. As this trend continued, the Internet, an international network of networks, was formed.

GETTING ON THE INTERNET

To use the Internet, you need a computer, a modem, a telephone line, communication software, and an Internet service provider.

Hardware and Software

Computer technology is rapidly changing. Efficient use of the Internet requires a fast and powerful computer. A fast modem is also needed. Fig. 6-2.

A **modem** (*modulator/demodulator*) changes the signals from your computer into signals that can be transmitted over telephone lines.

A modem installed inside the computer is called an *internal modem*. A modem in a separate outside box connected to the computer is called an *external modem*.

Modem speed is important. Early modems, which operated at 2,400 bits per second (bps), were once considered very fast. Modems today transfer information at much faster speeds.

▶ **Fig. 6-2** Computer technology changes quickly. The speed of the modem you use helps determine the speed of Internet access time.

In addition to the computer and a modem, *communication software* is required. This software connects the computer to the Internet. Many computers are sold with communication software installed. Software packages designed for the Internet are also available. The large on-line services supply at no charge the software needed to use their systems.

Internet Service Providers

In addition to large on-line services, thousands of smaller companies connect you to the Internet. These companies are called *Internet service providers (ISPs)*. Some ISPs charge by the hour. Others provide unlimited access to the Internet for a set monthly fee.

When choosing a service provider, consider several things. Be sure the provider has a telephone number in your area code (*local access*). With such a number, you will not pay long-distance telephone charges while you are on the Internet. Choose a provider that uses the latest communication software and offers free technical help when you have a problem. Compare the services offered by several companies. Try to speak with a few of their customers to find out if the provider's service is reliable.

INTERNET SAFETY AND APPROPRIATE USE

The Internet contains much valuable information, but it also has material that is offensive or inappropriate. This is why most schools have an *acceptable use policy*, which describes responsible use of the Internet in school.

Using the Internet Safely

Be safety conscious. Don't give out your full name, address, or phone number to people you chat with online. This is important even if you have chatted with a person several times and feel as if you "know" the person. If you receive a message over the Internet that makes you feel uncomfortable, tell your teacher immediately.

Netiquette

Being polite on the Internet is sometimes referred to as practicing *netiquette* (Internet etiquette). During conversations, typing in all capital letters is considered to be shouting. Remember always to use appropriate language.

Be patient. Sometimes the Internet may seem slow because many people are using it. Also, you may have to try several times to reach a particular site because that site is busy.

Explore

Design and Build an Internet Report

State the Problem

Use the Internet as a research tool to prepare a report on the development of the Internet and to explain to others how you prepared your report.

Develop Alternative Solutions

Discuss techniques used to research a topic on the Internet.

Select the Best Solution

Select the research technique that you think will be most effective.

Implement the Solution

1. Using the Internet, research the development of the Internet.
2. Use at least two search engines for your research.
3. Use at least three sources for your research.
4. Prepare two written reports. The first report should outline the results of your research (the actual content). The second report should detail how you carried out your research (the process).

Evaluate the Solution

1. Present your report on the process to several classmates. Ask what they liked about your report. Ask if they have any suggestions for improving it.
2. Give your report on the development of the Internet to a classmate. Ask your classmate to reword your report. Did your classmate find the material easy to understand?

Expressing Emotions

Some of the terms listed above help people express their feelings. Some people also try to express their feelings over the Internet using "faces" made up of two or three characters. For example, :-) is a "smiley face" that indicates a happy mood, and :-(indicates a sad or unhappy mood. (Look at the symbols sideways to see the faces.)

WORLD WIDE WEB

The *World Wide Web* ("the Web") is the part of the Internet that allows users to connect to computers all over the world. The Web is made up of millions of pages. Each page can include text and pictures. It might also contain sound and video.

A *web site* is an electronic home for the publication and collection of information. A web site can be any length. The first page of a web site is called the *home page*.

Web pages can be linked by the use of hypertext. **Hypertext** is text that provides a link to another web page. In other words, hypertext is not strictly "text." Hypertext contains hyperlinks.

Hyperlinks are links attached to text, buttons, or graphics. The mouse arrow will turn into a hand when it is placed on a hyperlink. When you click on a hyperlink, it takes you to another web page. The web page might contain text, graphics, or sound.

The World Wide Web uses a special address system called a **Uniform Resource Locator (URL).** Every Web site has its own URL. You have probably seen many of these. For example, the URL of the McGraw-Hill Companies is www.mcgraw-hill.com.

Text links are highlighted in a different color and are usually underlined. Many graphics also function as links. If you are not interested in the page a link takes you to, you can return to the previous page by clicking on "Back."

FASCINATING FACTS

The telephone company has added new area codes. Now new TLDs (top-level domains) are needed on the Internet. The planned TLDs are .firm (for firms), .shop or .store (for retailers), .web (for web activities), .arts (culture), .rec (recreation), .info (information services), and .nom (individual).

Web Browsers

Web browsers are the software packages that allow you to use the Internet for different kinds of communication. Web browsers allow you to use all of the features of the Internet. Fig. 6-3.

Electronic Mail

Messages sent over the Internet are known as **e-mail**. This is one of the most popular uses of the Internet.

Anyone who wants to send and receive electronic mail needs an e-mail address. An Internet e-mail address has three parts. Consider the address jdoe@tech.org. The first two parts are separated by the "@" symbol. The part before the @ is the *mailbox*. Usually this is the user's name (in this case jdoe). The part after the @ but before the period, or dot (.) is the domain. The **domain** is the name of the Internet provider. The third part, after the dot, has three letters that identify the *zone*. The zones used in the United States include:

.com	(commercial firms and on-line services)
.gov	(government offices)
.edu	(educational organizations, usually schools and colleges)
.mil	(military site)
.net	(networks)
.org	(nonprofit organizations)

To send e-mail, choose the MAIL feature on your browser. Click on "new mail." This will open a window with boxes in which you enter the address of the person who will receive your message. You can also enter the subject of your message. Type your message. When you are done, click "send" to send your message.

When you receive e-mail, you can decide what to do with the message. You can read it and then delete it, or you can file it for later use. If you want a permanent record, you can print the message. You can also reply to the message or forward it to someone else. Fig. 6-4.

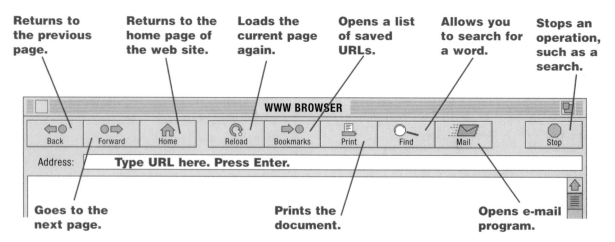

Fig. 6-3 The tool bar of a web browser screen.

> **Fig. 6-4** The tool bar of an e-mail screen.

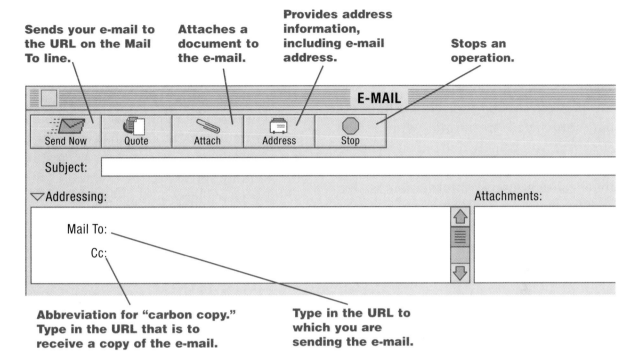

Sends your e-mail to the URL on the Mail To line.

Attaches a document to the e-mail.

Provides address information, including e-mail address.

Stops an operation.

Abbreviation for "carbon copy." Type in the URL that is to receive a copy of the e-mail.

Type in the URL to which you are sending the e-mail.

Searching the World Wide Web

The Internet can be used to gather information on almost any topic. The tools required to do this are called *search engines*.

Search engines do not actually search the Internet. Instead, they check a catalog of web sites maintained by a company. These companies continually update their information.

To access a search engine, start your web browser and enter the browser's URL. You can then begin a search by clicking on one of the subjects (words in bold type) or entering one or more keywords in the space provided. A *keyword* is any word or combination of words that you think will help the search engine find the information you need.

If the first keyword does not produce the results you want, try to think of another term. If the search engine finds too many sites, be more specific by entering several words. For example, if you choose "pets" as a keyword, the search engine will probably identify thousands of sites. If you are really interested in finding out if a particular breed of dog will make a good pet, use the name of the breed as the

Linking to COMMUNICATION

New Words. Language changes because of new discoveries and inventions. The use of computers and the Internet has created a need for a new vocabulary. Some call this new vocabulary "computerese." The word *netiquette* is one example. Interview your parents and some of your teachers to determine a list of new words or phrases generated by the electronic age. Compare your list with the lists of others in your study group. Write a definition for each of your terms.

keyword. This will narrow your search. Fig. 6-5.

Most search engines place the most important or most likely matches at the top of the list of retrieved sites. If your search produces thousands of sites, but the first ones listed look interesting, you may not need to refine your search. However, you usually don't need to bother to look at the last sites. The chance that they contain useful information is very small.

If you are not successful in finding the information you want, try one or more additional search engines.

Chat

Internet relay chat (IRC) is a real-time system that allows people to talk to each other by typing their messages. During chat sessions most people use nicknames, rather than their real names. Each "meeting room," or "channel," is devoted to a particular topic. To participate in a chat group, you may need to install additional software. However, the software is available free on the Internet.

After you install the required software, you can connect to a chat group. Spend a few minutes observing the conversation before typing your message. Remember to

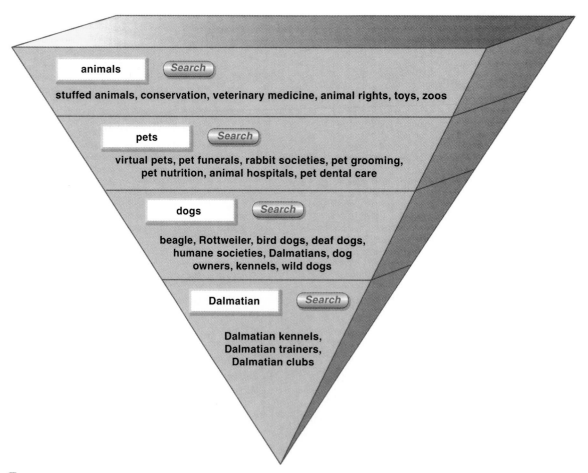

▶ **Fig. 6-5** By narrowing your search, you retrieve more specific results.

apply the appropriate use guidelines covered earlier in this chapter.

Newsgroups

Newsgroups are bulletin boards where messages are posted electronically. Anyone can read and respond to the messages. After a set number of days, the messages are removed to make room for new messages. Thousands of newsgroups handle millions of messages each day.

Most web browsers have a button that lets you search for newsgroups by subject. Once you find a newsgroup of interest, you can just read what others have said or you can reply. After you have read an article, it is removed from the list of articles you see. When you visit the site again, it will contain only articles you have not read. Fig. 6-6.

Most newsgroups have an FAQ (frequently asked questions) section. The FAQ area provides information about the site that can help you decide if you are interested in using it.

Mailing Lists

Mailing lists are discussion groups for people interested in a particular topic. E-mail sent to a mailing list is received by everyone on the list. Some web sites have their own mailing lists. Many companies use mailing lists to inform customers about new products and services.

▶ **Fig. 6-6** Newsgroups provide a convenient way to receive information electronically.

Explore

Design and Build a Timeline

State the Problem
Create a multimedia timeline presentation showing how communication has evolved. Use the Internet to conduct your research.

Develop Alternative Solutions
Most of our communication devices were invented over the last one hundred years. The telegraph, telephone, radio, and television are really very young technologies.

Determine the time period on which you will be reporting. Prepare sketches to show the rough layout of your timeline.

Select the Best Solution
Select the timeline model that you think will be most effective.

Implement the Solution
1. Research the time period in which you have an interest.
2. Decide how your research will be presented. Seek assistance if unsure.
3. Present your timeline to the class.

Evaluate the Solution
1. Was the subject matter of your report appropriate?
2. Did your report fulfill the mission, filling in the timeline as assigned?

Rotary Telephone. Combined the mouthpiece and the listening device in one unit. A later version is shown here.

Touch-Tone Telephone. Each key has its own tone. This allows user to obtain information by making choices from automated menus.

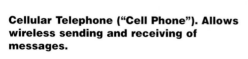

Cellular Telephone ("Cell Phone"). Allows wireless sending and receiving of messages.

IMPACTS

The Internet has brought about many changes. How many times have you written a letter, addressed the envelope, attached a stamp, and then dropped the letter into a mailbox? It might then take several days for your letter to be delivered. With e-mail, letters can be read shortly after they are sent.

Businesses have found that potential customers want to use the Internet to learn about the products and services they offer. Every day, additional commercial web sites are being set up.

Even though scientists have used it from the beginning, the Internet is changing the way they conduct research. This is because the Internet is now easier to use, more powerful, and provides a much greater database of information.

For many people, the Internet reduces research time. Students of all ages use it to gather information for school and personal use. Purchasing agents use the Internet to locate materials at the best possible price. Physicians use it to obtain information about products and procedures that will help their patients.

There is concern that our telephone systems will not be able to keep up with the demand for Internet usage. Many telephone companies are upgrading their facilities to provide additional phone lines. In the meantime, researchers are looking for alternative ways to transmit Internet traffic.

Another concern is that few laws govern use of the Internet. Some people and organizations want the Internet regulated. Others oppose such regulations.

THE FUTURE

Internet use will increase. More products and services will become available. Internet shopping will increase as procedures for protecting credit card information are improved. Competition among retailers on the Internet is already lowering the cost of products such as books and computers. Several airlines now use the Internet to offer bargain prices on flights.

Other devices will be used to access the Internet. For example, web televisions do not require a separate computer. In Japan, one popular television has a double-wide screen. This allows the viewer to watch a television program and "surf" the Internet at the same time. Some engineers predict that pocket-size Internet computers will be available in a few years.

Web telephone service may change the way long-distance calls are made. By purchasing inexpensive software, a microphone, and speakers, you may be able to place long-distance telephone calls without using a long-distance service. Long-distance providers are concerned about this. They have requested laws that would prohibit people from using the Internet for this purpose.

Small video cameras will make it practical to transmit video along with voice or e-mail messages. You will then be able to see the people you talk to in chat rooms.

It is clear that the Internet is bringing about many changes. Some people predict that we are at the beginning of a new revolution in communication technology. What do you think?

Apply What You've Learned

Design and Build a Home Page

State the Problem

A home page is the opening page of a web site. Web sites are used by government agencies, businesses, universities, individuals, and others to create a presence on the Internet. A Universal Resource Locator (URL) identifies each web site. The URL is like a phone number or a street address.

Why do you think so many people and organizations have established home pages? Will the number of such sites and the number of users continue to increase? Why do you think so?

In this activity you will be encouraged to create a web page for your class.

Develop Alternative Solutions

Decide on the purpose of your home page. For example, will your home page be informational, humorous, or educational? What information do you want to present on your home page? When you have decided on the information you want to present, you will need to decide on the text you want to include. Prepare a few designs for your home page.

Select the Best Solution

Select the home page design you think will be most effective.

> **Collect Materials and Equipment**
>
> You will need a computer with Internet access. You will also need software that will allow you to use HTML (Hypertext Markup Language). HTML is the language needed to prepare a web page. Some software helps you write HTML. Other software allows word-processed documents to be converted into HTML. Your instructor will recommend the approach you should take.

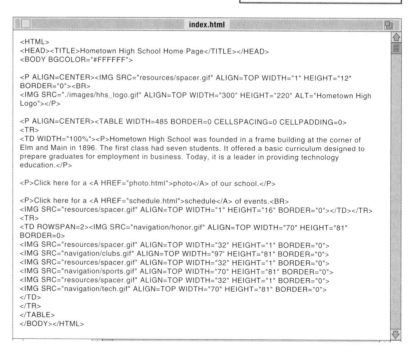

Fig. A. The HTML source code with the text of the home page in Fig. B.

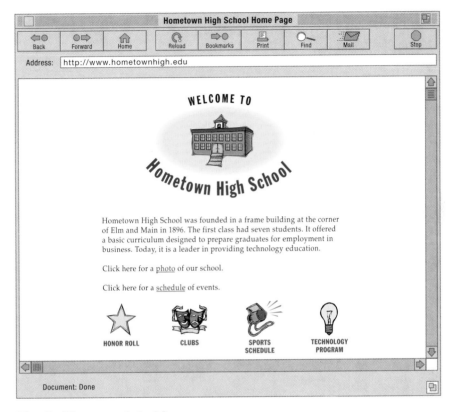

Fig. B. The completed home page.

Implement the Solution

1. Choose which, if any, web sites will be linked to your home page.

2. Create the entire home page (or the portion of it for which you are responsible).

3. Place the web site on a web server (web publishing service). Most providers will charge a fee.

Evaluate the Solution

1. Is the home page attractive?

2. Are the home page commands easy to understand?

3. Can the user easily locate the links?

4. Do the embedded links in the home page work properly?

5. Does the home page fulfill its purpose?

CAREERS IN
Internet Technology

INTERNET ELECTRONIC STOREFRONT OPERATOR

Provide your technical knowledge to Internet store owners by designing web sites, updating and adding information, and improving graphics and sound. This part-time position requires basic programming skills. Fax resume to: Internet Consulting Company at (312) 884-7878, ATTN: Jim Esop.

TECHNICAL SUPPORT SPECIALIST

Strong customer service skills needed to work in technical support. Must be able to work a flexible schedule in a team environment. College and related work experience helpful. Strong knowledge of PC configurations and Internet access required. Call Diane Jones to schedule an interview at (203) 856-0772.

LIBRARY TECHNICIAN

High-tech librarian position open. Requires Internet skills and computer programming experience. Conduct digital searches and teach visitors how to navigate the Internet. Degree in library science required. Send resume to: Seattle Public Library, 842 Main Street, Seattle, WA 90244.

WEB MASTER

Computer company has opening for creative web master. Will design web site to include company information, new products, and technical support. Broad computer hardware and software knowledge, programming skills, and graphic design skills required. Must work well with people from various departments to update and maintain web site. Competitive salary and excellent benefits. Please submit resume to: Supertech Computer Company, 112 Littleton Street, Round Rock, TX 52032.

WEB PAGE DESIGNER

Internet consulting firm has immediate opening for web page designer. Will assist clients in setting up Internet sites by providing design expertise and helping organize information. Responsible for writing text, scanning photos, building graphics, and creating underlying software code to run program. Requires creative talent and technical skills. Good writing skills a must. Excellent growth potential. Contact Internet Designers, Inc., 145 Ocean View Blvd., Coconut Grove, FL 30442.

INTERACTIVE ADVERTISING CREATIVE DIRECTOR

Large advertising firm has immediate opening for inter-active advertising director with knowledge in advertising and in building web sites. Should have an interest in technology. Will be involved in production and design. Fast-paced working environment. Please send resume to: Upfront Advertising, Inc., 1843 63rd Street, New York, NY 02301.

Linking to the WORKPLACE

Many of the career titles listed above can be found on the Internet at those web sites where employers list job openings. Working with your teacher or librarian, conduct an Internet search to locate an actual job listing. Write down the job title and the information provided. Share this information with the class.

Chapter 6 Review

SUMMARY

▶ The Internet is the world's largest computer network.

▶ To use the Internet, you need a computer, a modem, a telephone line, communications software, and an Internet service provider (ISP).

▶ The World Wide Web is the part of the Internet that lets users connect to computers around the world.

▶ Hyperlinks, which take you to another Web page, are an important feature of the Web.

▶ Search engines help you find information on the Internet.

CHECK YOUR FACTS

1. What is a computer network?

2. Identify the products and services needed to get on the Internet. Explain what each item does.

3. What is the purpose of a hyperlink?

4. Explain the parts of an e-mail address.

5. What is a mailing list?

CRITICAL THINKING

1. Select a topic you would like to learn more about. Briefly describe how the Internet could be used to learn more about it.

2. Explain why it is important to follow your school's acceptable use policy.

3. List the URLs for five Web sites you would like to visit. Discuss the type of information presented on each site.

Audio, Video, and Multimedia

OBJECTIVES

▶ explain the terms *audio*, *video*, and *multimedia*.

▶ identify three ways television signals can be received in homes.

▶ define sound.

▶ explain what multimedia is and give examples of how it is used.

KEY TERMS

amplitude modulation (AM)

audio

frequency modulation (FM)

hypermedia

multimedia

sound

UHF

VHF

video

Could this be you? The day begins with the sound of a clock radio. After listening to a few songs, you get out of bed and begin to prepare for another day at school. On the way to the school bus, you listen to several songs on a personal radio. You're glad that the news, sports, and weather took less than five minutes.

In your technology class, the teacher introduces a new topic by showing a video produced by NASA. You think about what it would be like to be an astronaut on-board an international space station. In biology class, you dissect a frog—on a computer screen. After school, you go for a run. Afterwards, you walk around the track a few times, listening to a song on your portable CD player.

ELECTRONIC COMMUNICATION

In this chapter you will learn about audio, video, and multimedia communication. Radio is an example of audio communication. Television combines audio and video communication. **Audio** refers to what we hear on a telecast or broadcast. A telecast is a broadcast by a television station. **Video** refers to the part of a telecast that you see. **Multimedia** is the combination of several forms of communication such as text, video, still photographs, and music. Many exciting games and educational multimedia programs are now available on CD-ROMs.

Radio and television have several things in common. They deliver information and entertain. Live broadcasts keep us informed about news stories and sporting events as they take place. Both radio and television are important providers of electronic communication.

Electronic communication processes begin with a message. The message is changed into a signal that is then *transmitted*. This signal travels on a *channel*, which can be the atmosphere or a cable. A *receiver* then changes the signal back into audio or audio and video information.

Sound is a form of energy produced by vibrations that act on the ear so that we can hear. When we speak, our vocal cords vibrate to create the words we want others to hear. These vibrations spread out in the same way that waves spread when you drop a pebble into a pond. When the sound waves reach other people, their eardrums vibrate and they hear the sound.

The distance between the peaks of any two successive (one after the other) waves is called the *wavelength*. The number of vibrations per second is the *frequency* of the sound. The size of the vibrations is called *amplitude*. Loud sounds have greater amplitude than soft sounds. Fig. 7-1.

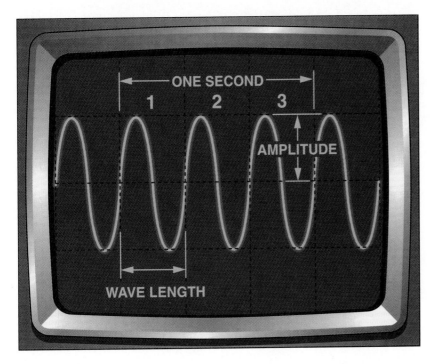

▶ **Fig. 7-1** Wave diagram showing amplitude and wavelength. It also shows frequency, which is three cycles per second.

RADIO

All radio stations have a studio and a control room. Usually they are adjacent rooms separated by a large window.

Radio Broadcasting

The *studio* is where the on-air performers work. Fig. 7-2. Radio studios are designed to broadcast live and recorded sounds. Studios are soundproofed to prevent outside noise from interfering with the broadcast. Disc jockeys, the other announcers, and guests create the live portion of a broadcast in the studio. They usually play compact discs (CDs) for the music. Commercials are typically prepared in advance and stored on tape.

Radio and television messages travel on *electromagnetic waves*. These waves travel at the speed of light—nearly one billion feet every second. This great speed makes it possible for an electronic message to reach its destination almost instantly.

Fig. 7-2 A broadcast is a message sent over the air by a radio or television station.

In the studio, a microphone picks up the sound created by the voices of the disc jockey and the announcers. The microphone changes the sound energy into electrical energy. Fig. 7-3. The signal is sent to an *audio console* in the control room. The console is operated by an engineer who combines live sound with recorded music and commercials. This process is called *mixing*. Mixing is a complex task that involves adjusting many controls to produce the *program signal* that travels through a cable to the transmitter.

The transmitter combines the program signal with carrier waves that "carry" the program signal away from the transmitter. Radio broadcasts can be transmitted by **amplitude modulation (AM)** (AMP-luh-tood MAHD-you-LAY-shuhn) or **frequency modulation (FM).** In AM radio transmission, the amplitude (strength) of the carrier wave changes. In FM broadcasting, the frequency of the carrier wave changes.

The radio waves from an AM broadcast go a long distance because they travel along the ground and bounce off the ionosphere and back to Earth. Radio waves from FM broadcasts do not bounce off the ionosphere. This is why FM

Linking to SCIENCE

Vibrations and Sound. When you speak, your vocal cords vibrate. These vibrations produce waves in the air around you and reach the ear as sounds. You can use a balloon to feel these vibrations. Blow up the balloon as fully as possible. Speak, hum, whistle, or shout, holding the balloon between your open hands. You will feel the vibration of the air on the balloon as you produce sound. What kinds of sound produce the strongest vibrations? ▲

▶ **Fig. 7-3** A microphone converts energy. It changes sound into electrical energy.

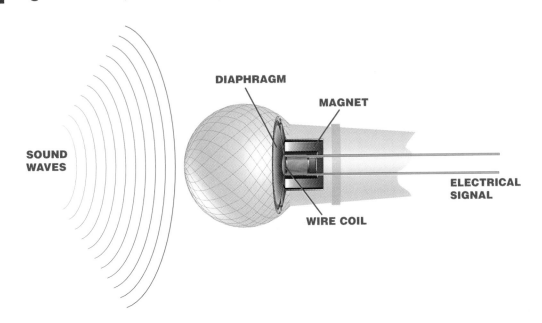

SOUND WAVES

DIAPHRAGM

MAGNET

ELECTRICAL SIGNAL

WIRE COIL

broadcasts do not travel as far as AM broadcasts. However, FM radio broadcasts are usually of better quality. They are affected less by static than AM broadcasts. Fig. 7-4.

The distance a radio broadcast travels also depends on the power of its transmitter. An AM broadcast of 50,000 watts has a range of several hundred miles. Powerful FM broadcasts of 100,000 watts have an effective range of 50 to 60 miles.

Every station has an assigned frequency. Frequency is measured in *hertz*, or vibrations per second. The AM frequency band goes from 535 to 1705 kilohertz. (One *kilohertz* equals 1,000 hertz.) The FM band goes from 88 to 108 megahertz. (One *megahertz* equals one million hertz.)

Radio Reception

We are surrounded by radio waves, even though we are not aware of them. To listen to a broadcast, we need a receiver to change the radio waves back into sound. Most radios can receive both AM and FM broadcasts.

The main parts of a radio are an antenna, a tuner, amplifiers, and one or more speakers. The antenna can be inside or outside the radio. It picks up radio waves from all nearby radio stations. The tuner sets the radio to the desired frequency (station). An amplifier then strengthens the signal for that station. Another amplifier strengthens the signal leaving the tuner. The volume and tone

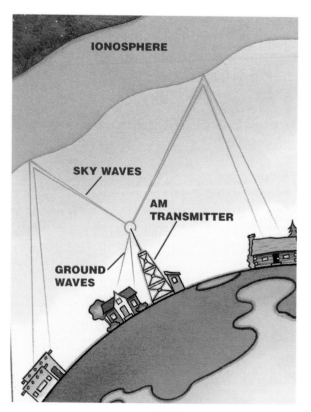

▶ **Fig. 7-4** FM broadcasts cannot be received beyond the horizon. AM broadcasts send out ground waves and sky waves. The sky waves bounce off the atmosphere and can travel long distances.

can be adjusted before the signal is sent to the speaker(s).

The speaker changes the electrical signal into sound. Figure 7-5 shows the basic parts of a speaker. The cone vibrates to create sound waves. These sound waves are nearly identical to those originally produced in the broadcast studio.

TELEVISION

Television emerged as an important communication system in the late 1940s when over-the-air broadcasting began. Over the air television broadcasting is similar to radio broadcasting. Electronic signals travel from a transmitter to an antenna, which sends the signals to a receiver in the TV.

Today most television stations still broadcast over-the-air, but nearly three of every four homes receive their television signals by cable or satellite. Satellite systems are particularly popular in remote areas. Both cable and satellite systems provide better reception and more programs than over-the-air broadcasting systems.

Television is also used in different ways by schools, businesses, and industry. *Distance learning* allows students in small, rural schools to participate in advanced classes offered in a larger school many miles away. Banks and stores use closed-circuit TV for security. *Video teleconferencing* has become a popular way of conducting business meetings. Telephone and satellite systems, combined with large video monitors, let participants at different locations meet face-to-face.

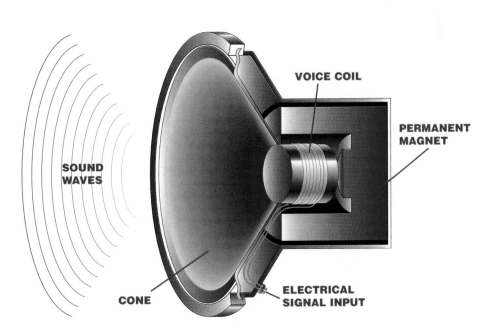

▶ **Fig. 7-5** Speakers change electrical signals into sound. The voice coil causes the cone to vibrate and produce the sounds we hear.

Explore

Design and Build a Storyboard

State the Problem

Create a storyboard for a television news show. The news story itself may be assigned or selected by the members of your team. Storyboards usually contain the following:

- Script. This is the text of what will be said. The script also indicates who will be talking.
- Scene. Tells where the action will be taking place. This is often sketched to show the placement of the "action."
- Sequence of events. Specifies what will be taking place and for how long.
- Camera directions. Specifies which camera will be used for each sequence. Specifies type of shot to select: close-up or not; narrow, normal, or wide field of vision; still or pan (moving) technique. Assume three cameras are available.

Develop Alternative Solutions

How might a producer develop a TV news sequence without using a storyboard? Would it be practical to tell everyone involved what is expected of them? Could everyone remember and understand purely spoken

FASCINATING FACTS

The images from the first television cameras were of poor quality. Actors had to wear black lipstick so their mouths would show up.

Producing a Television Program

Television production is exciting and challenging. Most of the shows on television are prerecorded. Some programs, such as the news and sporting events, are done live. Although the production of live and prerecorded TV programs are not exactly the same, the processes they require are similar.

instructions? Would it be useful to give all participants written notes describing the action to come? What advantages might a storyboard have over these techniques? Prepare several possible storyboard designs.

Select the Best Solution

Select the storyboard design that you think will be most effective.

Implement the Solution

1. Decide on your plan of action. Which part of the storyboard should be prepared first?
2. Prepare each part of the storyboard's requirements.
3. Compare and contrast all the parts. Do they merge smoothly? If not, some parts must be modified.
4. Arrange the papers on the cardboard or matboard.
5. Present your storyboard to the class.

Evaluate the Solution

1. Are there any questions from your classmates that would indicate that the storyboard is inadequate?
2. Could another group use your storyboard as is and successfully present a program?

All television programs are carefully planned. The *producer* hires *writers* to prepare a script. The producer also hires a *director* to turn the script into a TV program that viewers will find informative or entertaining. Many directors require their staff to produce storyboards. *Storyboards* include sketches that show what should happen in a scene and text to summarize the dialogue. They also identify the camera shots needed.

The producer and director usually work together to select the people who will appear in the TV program. The people who appear are referred to as *talent.* They include announcers, actors, newscasters, reporters, and hosts of game and talk shows. Selecting the right talent is important. A program that starts with a good script, but uses the wrong people, will not be successful. Fig. 7-6.

Many television programs must be rehearsed (practiced). Before the actual broadcast or taping session, the studio must be prepared. Microphones and lighting are set up and tested. Most shows are *shot* (recorded) using two or more cameras. Studio cameras are similar to home video cameras. However, they are much larger and have wheels so that they can easily be moved around the studio.

The *control room* is a busy place. Monitors show the image captured by each camera. A technician uses a *switcher* to select the camera specified by the director.

Here is an example of how the switcher is used. Imagine a news broadcast with two anchors. As one anchor finishes a story, the director might say, "Camera two." A minute later, as the other anchor reads another story the director might say, "Camera one." After a few stories, a technician uses the switcher to *cut* to a commercial. While all of this is happening, engineers make sure that the picture and sound quality are satisfactory.

Television Cameras

Television cameras use a lens to collect light from the scene being shot. Mirrors inside a color camera separate the image into three images—blue, red, and green. Sensors then create an electronic signal for each of the three colors.

Two types of sensors are used in television cameras. Some have a tube called a *vidicon*. Others use an electronic part called a *charge-coupled device* (*CCD*). Although they work differently, both systems change the color image that hits them into an electronic signal. The signal is then amplified and sent to an encoder. The encoder combines this signal with several others. These signals and the audio signal go to the transmitter, which produces the broadcast signal. Fig. 7-7.

▶ **Fig. 7-6** Actors and announcers provide the input for a television program. The producer plays a part in deciding which camera shots will be used.

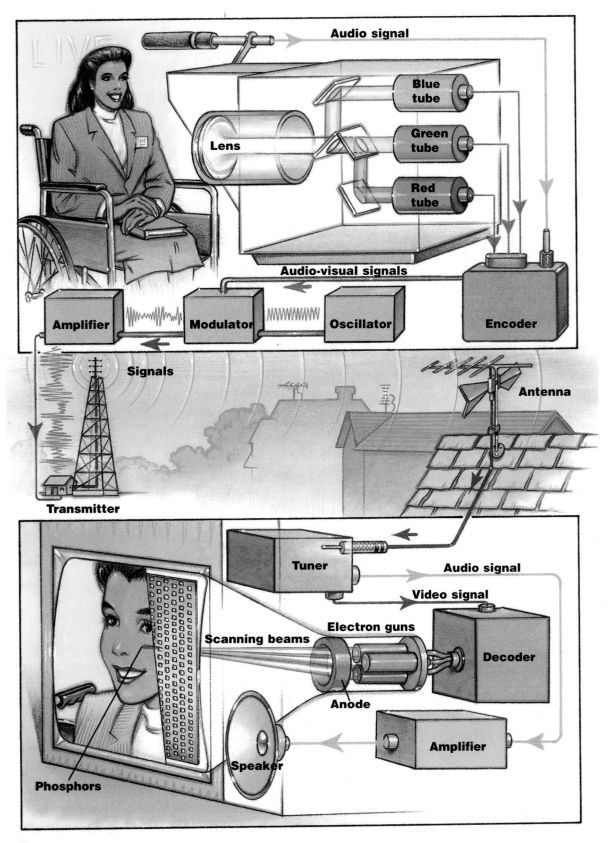

▶ Fig. 7-7 This systems diagram shows how a color television signal is produced and transmitted.

Over-the-air television transmission is similar to AM and FM radio transmission. The frequency a station uses for broadcasting is called its *channel*. Channels 2 through 13 are **VHF** (very high frequency) channels. VHF signals have a frequency between 54 and 216 megahertz. Channels 14 to 69 are **UHF** (ultra-high frequency) channels. They have a frequency between 470 and 806 megahertz.

Over-the-air television broadcasting has an effective range of up to 150 miles. Cable, microwave, and satellite systems carry TV signals over longer distances. The networks often use cable to send their programs to local stations. Also, most communities have cable companies that install cables to carry the signals into homes.

Microwave towers throughout the country relay signals from one tower to the next. At each tower, the signal is amplified before being sent to the next tower. When the microwave signal reaches the TV station, that station changes the microwave signal into a television signal.

Satellites also relay signals around the globe to television stations, cable companies, and homes, which use satellite dish antennas to receive the signals. Several satellite TV networks exist. Some networks rent the receiving equipment; other providers require the user to buy it.

Television Reception

Televisions receive the signals they need from an antenna, cable, or satellite system. The tuner amplifies the signal for the selected channel. The speaker changes

> ## Linking to MATHEMATICS
>
> ***Speed of Electromagnetic Waves.*** A satellite positioned 22,300 miles above Earth circles Earth at the same rate as Earth's rotation. Thus, it seems to hover over one fixed spot on Earth. This satellite will receive signals from Earth and beam them back to another station on Earth.
>
> If the electromagnetic waves travel at the speed of light (186,000 miles per second), how long will it take for a radio signal to reach the satellite from a station on Earth?

audio signals back into sound. The video signals produce the picture.

The *neck* (back) of a color picture tube contains three *guns*, one each for blue, red, and green signals. Each gun directs a beam at the screen in a pattern that reproduces the images originally captured by the camera. The screen is coated with dots of blue, red, and green *phosphors* that glow when struck by the beam of their corresponding color. The glowing dots blend to produce the picture we view. Fig. 7-8.

Impacts of Television

Like many other technologies, television has positive and negative impacts. Positive impacts of television include the information and entertainment it provides. Some of the negative impacts affect people your age. Young people who spend too much time watching television may neglect their homework. Also, many parents are concerned that some programs show too much violence.

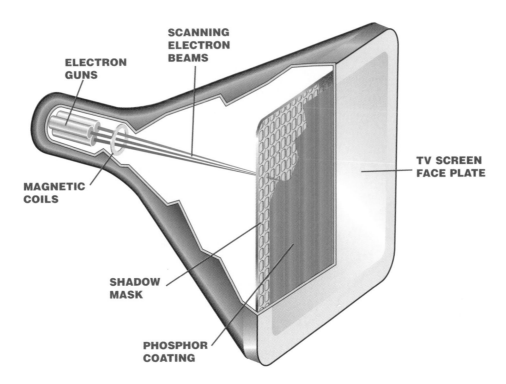

ELECTRON GUNS

SCANNING ELECTRON BEAMS

TV SCREEN FACE PLATE

MAGNETIC COILS

SHADOW MASK

PHOSPHOR COATING

Fig. 7-8 A color television picture tube. Red, blue, and green beams directed at the screen cause the phosphor coating to glow and produce the television picture we see.

Television in the Future

Cable networks will continue to add channels. Interactive television will become more common. It will encourage people to use their televisions for other services, including the Internet, pay-per-view programs, and shopping. Satellite networks will continue to increase in popularity as large satellite receivers are replaced by smaller receivers.

High-definition television (HDTV) is a new system. It is digital (based on the 0s and 1s used by computers) and produces incredibly sharp pictures and sound.

VIDEO RECORDERS

Video recorders, or videocassette recorders (VCRs), have changed the way

we use television. The VCR makes it easy to watch prerecorded tapes. A VCR can also be used to record a program for later viewing.

Videotape recorders capture the electronic signals from a television or camera. A recording head acts as an electromagnet to create magnetic patterns. These patterns represent the original sound and picture on the iron oxide coating of the tape.

During playback, a playback head changes the magnetic patterns on the tape into electrical signals. This signal is converted back into pictures and sound by a television. Fig. 7-9.

Many people use their VCRs to watch their own videos. Video cameras are now smaller, less expensive, and easier to use than they were just a few years ago.

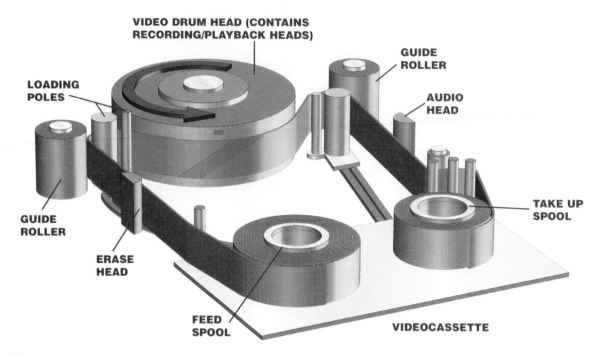

VIDEO DRUM HEAD (CONTAINS RECORDING/PLAYBACK HEADS)

GUIDE ROLLER

LOADING POLES

AUDIO HEAD

GUIDE ROLLER

ERASE HEAD

TAKE UP SPOOL

FEED SPOOL

VIDEOCASSETTE

▶ **Fig. 7-9** Video recorders change sound and video signals into magnetized patterns on tape. During playback the stored patterns are changed back into electrical signals. When you start the VCR, loading poles take tape out of the plastic cassette and bring it into contact with the record/playback heads. The tape is then guided back into the cassette.

MULTIMEDIA

Multimedia combines of several kinds of communication, such as text, pictures, sound, video, and animation. Most multimedia products are based on the use of a CD-ROM. Usually, these products are *interactive*. This means that the user can control parts of the action or decide how the program is used. CD-ROM programs allow you to go forward or backward in the program at any time.

The word *multimedia* can also refer to a combination of communication systems, such as cable television and telephone. In some areas, cable TV and telephone companies are working together to deliver new forms of entertainment, including video games and video-on-demand.

CD-ROM

CD-ROM is short for "compact disc read-only memory." It was developed from the music CD. In addition to music, it can store many other kinds of information, including video and animation.

Most home computers now include a CD-ROM drive. Popular multimedia software includes encyclopedias, atlases, magazines, books, and games. An important advantage of a CD-ROM is that it can hold a large amount of information. A single CD-ROM can hold more information than two complete encyclopedias. Fig. 7-10.

CD-ROM programs use hypermedia to make them interactive. **Hypermedia** provides links to other parts of the program using hypertext and hotspots.

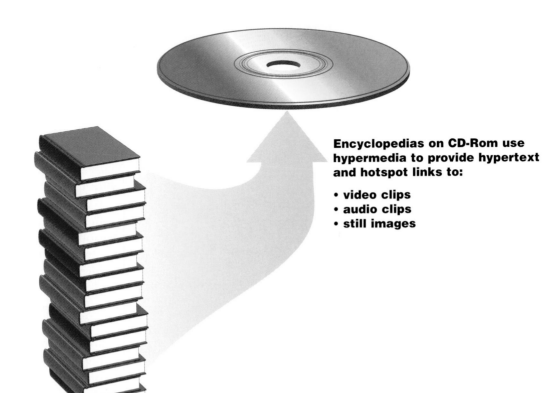

Encyclopedias on CD-Rom use hypermedia to provide hypertext and hotspot links to:

- video clips
- audio clips
- still images

▶ **Fig. 7-10** Some encyclopedias are now on CD-ROM. What are some advantages and disadvantages of presenting information on CD-ROM rather than in a book?

Hypertext allows you to choose the information you want.

To use *hypertext*, you select a particular word by pointing and clicking with a mouse. You can then view a list of pages where the term is used and go to one of the pages instantly. Some multimedia programs use words in a different color to suggest visiting other parts of the program.

Hotspots are pictures or buttons that take you to another part of the program. Hypertext and hot spots are similar to the links used on the World Wide Web.

Linking to COMMUNICATION

Abbreviations. An abbreviation is a shortened form of a word or phrase that is used to represent that word or phrase. We use abbreviations for convenience. In this chapter many abbreviations are used—TV, CD, AM, FM, VCR, VHF, UHF, and HDTV. Abbreviations using only one syllable or a portion of the word are usually written in lowercase followed by a period, such as "fig." for "figure."

Abbreviations using the first letter of each word in the phrase are written in upper case letters with no period, such as "CD" for "compact disc." With your study group, brainstorm abbreviations used in your technology class and textbook. Prepare a list of correctly written abbreviations followed by their longer forms. ▲

Explore

Design and Build a Megaphone

State the Problem
A megaphone is a cone-shaped object used to enhance sound. Your instructor may assign an approximate size for your megaphone.

Develop Alternative Solutions
1. Decide on the size and shape of the megaphone you will be making.
2. Select the materials you wish to use.
3. Note that the large and small ends of the megaphone are variable for any given design. Prepare several designs for shapes of different sizes.

Select the Best Solution
Select the design that you think will be most effective.

Implement the Solution
Form the megaphone. Add a handle for larger models.

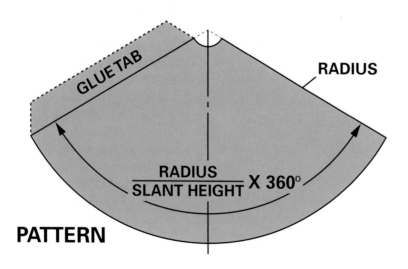

PATTERN

$$\frac{\text{RADIUS}}{\text{SLANT HEIGHT}} \times 360°$$

GLUE TAB

RADIUS

> **Collect Materials and Equipment**
>
> various cone materials (cardboard, sheet metal)
> various fastening materials (tape, metal fasteners)
> Megaphones may be made of almost any material. However, some materials are better suited for the purpose. Some characteristics to look for in materials are: strength, weight, availability, cost, appearance, and the ease with which it can be formed into a cone. Your instructor will suggest what materials are available to you in the classroom. Your teacher may also discuss materials you might wish to bring to class.

MEGAPHONE

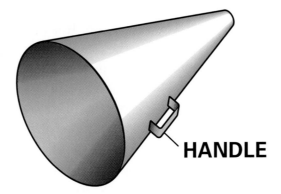

HANDLE

Evaluate the Solution

1. Set up the megaphone and a sound source (such as an audio tape recorder) so that the small end of the megaphone faces the sound source. Position a classmate a certain distance from the sound source (perhaps 20 feet or so). The classmate is to listen for the sound.

2. Play the tape (ordinary speech will serve well) at a volume just loud enough for the classmate to hear it.

3. Remove the megaphone and play the same tape at the same volume. Can your classmate still understand the words?

4. Do some megaphones seem more effective than others? Why might this be so? Do you think the type of material used for the cone affects the megaphone's amplification?

Multimedia Video

Video information is normally recorded in analog form. Before the video can be used on a CD-ROM, it must be converted to digital form. This is done by connecting a cable from the video recorder to a computer. Software and special hardware in the computer convert the analog information into digital information that can be stored on the computer's hard disk.

The quality of CD-ROM video is not as good as that of the original videotape. Computers and video hardware are not yet capable of producing high-quality images. Because of these problems, image quality is affected by the size of the video window and how many colors are used for the images. Also, the number of times the image is updated each second is very important. Television uses 30 frames per second. Multimedia video uses about 15 frames per second.

Digital Video Editing

An important advantage of digital video is that it can be edited on the computer. Each frame can be viewed, deleted, copied, or moved with the click of a mouse. It is also easy to access any part of the video almost immediately. This is because hard drives can find a specific address much faster than videotapes can be wound and rewound. Another advantage of digital video is that there is no loss of quality during the editing process. When the editing is complete, it is easy to make additional copies of the video. Fig. 7-11.

▶ **Fig. 7-11** A video editing station.

Digital video takes up a large amount of space on a CD-ROM. To reduce the space required for video clips, a technique called *video compression* is used. Special software reduces the space needed to store the data. During playback, the data is decompressed. Compression makes it possible to put about 60 minutes of video on a single CD-ROM.

IMPACTS OF MULTIMEDIA

Our age is the Information Age. The main characteristic of the Information Age is the development of new methods of communication. The communication advances of the last twenty years have changed the ways in which we learn, work, and entertain ourselves. Information is now easier to obtain. Some have said that this has led to a decline in reading skills.

THE FUTURE OF MULTIMEDIA

In the future we will see great changes in audio, video, and multimedia. Multimedia will continue to become a more familiar and important form of communication. People who own video recorders will make sure their next computer has a CD-ROM drive. CD-ROM drives will continue to increase in speed. As more CD-ROMs are sold, their cost is expected to drop. The Internet (see Chapter 6) will make it possible to access multimedia from many web sites that currently use only text.

▶ **Fig. 7-12** Multimedia allows this photo researcher to access audio and video clips on her laptop.

Apply What You've Learned

Design and Build a Tin Can Telephone

State the Problem

Design and build a tin can telephone to carry sound over a distance.

Develop Alternative Solutions

How may the range of spoken (oral) communication be extended? How does sound travel from mouth to ear? Does sound travel through solid materials? (If you put your ear on the desk, can you hear sound?) Can fluids carry sound? (Can you hear under water?) Do you think sound can travel through a string?

Select the Best Solution

Select the tin can telephone as a project that will demonstrate the movement of sound.

Implement the Solution

1. Acquire two cans. Make sure the cans are clean. Remove one end of each can. (Caution: tin cans can have sharp edges.) If the cans being used have sharp edges, consult your teacher. (Tape can be placed over the edge.)
2. Create a hole in the end of each can. Use a punch, an awl, or a long nail.

> **Collect Materials and Equipment**
>
> 2 cans (large enough to fit comfortably around the ear). The average soup or vegetable can is about the right size.
> string (You will need about 20 feet. The string should be strong enough to be pulled tightly without breaking.)
> hammer
> punch, nail or awl

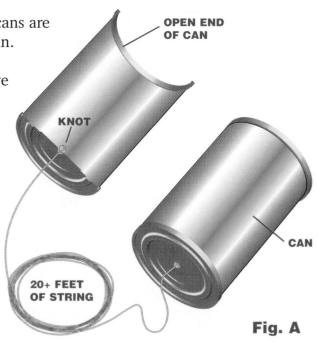

OPEN END OF CAN

KNOT

CAN

20+ FEET OF STRING

Fig. A

3. Place the string (20 feet or more) through the hole. Tie a knot on the end so that the string cannot be pulled out. (See Fig. A.)
4. Give each can to a classmate.
5. Ask the two classmates to walk away from one another until the string becomes taut.
6. Ask one person to place the can over his or her ear. The listener might want to cover the other ear to muffle sounds not coming from the can.
7. Ask the other person to speak into the can.
8. Reverse the roles. Ask the listener to speak and the speaker to listen.

Evaluate the Solution

1. Ask several teams of students to test your tin-can telephone system. Prepare a script for every team. Every team should speak into the system using the same script. Members of each team should try to speak at the same voice level. Does the system work for everyone testing it?
2. Can the system be improved by using a material other than the string you have selected? Would a lighter or heavier weight string material other than string transmit sound more effectively? For example, would fishing line or wire work as well?
3. Are the open ends of the cans dull enough to be safely placed over the ear?

CAREERS IN
Audio, Video, and Multimedia

BROADCAST TECHNICIAN

Broadcast technician needed at local radio station. Duties will include installing, operating, and maintaining a variety of electronic broadcasting equipment. Two years of technical training and the ability to work well with people are required. Fax or send resume to: Werning Communications, 8800 Gulf Drive, Coral City, FL 22465

VIRTUAL REALITY DEVELOPER

Software programmer needed at small virtual reality development company. Must be logical thinker and able to work as part of a team. Degree in computer science required. Come join our company in this competitive field. Salary and benefits. Send resume to: Virtual Designs, 1200 Howard Street, Redwood City, CA 99824.

ILLUSTRATOR

Advertising agency needs illustrator to draw storyboards for presentations to clients for proposed television commercials. Knowledge in computer design techniques helpful. Creative and talented artist should apply to: Lashley Advertising, Inc., 3535 East Wacker Drive, Chicago, IL 30060.

ART DIRECTOR

Telecommunications company needs experienced art director to research and develop concepts and supervise staff engaged in all facets of creative layout and design. Management experience required. Candidates interested in working for a growth-oriented company should send resume to: Human Resources, 9000 Clinton Road, Brooklyn, CA 99878.

MULTIMEDIA PRODUCER

Multimedia company seeks multimedia producer to make assignments and oversee writers, graphic artists, and programmers in developing computer programs. Responsible for production and budgeting. Must have previous experience as production assistant. Please send resume to: Multimedia Makers, 9856 Makenzie Street, Tulsa, OK 66982.

FILM/VIDEO TAPE EDITOR

Combine your artistic ability with technology to perform digital film editing of educational films and videos. Previous experience as editing room assistant required. Submit resume to: Educational Publishers, Inc., 1818 North Lane Avenue, Hartford, CT 36854.

Linking to the WORKPLACE

The telephone book can be used to locate employers when you are looking for a job. Using your local Yellow Pages, locate potential employers for the jobs listed above. Look under the following headings: audio-visual production services, audio-visual consultants, artists, graphic designers, motion picture production studios, multimedia producers, radio and television stations, videotape editing services, video production services, and visual production services.

Chapter 7 Review

SUMMARY

▶ Audio is something we hear. Video is something we see. Multimedia combines several forms of communication.

▶ In electronic communication, a message is changed into a signal, which is then transmitted. The signal travels on a channel. A receiver changes the signal back into information.

▶ All radio stations have a studio and a control room. Mixing combines live and recorded sounds to produce the signal that is transmitted.

▶ Television broadcasts are transmitted to homes by over-the-air transmission, by cable, and by satellite.

CHECK YOUR FACTS

1. Explain the terms *audio*, *video*, and *multimedia*.

2. Briefly explain video switching.

3. Identify three ways television signals can be received in homes.

4. Define sound.

5. Identify the main parts of a radio.

6. Explain multimedia. Give five examples of its use.

CRITICAL THINKING

1. Explain how the activities in radio and television control rooms are similar.

2. Write a one-minute audio script to promote an upcoming event.

3. Imagine that you were not allowed to watch TV for a week. How would your lifestyle be affected?

4. Explain the term *interactive*. What makes a multimedia program interactive?

Production Technologies

CHAPTER 8 *Manufacturing*

CHAPTER 9 *Structures*

What do manufacturing and building construction have in common? Both of them use a system to produce the desired output. In manufacturing, the output is the manufactured product. In building construction, the output is the finished building.

The chapters in Section 3 will help you understand how a certain set of procedures can bring order to a process. This section discusses the basic stages in manufacturing and the basic parts of the system used to build structures.

The Secret Life of a Tower Crane

Society depends on a sound infrastructure. We need highways that are in good repair and structures that are well built. Tower cranes, like pigeons, are a familiar part of the urban construction scene. Yet who knows where they come from or where they go?

The upright mast and outward-reaching jib of a tower crane are made of steel. The design combines strength and light weight. Slung under the jib is a hook that can travel back and forth, as well as up and down. This hook can deliver a lifted load to any point under the crane's reach.

How a Tower Crane Climbs

The ascent of a tower crane remains a mystery to most city dwellers. The tower crane is sometimes set in a central stairwell or elevator shaft. This position offers stability and reach in all directions.

With this type of crane, the mast does not get taller (although some masts have a telescoping inner structure that can almost double their height). Instead the mast raises itself from floor to floor by climbing a set of ladders mounted within the building. The crane pushes itself up, rung by rung, using a hydraulic jack within the base of the mast.

Getting Down

Other climbing cranes go up the side of a building. In this case, braces are attached every few floors and new sections of mast are added at the base. The exterior-climbing crane is easier to take down. The stairwell-mounted crane may take weeks to dismantle. In some cases a derrick must be constructed on the roof to lower the pieces of the crane.

Linking to the COMMUNITY

Observe a local construction site. Identify ways in which equipment use and building operations have been designed to minimize neighborhood noise, traffic, and safety problems.

CHAPTER 8

Manufacturing

OBJECTIVES

▶ describe the difference between natural and synthetic materials.

▶ name primary and secondary processes.

▶ list the basic steps in the manufacturing system.

▶ describe the three types of production systems used in manufacturing.

▶ describe how computers are used in computer-integrated manufacturing.

KEY TERMS

computer-aided design/
  computer-aided
  manufacturing (CAD/CAM)
continuous production
custom production
industrial materials
job-lot production
manufacturing
primary processes
prototype
raw materials
secondary processes

Think about your plans for today. What will you wear? What will you eat? Where will you go? What will you do for entertainment? Your clothing, most of your favorite foods, the vehicles that take you places, and the television you watch all have something in common. Can you guess what it is? They are all products of manufacturing.

The word "manufacture" comes from the Latin words *manu* (hand) and *factus* (to make). Together they mean "made by hand." However, most products that we use today are not made in this way. Times have changed. The meaning of manufacturing has also changed.

MATERIALS IN MANUFACTURING

Manufacturing is the changing of materials into usable products. Most manufacturing takes place in factories. In this chapter, you will learn about some of the important materials and processes that are part of manufacturing today. You will also learn how manufacturing affects our lives and how it will be changing in the future.

One of the seven resources of technology is materials. Many different materials are used in manufacturing. *Natural materials* are found in nature. Wood, clay, metal ores, and oil are natural materials. *Synthetic* (sin-THET-ik) *materials* are made by people. Plastics and ceramics are important synthetic materials. Fig. 8-1.

Wood

Wood is either a hardwood or softwood. *Hardwood* comes from trees that lose their leaves in the fall. Oak, maple, and walnut are hardwoods. Most furniture is made from hardwood. *Softwood* comes from trees that keep their leaves or needles all year. Softwoods include pine, fir, and spruce. Wood from these trees is used for construction lumber.

Metals

Metals are among the most important materials used in manufacturing. Metals can be ferrous or nonferrous. *Ferrous* metals contain iron. *Nonferrous* metals do not. Natural metals include iron, aluminum, copper, zinc, tin, and lead.

▶ **Fig. 8-1** The first bicycles were made by hand. However, most bicycles today are mass produced in factories. Can you name some of the natural and synthetic materials that might have been used in manufacturing this racing bicycle?

Most metals used today are alloys. An *alloy* is made by combining two or more metals or a metal and a nonmetal. This is done to give the metal characteristics such as hardness and strength. For example, steel is an alloy of iron.

Plastics

Hundreds of different plastics are used to manufacture a variety of products. All plastics fall into one of two categories— thermoplastics or thermoset plastics. *Thermoplastics* are heated and then molded to shape. Later, they can be reheated and remolded. The acetate sheets used in packaging are thermoplastic. *Thermoset plastics* can be heated and shaped only once. Thermoset plastics are used to make fabric, dinnerware, and automobile parts.

Ceramics

Most ceramics are synthetic materials. They are made by *firing* (baking) clay, sand, and other natural substances at very high temperatures. Ceramic materials are used to make bricks, bathroom fixtures, and cookware. They are even used to make the insulating tiles on the Space Shuttle.

Other Materials

A *composite* is a material that is made by combining two or more materials. The materials that form a composite are not changed. They work together to create a new material with desirable qualities. Commonly used composites include concrete and plywood. Concrete is made from cement, sand, and gravel. Plywood is

FASCINATING FACTS

Silicon chips contain silicon—a material found in sand. Silicon is purified and made into a three-foot-long crystal. The crystal is then cut into tiny wafers with a diamond-tipped saw. These wafers are polished to perfection. Chips are then photoetched onto the wafer. The chips are used in many products, including wristwatches and computers.

made from sheets of wood and glue. Other composites are used to make boats, golf clubs, tennis rackets, and airplanes.

Textiles, leather, and rubber are also important materials. Textiles can be made from natural or synthetic materials. Wool and cotton are natural materials. Polyester is a synthetic material. Textiles are used to make clothing, carpeting, and many other products.

Leather is made from animal skins. Most leather is used to make shoes. Leather is also used to make clothing and sporting goods.

Rubber is made by adding a variety of chemicals to a fluid called *latex* (LAY-tex). Some latex comes from rubber trees. Latex can also be made synthetically. Rubber is used in tires, drive belts, gloves, and many other products.

PROCESSING MATERIALS

Primary Processes

Raw materials are materials as they occur in nature. For example, a tree is raw material. Most raw materials cannot be used until after they are processed into industrial materials. **Industrial materials** are materials that are used to make products. Lumber is one example of an industrial material.

Primary processes are processes that change raw materials into industrial materials. There are three kinds of primary processes: mechanical, thermal, and chemical.

Several different processes are required to change most raw materials into industrial materials. At sawmills, machines remove bark from tree logs and cut the logs into lumber. Smaller saws cut the lumber to length. These are *mechanical* processes. Next the lumber is graded according to quality and dried outdoors or in a kiln (oven). Drying is a *thermal* process. Fig. 8-2.

Metal ores are mined, crushed, heated, and changed *chemically* when other materials are added.

▶ **Fig. 8-2** Wood is a raw material. To be usable in construction, it must be processed into lumber. The cut tree trunks are transported to a sawmill. There the bark is removed from the trunks. The trunks are then sawed into lumber. After being dried, the lumber is ready for sale.

Secondary Processes

Secondary processes are processes that turn industrial materials into finished products. These processes include forming, separating, combining, and conditioning.

Forming

Forming is the process of changing the shape of a material. Fig. 8-3. The material itself is not changed by adding or taking something away. Forming is done in several ways:

- *Rolling* is the process of squeezing the material between rollers. Sheet metal is made in this way.

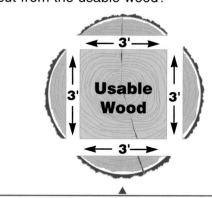
▶ **Fig. 8-3** Forming processes were used to manufacture many of the items you use daily. Can you name some items that may have been manufactured using forging, casting, rolling, and extrusion?

FORGING

CASTING

- *Casting* involves pouring or forcing softened material into a mold (hollow form).
- *Forging* is the process of hammering or squeezing material into shape. Usually dies (molds, patterns, or forms) are used to shape the material. Most wrenches are shaped by forging.
- *Stamping* is the process of squeezing sheet metal between dies to give it shape. Stamping is used to shape some automobile body parts.
- *Extrusion* (ex-TRUE-shun) is the process of pushing the material into shape. The shape of the opening in the die determines the shape of the material.

Separating

Separating is the cutting of materials to size and shape. Some material is usually lost. Heat, light, chemicals, and even water can be used for separating.

Shearing is separating part of a solid material from the rest of the material. No material is destroyed. Thin materials like paper and sheet metal can be cut by shearing. Material can also be separated by cutting. Usually, cutting a material involves *chip removal*. Common chip removal processes are shown in Fig. 8-4.

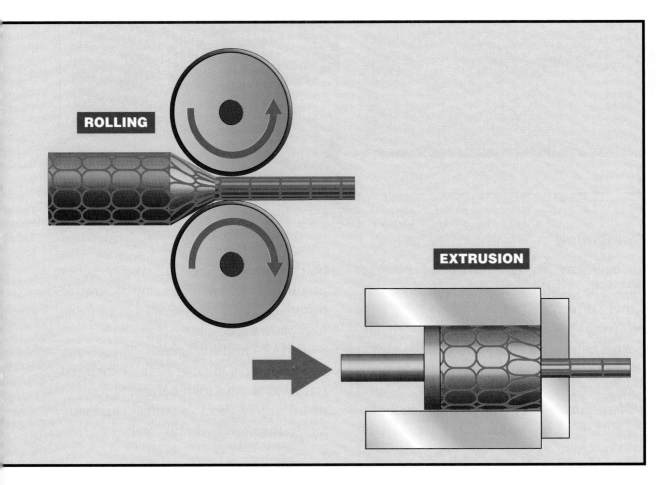

▶ Fig. 8-4 Wood is shaped by tools that produce chips. The chips are sawdust and wood shavings. Do you note any relationship between the speed of the operation and the size of the chips?

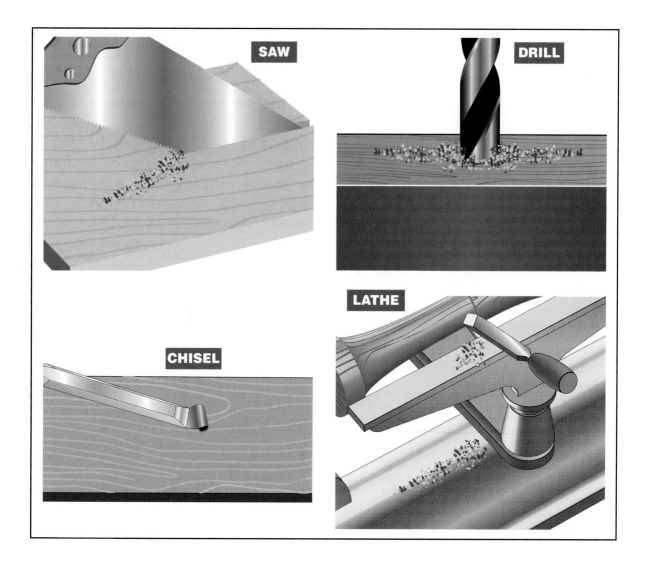

Combining

Combining is the process of putting materials together. Combining is done in a variety of ways:

- *Mixing.* Materials may be mixed together to form new materials. Food products such as cake mixes and soup mixes are made in this way.
- *Mechanical fasteners.* These include nails, screws, staples, and nuts and bolts. Automobiles and many other products are assembled using removable fasteners. Doing this makes it possible to replace parts.
- *Soldering* and *brazing.* These are processes that involve heat. A material is melted along a joint between pieces of metal to join them together. The metal pieces themselves do not melt.

- *Coating.* This process uses one material to cover another. It may be done to decorate or protect the covered material. Some coatings do both. Painting is the most common coating process. Fig. 8-5.

Conditioning

Conditioning is a process that changes the internal structure of a material. The three types of conditioning are thermal, chemical, and mechanical.

Thermal conditioning uses heat. Hardening, tempering, and annealing are thermal processes used to condition metals. *Hardening* makes metal products more wear-resistant. However, hardened metals are usually brittle. In *tempering*, hardened metals are reheated to remove brittleness and make them tougher. Tempering helps steel-cutting tools keep their sharp edges. *Annealing* is a softening

process. Steel must often be annealed. It can become too hard during processing. Annealing also makes metal less brittle.

▶ **Fig. 8-5** Coating is one of the final processes in manufacturing. The paint job on this car will be carefully inspected. Inspection is part of the quality control process.

Chemical conditioning uses a chemical reaction. *Vulcanization* (vul-cuh-na-ZAY-shun) is a chemical conditioning process used to make rubber durable. A substance such as sulfur is added to the rubber. Heat and pressure are then applied.

Mechanical conditioning uses force. Forging is an example of mechanical conditioning. Hammering a piece of metal will cause it to change. Usually this is done to make the metal harder.

THE MANUFACTURING SYSTEM

All manufactured items are either consumer products or industrial products. *Consumer products* are "used up." Televisions, cereal, and books are consumer products. Lumber, steel beams, and machine tools are *industrial products*. They are used in the making of other products. Products may vary greatly.

Explore

Design and Build a Mass-Produced Item

State the Problem
Your class has been asked to mass-produce a small wooden product. Your instructor will help you select and design the actual item. Assume that the product is made of several pieces. Your task is to organize the team members so that they will be able to manufacture the product by mass production.

Develop Alternative Solutions
Identify the item you will be manufacturing. Identify the processes needed to manufacture the item. Identify the order in which the individual processes are to be performed. Determine the order in which each part of each process is to be performed. Identify the individuals who will perform each operation. Remember that you will also need inspectors to ensure quality control. You may come up with several different plans. Write out each plan so that it is clearly understood.

Select the Best Solution
Select the plan that you think will be the most effective.

Collect Materials and Equipment
wood (perhaps pre-cut)
hand tools, as needed
machine or power tools, as available (used under the supervision of an adult)
fasteners (chemical or mechanical)

However, the steps used to produce different products are basically the same. Fig. 8-6.

Design and Engineering

To sell products, manufacturers must design products that people will buy. As time passes, the needs and desires of customers change. Manufacturers must continue to improve current products and design new ones.

Implement the Solution

1. Make working drawings for your item (Sketches might be acceptable. Ask your instructor.)
2. Determine how each part of the product will be made. Each part may require more than one step or process.
3. Assign jobs and materials.
4. Make a prototype (PRO-tow-type). A prototype is the first working model of the item.
5. Begin mass production. Assign one or more people as quality control inspectors. They will need to check progress for a minimal acceptable standard.
6. Assemble the item and complete finishing operations.

Evaluate the Solution

1. Were the job assignments clearly explained?
2. Was the order in which the tasks were done effective?
3. Did the pieces fit together as planned?
4. Did the product meet your expectations?
5. How might this product be improved if it were to be made again?

▶ **Fig. 8-6** The main stages in the manufacturing process. The product design department (top left) determines the materials needed. The purchasing department orders the materials. The product is manufactured by the production department and checked by the quality control department. After being packaged, the product is sold by the sales department.

Engineers design new products and determine the best ways to produce them. To test their ideas, they often build prototypes and small-scale production systems. A prototype is a full-size model of a product. It looks and works like the actual product.

Purchasing

Items needed for production must be ordered. A sawmill needs raw materials—logs. A manufacturer of metal products requires industrial materials such as sheet metal. Automobile manufacturers need finished products such as tires, batteries, and window glass to produce their own finished products—automobiles. Purchasing agents work to obtain the correct items at the best price.

Production

Both primary and secondary processes are used in manufacturing. Many different processing techniques may be needed to produce one type of product. Products with a number of different parts, such as appliances and computers, must also be assembled.

Production Systems

The three important types of production systems are custom, job-lot, and continuous. The production system used depends upon the kind of product and the number of products needed.

In **custom production**, products are made to order. For example, a cake made according to a buyer's choice of size, flavor, shape, and icing is a custom-made product. A custom-made product need not be a small item. A $300-million cruise ship designed for 2500 passengers is also a custom-made product.

In **job-lot production**, a specific quantity of a product is made. For example, suppose a manufacturer received an order for 10,000 specially designed skateboards. Job-lot production would be

Linking to COMMUNICATION

Research Recycling. Determine what materials you recycle. Research the use of these recycled materials in new products. As a research tool, use the media center in your school. Organize your information in a chart or table. Be prepared to explain your information to the class.

used. Many job-lot manufacturers make products on a seasonal basis. A company making skateboards used mostly in the summer might also make sleds for use during the winter.

In **continuous production**, products are mass produced, usually on an *assembly line*. Continuous production is also called *mass production*. Products move from one workstation to the next. Some mass-production systems make thousands of identical items each day. Fig. 8-7.

▶ **Fig. 8-7** Continuous production depends on the smooth transfer of materials from one workstation to the next.

Other manufacturing systems turn out products that are not identical. Modern manufacturing techniques allow flexibility. This is true in automobile manufacturing. Vehicles that vary in color and engine size can be produced on the same assembly line.

Quality Control

Products must meet the quality standards set by the company. The system used to check quality is called *quality control.* Automated devices and trained inspectors may be used. Materials are checked as they come into the factory. The quality of the work and of the products being produced are checked throughout production.

Distribution and Sales

Products must be prepared for shipping and then transported to customers. Industrial materials are usually shipped to other manufacturers. Most consumer

Explore

Design and Build an Assembly System

State the Problem

Most manufactured products are made of many different components. An automobile typically requires the assembly of more than 35,000 parts. A ballpoint pen might have five or more parts. Work as a member of a two-person team to develop a system for assembling a simple product such as an inexpensive retractable ballpoint pen.

Develop Alternative Solutions

1. Work as a member of a two-person team to disassemble all ten pens.
2. Each team member should assemble and then disassemble one pen.
3. Separate the parts and make a pile for each of the individual parts.
4. Work as a team member to decide how to position the materials on the work surface.
5. Two team members should complete the assembly of five products. Five products should be assembled using an assembly line process. One team member should use the stopwatch to time how long it takes to assemble the

Collect Materials and Equipment
10 identical retractable ballpoint pens
1 stopwatch
1 video camera

products are packaged. Packaging must protect the product. It should also attract attention on a store shelf.

To promote sales, manufacturers must advertise their products. For consumer products, television, newspaper, and magazine ads are used.

Salespeople are employed by many companies. The salespeople are paid according to their sales. The more products they sell, the more money they are paid.

COMPUTERS IN MANUFACTURING

Computers play an important role in manufacturing. They are used to design products, to control processes, to control machines, and to manage businesses.

Computer-aided design/computer-aided manufacturing (CAD/CAM) is a combination of two separate systems. This means that a computer can be used to design a product. That same computer

five products using both processes. If you are working as part of a two-person team, a volunteer will need to time the process. Which process takes less time?

6. Complete three to five trial runs using each process.

7. Record the results of each trial run on a chart of your own design.

Select the Best Solution

As a team member, discuss the most efficient method for assembling the products. Make a decision regarding the production method. Prepare for a production run that will be timed and videotaped.

Implement the Solution

Time and videotape the production run. View the videotape. Determine if the process can be improved. If necessary, make another production run. As before, time and videotape the run.

Evaluate the Solution

Write a brief script outlining the production process and the team decision-making process. Using this script, narrate the videotape. Present the videotape to the class.

might then be used to also control the machines used to manufacture the product.

Computer-numerical control (CNC) machines are an important part of CAM systems. They are programmed to perform a series of operations over and over. CNC mills and lathes are used for cutting, drilling, and turning. Some of these machines can produce thousands of identical parts each day. When production of a particular part is complete, the CNC machine is reprogrammed to make a different item.

In *computer-integrated manufacturing (CIM)*, computers monitor and control every aspect of manufacturing. Computers link design and production operations with purchasing, accounting, inventory, shipping, sales, and payroll.

IMPACTS

Economy

Manufacturing is an important part of most modern economies. It accounts for nearly 20 percent of the gross domestic product in the United States. The *gross domestic product (GDP)* is the total value of goods and services produced each year. Today, however, many companies have their products made in foreign countries. Labor costs are often lower.

Society

We benefit from manufacturing in many ways. New products make life easier and increase our leisure time. New medical products can improve the quality of life for people of all ages.

The number of workers and the skills that manufacturers require are constantly changing. Every year some jobs become obsolete and new types of jobs are created. In many industries, the need for unskilled workers is declining. Automated equipment is used instead. At the same time, the need for skilled technicians is rapidly increasing. These technicians are needed to set up and maintain complex equipment.

Safety

Most factories are safe places to work. To reduce injuries, workers are required to wear safety equipment. Safety glasses, hard hats, and steel-toed shoes are commonly required.

Workers who perform the same task all day can develop repetitive motion injuries. To reduce the risk of this happening, workers are often rotated through several different jobs each day.

Table 9-A shows colors that signal safety messages.

The Environment

Many factories use materials that can harm the environment. These materials must be handled properly. Strict government regulations have been passed. Industries have cooperated with the government to reduce air and water pollution.

Recycling

Manufacturers can reduce their costs by reprocessing materials. Automobiles and other products are being designed so that they are easy to disassemble and recycle. Several steel companies have developed systems to produce high-quality industrial materials from scrap. Fig. 8-8.

Consumers can help by purchasing products made from recycled materials. Participating in community recycling programs is also helpful. These programs help preserve raw materials. They also provide income to the community. Paper, glass, aluminum, and plastic can be sold. Disposal costs are reduced.

Table 9-A. Safety Colors	
Color	**Meaning**
Red	Danger or emergency
Orange	Be on guard
Yellow	Watch out
White	Storage
Green	First aid
Blue	Information or caution

THE FUTURE

Manufacturing is continually changing. One important trend will be an increased emphasis on *teamwork*. Teams will be given the responsibility to make important decisions. Team members will plan assembly operations. They will also decide who to hire when a new worker is needed. Teams will also be asked to design new products, develop cost-saving techniques, and maintain their equipment.

Automation will continue as manufacturers seek to reduce costs and increase product quality. However, installing machines does not always mean

▶ **Fig. 8-8** Recycling expands the quantity of available materials for manufacturing.

that workers will be replaced. Some companies will design machines to make work tasks easier. Workers will be able to work more efficiently. Productivity will improve.

Manufacturers will continue to use materials creatively. The use of lighter materials such as plastics, aluminum, and composites will increase. This will reduce the cost and weight of products. More recycled materials will be used.

New products will be developed more frequently. Ongoing education and retraining programs will be offered for all employees. Continuously improving programs and practices will help manufacturers compete in the *global economy*.

Apply What You've Learned

Design and Build Product Packaging

State the Problem

Imagine that a candy manufacturer is willing to donate a large quantity of wrapped mints to your school. In return, you have been asked to design a unique and attractive package that will hold ten mints.

Product packaging is important. It is one of the last steps in the manufacturing process. There are several reasons for product packaging. The primary reason is to provide protection. Very few small products are shipped without packaging. This is because they generally require protection. A package must offer enough protection to keep the product intact. For example, if candy mints are being packaged, the package must be strong enough to keep the mints from being broken during normal handling and shipment.

Have you ever noticed the number of small items that are sold in the area around a supermarket cash register? Protection is not the only reason for packaging. There are other reasons. Because many smaller items are bought regularly or impulsively, packaging is used to draw the buyer's attention. To attract attention, a package must have an eye-catching design.

<table>
<tr><th colspan="2">Collect Materials and Equipment</th></tr>
<tr><td colspan="2">individually wrapped circular mints
paper of assorted colors and weights
lightweight card stock of assorted colors and weights
plastic wrap
tape
color markers
color pencils</td></tr>
</table>

Fig. A

Such a design may attract a first-time buyer. Of course, the design will be familiar to the repeat customer. Because the design is familiar, it allows the repeat buyer to easily find the item he or she is looking for. In this way, an attractive package will help build new and repeat sales for a business.

Develop Alternative Solutions

1. Remember that the manufacturer needs a new product package that will encourage people to buy the mints. Disassemble several small boxes. This will show you how some packages are constructed. You might disassemble a paper clip box, a toothpick box, and a fast-food carton. Note how they are constructed. A package can be made strong not only through materials, but also through design. Remember that some shapes are stronger than others. Remember also that packaging material adds weight. This increases shipping costs. Keep this in mind as you choose packaging materials.

2. Sketch several possible package designs. In sketching the designs, remember the reasons for packaging. Remember that the package material will need to form a protective shape around the product. To provide this shape, it will need to be folded or cut into pieces. If the material is to be folded, make sure that you show the fold lines on your design. If the material is to be cut to form the package, make sure that you show the cut lines. Show also how the parts are to be joined together. Specify what material (i.e., glue or tape) will be used to join the parts.

3. Using your sketches, create three prototype packages. Use the markers and pencils to provide written information about the mints and to add color. Examples of various package designs are shown in Fig. A.

Select the Best Solution

Show the three prototype packages to several classmates. Ask for suggestions that will improve your designs. Select the design that you think will be most effective.

Implement the Solution

Using your chosen design, construct a final appearance model of the mint package.

Evaluate the Solution

Prepare a brief written or oral presentation. Explain why your package is unique. Point out features that will increase sales of the company's mints.

CAREERS IN
Manufacturing

QUALITY ASSURANCE TECHNICIAN

Rapidly growing manufacturer has position in quality assurance. Will assist engineer in implementation of quality-control program that ensures continuous production of products. Also responsible for conducting inspections and testing. Individual with degree and exceptional communication and computer skills should submit resume to: Stilson Manufacturing, Inc., 6121 Huntley Road, Marietta, CA 99804.

INDUSTRIAL ENGINEERING TECHNICIAN

Manufacturer of water processing systems seeking technician to join our growing team. This position will perform statistical studies, analyze production costs, and assist engineers in developing cost-effective operations. Applicants must have a two-year degree in engineering technology. We offer a unique work environment with a competitive salary. Submit resume to: Kinetic, Inc., 10854 Kingsman Road, Newbury, NJ 10904.

SALES REPRESENTATIVE

Manufacturer of pressure-sensitive tapes is seeking aggressive, personable individual to sell to established accounts and to generate new business. Communication and organizational skills a must. College degree is required and industrial sales experience will be a definite plus. Salary and benefits plus commission program. For prompt consideration, please fax your resume to Tees Tape, Human Resources at (602) 355-1614.

MANUFACTURING TECHNICIAN

Machine tool manufacturer is looking for a technician with the desire to take on challenging projects and follow them through from start to finish. Work closely with engineers and operators to integrate new machinery into production lines. Two-year associate's degree in electronics is required. Good technical writing skills and an understanding of mechanics and hydraulics needed. Send resume to: Power Manufacturing Company, 7711 Pleasant Valley Road, St. Louis, MO 43635.

MECHANICAL DRAFTER

Engineering firm is seeking drafter with good conceptual design skills. Will prepare detailed working drawings of mechanical devices showing dimensions and tolerances. CAD training a must. Send resume to: Human Resources, Osborn Engineering, 1300 East 85th Street, Youngstown, OH 64088.

Linking to the WORKPLACE

Identifying Materials. Manufacturing companies are often organized around the materials they use to make products. These materials include steel, aluminum, plastic, glass, fabric, rubber, petroleum, and wood. Often, a variety of manufacturers contribute to making a single product. List the various parts of an automobile. Identify the material used to manufacture the parts you have listed. Each of these materials resulted from the processing of a raw material. You can see that manufacturers often depend on hundreds of other companies for materials.

Chapter 8 Review

SUMMARY

▶ Natural materials are found in nature. People create synthetic materials.

▶ Primary processes are processes that change raw materials into industrial materials. Secondary processes are processes that turn industrial materials into finished products.

▶ The manufacturing system includes design and engineering, purchasing, production, and distribution and sales.

▶ Three types of production systems include custom, job-lot, and continuous. The type of product and the number of items required determine which process is used.

▶ Manufacturing has positive and negative impacts on the economy, on people, and on the environment.

▶ Manufacturing trends include the use of teams, increased automation, creative use of materials, and improvement of programs and practices.

CHECK YOUR FACTS

1. Describe the difference between natural and synthetic materials. Give two examples of each.

2. What is a composite?

3. What are primary processes?

4. What are secondary processes?

5. Name three ways in which forming can be done.

6. Name the three types of conditioning.

7. List the basic steps in the manufacturing system.

8. Describe the three types of production systems used in manufacturing.

9. How are computers used in computer-integrated manufacturing?

10. What is the gross domestic product?

CRITICAL THINKING

1. Describe how thermoplastics and thermoset plastics are different.

2. Make a series of sketches to show two examples of each of the four secondary processes discussed in this chapter. Label the processes shown.

3. A company manufactures artificial Christmas trees for sale in winter. Another product is needed for spring and summer sales. What product could the company make? What kind of manufacturing system would probably be used?

4. Why is teamwork important in manufacturing?

5. Describe some ways in which manufacturing is expected to change in the future. What changes might occur that are not discussed in this chapter?

Structures

OBJECTIVES

▶ identify the four parts of a system.

▶ describe the loads and forces that act on structures.

▶ be able to explain the difference between a static load and a dynamic load.

▶ identify the main parts in the system used to build a house.

KEY TERMS

building codes

force

foundation

load

plans

smart materials

specifications

structural member

structure

superstructure

Imagine that a new house is being built across the street from you. The house is nearly finished. You and a friend have been eyeing a small pile of scrap materials in front of it. Finally, you get up the courage to ask the builder if you can have the lumber.

"Yes," the builder answers. He says that the scrap materials are yours. In the pile, you find four sheets of plywood, twelve 2 x 4's, a small amount of vinyl siding, and a can of nails.

You want to build something, but what could you build? A tree house? A doghouse? A shed? Each of these is a structure. This chapter will give you some ideas about how a structure is planned and built. For our example, we'll use a small house.

BUILDING IS A SYSTEM

As you read this chapter, think of the building of a house—or any structure—as a system. A *system* is an organized procedure for doing something. There are four parts to any system: inputs, processes, outputs, and feedback.

- *Inputs* are activities and items that go into a system.
- *Processes* are actions that accomplish a result.
- *Outputs* are results.
- *Feedback* is the comments on the outputs.

There are inputs, processes, outputs, and feedback at every stage of a building system. As we discuss the building of a house, try to identify each of these parts of the system.

WHAT IS A STRUCTURE?

A **structure** is something that is constructed, or built. Structures are made by joining parts to meet a certain need or perform a certain task. There are natural structures and human-made structures.

Examples of natural structures include a spider web, a bird's nest, and a wasp nest. Fig. 9-1. Examples of human-made structures include houses, high-rise buildings, and bridges. Fig. 9-2. What

▶ **Fig. 9-1** Structures exist in nature. These natural structures provide us with design ideas. What are these natural structures designed to do? How does the design meet each structure's purpose?

▶ **Fig. 9-2** Many human-made structures are buildings for specialized uses. We live in residential structures such as homes and apartments. Goods are manufactured in factories, which are examples of industrial buildings. Civil construction projects, such as courthouses and public libraries, meet public needs. Stores, restaurants, and office buildings are examples of commercial structures. Bridges, piers, and highway interchanges help serve our transportation needs. Dams can provide electricity. What other types of structures do you see here? What purposes do those structures serve?

similarities do you see between natural structures and human-made structures?

The design of any structure depends on its use. For example, a dam used to control the Colorado River must be rigid. A hot-air balloon must be flexible. A television-signal transmitting tower must be tall.

Figure 9-2 shows several human-made structures. Can you see that human-made structures have specific uses? For example, people *live* in houses. They *work* in factories and office buildings. This was not always the case. Hundreds of years ago, most people lived and worked in the same place. We now have buildings for specific uses because the range of human activities has expanded.

Think about why the buildings in the neighborhood of your school were built. When they were built, what human needs were they meant to satisfy? Do they still satisfy those needs?

FORCES ON STRUCTURES

A **force** is a push or a pull that transfers energy to an object. Forces on a structure can be external or internal. *External forces* are those that one body (substance) exerts on another. They are applied forces, or forces acting *upon* a structure. *Internal forces* are those that one *part* of a body exerts on an adjacent or adjoining *part* of the same body. They are forces acting *within* a structural material.

The internal forces that act within structural materials include tension, compression, shear, and torsion. A material that is being pulled is in *tension*. A material that is being pushed is in *compression*. A material that is being pushed in opposite directions along *adjacent planes* (bordering surfaces) is being subjected to *shear*. A material that is being twisted is being subjected to *torsion*. Fig. 9-3.

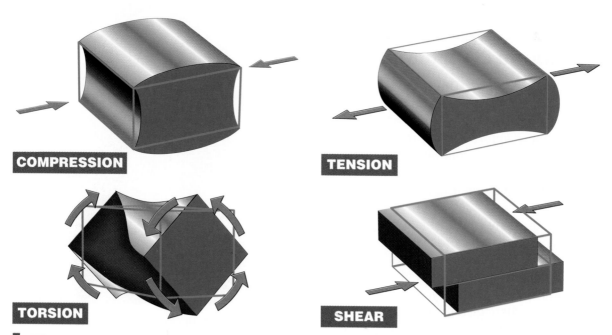

COMPRESSION

TENSION

TORSION

SHEAR

▶ **Fig. 9-3** The four types of internal forces: **(A)** compression is the force that tends to shorten an object or even crush it, **(B)** tension is the force that tends to stretch an object or even pull it apart, **(C)** torsion is the force that tends to twist an object along its axis, **(D)** shear is the force that tends to push adjacent parts of a material in opposite directions.

A **load** is an external force on an object. A load on a structure can be a weight of some sort. It can also be force caused by wind pressure or water pressure. Structures must be built to withstand loads. There are two types of loads: static and dynamic.

A *static load,* also called a *dead load,* changes slowly or not at all. The materials used to build a structure are part of this kind of load. For example, the bricks in a building are part of its static load. The twigs in a bird's nest are part of its static load. Can you think of other examples?

Dynamic loads, or *live loads,* move or change. A car crossing a bridge and oil flowing through a pipeline are examples of dynamic loads. Wind blowing on a building and waves pounding on a seashell are also examples of dynamic loads. Fig. 9-4.

STRUCTURAL MATERIALS

The materials used to build a structure are subjected to various loads and forces. The materials will help determine the structure's strength, cost, and appearance. Commonly used structural materials include steel, wood, brick, concrete, aluminum, and plastic.

It is important to choose the right material for each part of a structure. You must also choose the way the material will be used. Before choosing the materials for a structure, you should learn how they will react to loads and forces. Testing is a good way to gather this data.

▶ **Fig. 9-4** Both static and dynamic loads must be considered when designing a structure. Can you identify the static and dynamic loads on this suspension bridge?

Linking to COMMUNICATION

Word Meanings. The same words can be used in different subjects to mean different things. Review the meaning of *static* and *dynamic* as they refer to loads in technology. Ask your language arts teacher to help you understand *static* and *dynamic* as they refer to characters in stories.

Are there any similarities in the meanings of these words when talking about loads and when talking about characters? Give examples of static and dynamic loads. Explain your choice of examples. Also, give examples of static and dynamic characters. Explain your choice of examples.

STRUCTURAL MEMBERS

A **structural member** is a building material connected to another structural member to make up the frame of a structure. Wooden studs, joists, and rafters, discussed later in this chapter, are structural members typically used to frame houses. Steel beams and columns are structural members typically used to frame towers, bridges, and large buildings.

Horizontal structural members, or supports, are known as *beams*. The top and bottom surfaces of a beam are subjected to the greatest internal forces. Fig. 9-5. Beams can be strengthened by giving them shapes such as those in Fig. 9-6. What other shapes would add strength?

Vertical structural supports are known as *columns*. Columns must have high compression strength to support the weight of a structure.

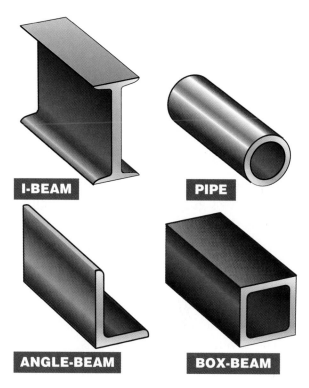

I-BEAM PIPE

ANGLE-BEAM BOX-BEAM

▶ **Fig. 9-6** These structural members are used in the framework of many different kinds of structures.

▶ **Fig. 9-5** This greatly exaggerated illustration shows what happens when you place a weight on a beam.

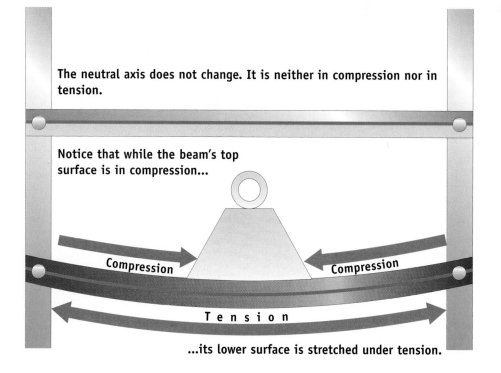

The neutral axis does not change. It is neither in compression nor in tension.

Notice that while the beam's top surface is in compression...

Compression   Compression

T e n s i o n

...its lower surface is stretched under tension.

Bridges

A bridge extends a roadway across a land obstacle or over water. On land, bridges are used to cross gullies and ravines, as well as highways and rail lines. Bridges also are used to cross streams, rivers, and bays.

The design of a bridge depends on the obstacle being crossed and the load the bridge will carry. For example, a simple rope suspension bridge might be used to allow people to cross a mountain ravine. However, a suspension bridge such as the Golden Gate Bridge is needed to carry traffic across San Francisco Bay. The obstacle being crossed will determine the length of the bridge.

The seven basic bridge designs are shown here. Of these designs, the beam bridge is most commonly used.

BEAM BRIDGE

This bridge is built from steel or concrete beams, or girders. The beams provide horizontal supports on which the concrete roadway rests.

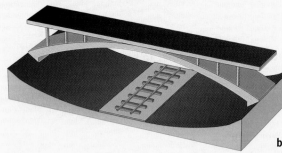

ARCH BRIDGE

The load on the roadway of this bridge is carried by the arch. The arch is supported at each end by a support called an abutment. Here the roadway is shown above the arch. However, the roadway can also be below the arch.

CABLE-STAYED BRIDGE

Cable-stayed bridge. This bridge supports the roadway by cables that run from towers to the roadway. In this way it is similar to a suspension bridge. Most cable-stayed bridges have been built outside the United States. American engineers are not convinced that the bridges have the strength and durability needed for use in the United States.

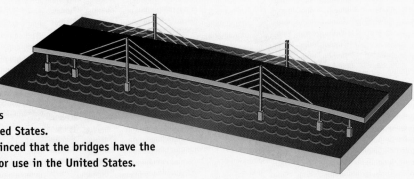

SUSPENSION BRIDGE

Tall towers on both sides of the roadway support the main cables. These cables run the entire length of a suspension bridge. These main cables are anchored in concrete at each end. Smaller cables are suspended from the main cables. These suspended cables support the roadway.

MOVABLE BRIDGE

This bridge is usually used to span canals and rivers that carry heavy boat traffic. Also called a lift bridge, this bridge has a section of roadway that can be raised to allow large ships to pass.

TRUSS BRIDGE

This bridge uses trusses to carry the load of the roadway. A truss is a framework formed from triangles. The truss may be placed above or below the roadway. A truss bridge is strong and economical to build. A truss may be used in other bridge designs to add strength.

CANTILEVER BRIDGE

A cantilever is a beam that extends from each end of the bridge. A cantilever does not reach all the way across the bridge. Cantilevers are connected in the middle of the bridge by a part called the suspended span.

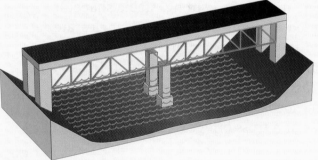

STRUCTURAL SHAPES

The triangle is a strong, stable shape. It is used in many structures. Triangles are used in the construction of towers and in domes. Triangles are also used in trusses. A *truss* is a triangular framework that can carry loads. Perhaps you have seen a roof truss or a truss bridge.

The *arch* is another common structural shape. An arch transfers the load it carries to the earth below. The Romans used masonry to build arch bridges. Today we use concrete and steel.

BUILDING A RESIDENTIAL STRUCTURE

A house is a residential structure. As mentioned, a system is used to construct a residential structure. You may remember that a system requires inputs, processes, outputs, and feedback. Try to identify these as we present a short outline of house construction.

Site Selection

Choosing the right location for a structure is important. A family with three children might want to build their home where there are other families with children. A retired couple might want a small home near stores and public transportation. A person planning a summer cottage might want to build it in the country on a lake or hilltop. The site will help determine the plans.

Plan Preparation

Before a structure can be designed, its desired features must be identified. An *architect* (ARC-uh-tect) is a person trained in building design. An architect designing a new home will ask questions such as:
- Who will live in the house?
- How big should the house be?
- How big will the lot be?
- Will the building site be level?
- Which rooms will be used the most?

- How many bedrooms will be needed?
- When must the house be completed?
- How much money can be spent?

Based on the information gathered (input), the architect can make up a set of plans. Plans are also called working drawings or blueprints. **Plans** are drawings that show the builder how to construct the structure. The finished plans (output) may change as the architect receives comments (feedback) from customers. He or she may then act on their comments to produce a new set of plans (output). Can you see that a process is being used? The process involves input, output, and feedback. A similar process occurs at each stage of building construction.

Small sections of the drawings required for most residential projects are also given. These include:

- A *site plan*. This shows the location of the building(s) on the lot. It also shows sidewalks, driveways, utilities, and streets. A site plan is also called a *plot plan*.
- *Floor plans*. These show the arrangements of rooms as viewed from above. A separate drawing will be made for each floor level. Fig. 9-7.

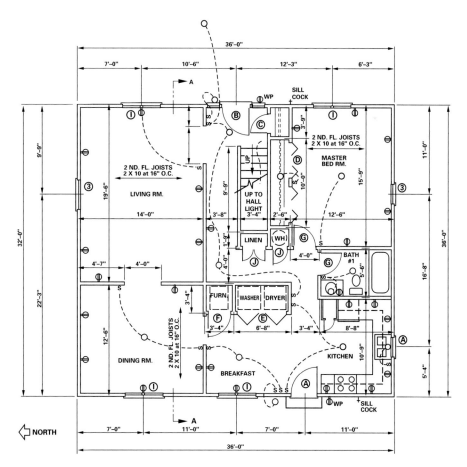

▶ **Fig. 9-7** The floor plan of a house.

- *Elevations*. These show the outside of the structure. A separate drawing (or elevation) will be made for each side of the structure. Fig. 9-8.
- *Detail drawings*. These show items that must be shown more clearly than they appear on the floor plan or elevations. Fig. 9-9.
- *Section drawings*. These show cross-sections of the inside of a structure. Fig. 9-10.
- *System drawings*. These show plumbing, electrical, and heating and ventilating systems.

A set of specifications must also be prepared. **Specifications** are written details about materials and other project-related concerns.

Fig. 9-8 An elevation drawing.

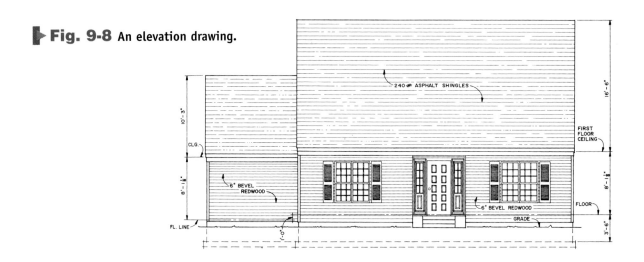

240# ASPHALT SHINGLES
16'-6"
FIRST FLOOR CEILING
10'-3"
CLG.
8'-1 1/8"
6" BEVEL REDWOOD
6" BEVEL REDWOOD
FLOOR
GRADE
FL. LINE
1'-0"
3'-6"
EAST ELEVATION
SCALE 1/4" = 1'-0"

FIN. SEC. FL.
HANDRAIL
2'-10"
FIN. FIRST FLOOR
15 RISERS AT 7 3/16" = 8'-11 3/16"
14 TREADS AT 10" = 11'-8"
STAIR DETAIL
SCALE 3/8" = 1'-0"

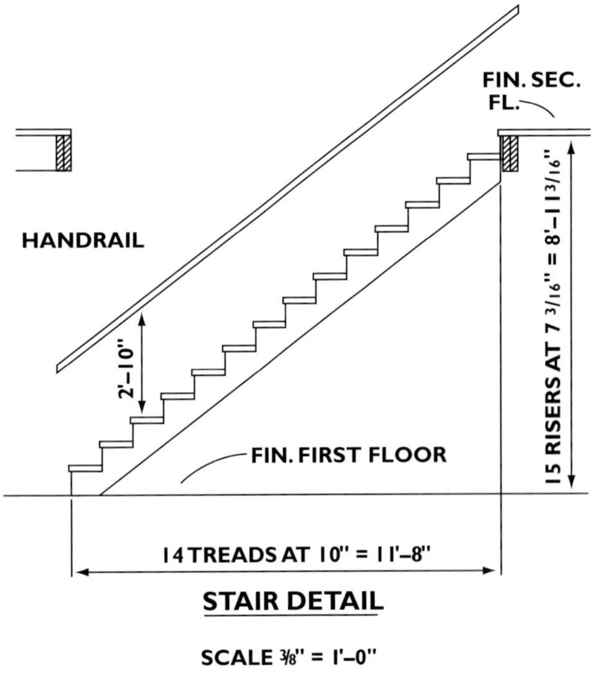

Fig. 9-9 A detail drawing.

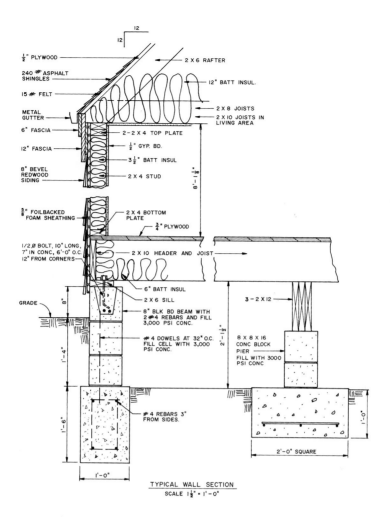

240 # ASPHALT SHINGLES

½" PLYWOOD

2 X 6 RAFTER

12" BATT INSUL.

15 # FELT

2 X 8 JOISTS

2 X 10 JOISTS IN LIVING AREA

METAL GUTTER

6" FASCIA

2 – 2 X 4 TOP PLATE

12" FASCIA

½" GYP. BD.

3½" BATT INSUL

8" BEVEL REDWOOD SIDING

2 X 4 STUD

⅝" FOILBACKED FOAM SHEATHING

2 X 4 BOTTOM PLATE

¾" PLYWOOD

1/2 Ø BOLT, 10" LONG, 7" IN CONC, 6'-0" O.C. 12" FROM CORNERS

2 X 10 HEADER AND JOIST

GRADE

6" BATT INSUL

2 X 6 SILL

3 – 2 X 12

8" BLK BD BEAM WITH 2 #4 REBARS AND FILL 3,000 PSI CONC.

4 DOWELS AT 32" O.C. FILL CELL WITH 3,000 PSI CONC.

8 X 8 X 16 CONC BLOCK PIER FILL WITH 3000 PSI CONC

4 REBARS 3" FROM SIDES.

2'-0" SQUARE

TYPICAL WALL SECTION
SCALE 1½" = 1'-0"

▶ Fig. 9-10 A section drawing.

Linking to MATHEMATICS

Drawing to Scale. A floor plan obviously cannot be drawn to actual size, so it is drawn to scale. Various scale dimensions can be used, but a popular scale is 1/8" = 1'. That means that a line 1/8" long represents 1' of actual house. If you draw a line 2 3/4" long, how many feet of house does this represent?

▲

Site Preparation

Actual construction of a structure begins with *site preparation*. The site is surveyed to establish property lines. Next, trees and debris that might interfere with construction are removed. Then, the position of the house is laid out. Refer to Fig. 9-11. It shows the basic parts of a foundation and superstructure.

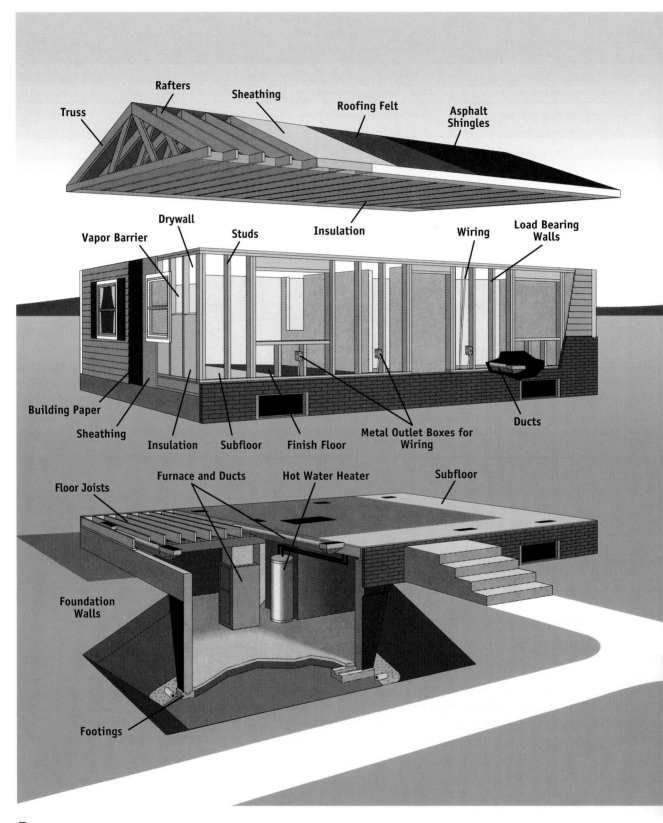

Fig. 9-11 The basic processes for building are the same for most types of structures. Note the parts of this building.

Building the Foundation

The **foundation** of a building is the part of the structure in contact with the ground. The foundation is also known as the *substructure*. The two main parts of the foundation are the *foundation walls* and the *footings*. The foundation walls are built on top of the footings. The foundation walls support the **superstructure**, which is that part of a structure above the foundation.

Some houses are built on a poured slab of concrete. Such houses do not have basements.

Building the Superstructure

The frame of the superstructure provides its basic support. Most single-family homes have a wood frame. Important structural members include studs, joists, and rafters. *Studs* are upright members that help form the walls. *Joists* are horizontal members that help form the floors and ceilings. *Rafters* help form the roof.

Framing, or building the frame of the structure, typically begins with the floor. Floor joists are usually covered with plywood to create floors. The plywood is then covered with a finished floor.

Exterior (outside) walls enclose a structure to protect it from the weather. Walls that support a structure are called *load-bearing walls*. Walls that divide a building into rooms, but that are not load bearing, are called *partitions*.

There are two ways to frame the roof and ceiling of a house. Some homes use ceiling joists and rafters. Others have trusses, which serve as ceiling joists and rafters.

The roof and exterior walls are covered with sheathing. *Sheathing* is a layer of material placed between the framing and the finished exterior of a structure.

To enclose a structure, the roof, doors, windows, and siding are added. *Utility systems* are then added in two stages. Utility systems supply the services that allow someone to live comfortably in a building. Utility systems include electrical, plumbing, and climate-control systems. Pipes and wiring are installed just after the building is enclosed. This is called *rough work*. After the walls are covered, the remaining utilities are installed.

Finishing the Structure

Insulation helps keep a heated house warm in the winter. It keeps an air-conditioned house cool in the summer. After the rough work is complete, insulation is installed.

After being insulated, the ceilings and interior walls are covered. *Drywall* is the most common material used to cover ceilings and walls. Ceilings are covered first. To complete the drywall job, joint compound is applied to the nail holes and seams. Tape is pressed into the joint compound and additional joint compound is applied. After this has dried, two additional coats of joint compound are applied and lightly sanded.

Explore

Design and Build a Testing Station

State the Problem
Determine how the strength of a material can be affected by changing its shape.

Develop Alternative Solutions
Refer to the beam- and column-testing setups shown here. Prepare a table that will allow you to record the results of your tests. Decide on the items that should be represented in the table. Prepare some designs for the table.

Select the Best Solution
Select the table that you think will provide the best test record.

Collect Materials and Equipment
4" x 6" index cards
masking tape
set of standard weights
ruler
string

BEAM TESTING

COLUMN TESTING

Molding is then added around the doors, windows, and floors. Cabinets are installed. The interior is painted. Finished flooring, such as carpeting or tile, is installed. Exterior finishing, which includes painting and landscaping, is then done.

CONTROL SYSTEMS

Control systems help ensure that a structure of acceptable quality has been completed on time at the expected cost. The systems used to control the quality of residential construction include zoning

Implement the Solution

1. Prepare the table in which you will record test results.
2. Make each of the shapes shown using a single index card for each shape. Hold each beam together with small pieces of masking tape. Use no more than 2 linear inches of tape per beam.
3. Find the force required to buckle the beam by testing, as shown.
4. Record the test results in the table.
5. Create smaller beams of the same shape and length.
6. Test the beams for strength as before.
7. Again record the results in the table.
8. Conduct similar tests to determine the strength of columns in these same shapes.
9. Record your results in a different table.

Evaluate the Solution

1. Which shapes were strongest during the first beam test?
2. Did reducing the size of the beams affect strength?
3. Compare the results of the beam and column tests. Which shape(s) performed well in both tests?

laws, building codes, building permits, and building inspections. The systems used to control the building costs include written contracts and agreements between the builder and the client.

Zoning Laws

Most communities have *zoning laws*. These regulate the kinds of structures that can be built in each part of the community. Zoning laws also usually specify minimum property size. They also specify how close a structure can be to the property line.

Building Codes and Building Permits

State and local governments enact **building codes.** These specify the methods and materials that can be used for each aspect of construction. Before construction can begin, most communities require a *building permit* that approves the structure for construction.

Building Inspections

In many communities, a building inspector will visit the site periodically. The inspector will make sure that proper construction methods are being used. The inspector usually checks footings, framing, and electrical and plumbing systems.

Before an owner accepts a building from the builder, the owner makes a final inspection. Needed corrections are written on a *punch list*. When all the corrections on the punch list have been made, the structure is considered finished.

IMPACTS

Structures benefit people in many ways. Houses and apartments provide shelter. New roads and airports improve our transportation system.

Construction creates employment opportunities for many people. Architects and drafters design new structures. Decorators suggest ways to improve existing buildings. New stores and office buildings create jobs for construction workers, as well as permanent jobs. Today, construction workers are the largest group of skilled workers in the United States.

Construction can also have negative impacts. The noise and debris created during the construction of a structure can be annoying. A new shopping mall may create traffic problems. Accidents and injuries occur during construction activities. These can be kept to a minimum by observing safe work habits.

FASCINATING FACTS

Adobe bricks are often made of clay, straw, and water. These materials are kneaded together and pressed into a mold. After being baked in the sun for a few weeks, they are often finished with a coat of moist adobe or lime. Adobe is a good building material in dry climates. It is fireproof and an excellent insulator.

THE FUTURE

Advances in technology will continue to change the way structures are designed, built, and used. Computer-aided drafting (CAD) systems make it possible to design structures more quickly and at less cost. The use of manufactured components such as roof trusses and floor joists improves quality. It also reduces construction time. Factory-made houses will continue to improve in quality and increase in popularity. Fig. 9-12.

In just a few years, we can expect to see more smart buildings and even some smart materials. *Smart buildings* are buildings in which computers control lighting, heating, air-conditioning, and security systems. Appliances will be controlled so that they operate when utility rates are lowest. **Smart materials** are materials that have built-in sensors to warn of unsafe conditions. For example, after an earthquake, the materials in a bridge would change color to show that the structure had become unsafe.

▶ **Fig. 9-12** Factory-made houses are assembled from manufactured components.

Apply What You've Learned

Design and Build a Small Structure

State the Problem

A structure must be able to support the weight of its building materials. A structure must also be strong enough to support furniture, people, vehicles or other weight depending on the use of the structure. A structure must be strong enough to resist forces such as wind, snow load, and vibration.

Design a small structure capable of supporting as much weight as possible. The structure must meet the following specifications:

- The structure cannot exceed the weight limit set by your instructor (e.g., 500 grams).
- The structure should be no larger than the size specified by your instructor (e.g., must fit within a 6-inch cube).
- The structure should have a flat surface on top and bottom to allow for easy testing.

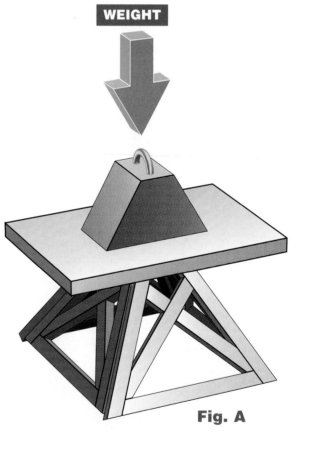

WEIGHT

Fig. A

Develop Alternative Solutions

Design and draw a small structure that could be made from the materials provided. You will probably need to prepare several design sketches. One possible design is shown in Fig. A. This design is presented only to give you a general idea of what will be needed. Don't copy this design.

Select the Best Solution

Select the design you think will be most effective.

Implement the Solution

1. Cut the wood pieces to length and assemble the structure.
2. Weigh the structure.
3. Test the structure to determine how much weight it will hold. Your instructor will assist you.
4. Compute the weight-to-strength ratio. The more weight a structure can hold (per unit of weight of the structure), the stronger and more efficient its design.

Evaluate the Solution

1. Is the structure within the specified weight limit?
2. Did the structure have a weight-to-strength ratio as good as or better than the average of the other structures built by the class?
3. Was it possible to determine exactly where the structure failed as the load was applied? (Filming the test with a video camera and playing the film back frame by frame, will provide the answer.)

CAREERS IN
Structures

CONSTRUCTION ELECTRICIAN

Commercial construction company needs trained electrician to install electrical systems in office buildings. Must ensure that work conforms to state and local building codes. Contact American Construction at (312) 466-0902 for additional information.

INTERIOR DESIGNER AND DECORATOR

Growing downtown firm with wide ranging interiors practice seeks interior designer with project experience and solid computer-assisted design background. Artistic talents and attention to detail required. Knowledge of colors and textures important. Send resume to: Box BAO13, The Design Firm, 1905 North Superior Avenue, Cleveland, OH 44114

CARPENTER

Full-time needed immediately. Commercial construction company offers year-round work. Drywall finishing a must. Experienced only need apply. Benefits. Call (312) 656-9353 for details. Office open on Sunday.

CIVIL ENGINEERING TECHNICIAN

Engineering firm has entry-level position for a technician with a strong working knowledge of CAD. Associate's Degree with emphasis in civil, highway, or environmental plan preparations required. Submit resume to: MRS Consultants, Inc., 6500 Bush Blvd., Topeka, KS 45987

SURVEY TECHNICIAN

Entry-level position for technician to work on surveying crew. Dependable individual with strong work ethic needed. Experience helpful but not required. No phone calls accepted. Send resume to: Hoffman-Metz, Inc., P.O. Box 343, Richmond, VA 42286.

ARCHITECTS

Immediate opening for registered architect to manage medium-sized retail and commercial project from initial planning through construction phase. Bachelor's degree required with computer-assisted drafting (CAD) knowledge preferred. Send resume or call: Stephen Karls, Cedar Architectural, Inc., 1567 Merriman Road, Atlanta, GA 30030, (404) 836-9972.

Linking to the WORKPLACE

There was a time when job ads in the newspaper were divided into "Help Wanted—Male" and "Help Wanted—Female." Federal laws now require that people be hired based on their qualifications, not their gender. There are still jobs held mostly by women or by men, but that is changing. Some women climb telephone poles and some men are secretaries. Are the careers you are considering held mostly by men, women, or both?

Chapter 9 Review

SUMMARY

▶ A system is an organized procedure for doing something. There are four parts to any system: inputs, processes, outputs, and feedback.

▶ A structure is something that is constructed, or built.

▶ A load is an external force on an object.

▶ The internal forces that act on structures include tension, compression, shear, and torsion.

▶ Beams, columns, studs, joists, and rafters are structural members.

▶ Building a structure involves building a foundation, or substructure, erecting the superstructure, and finishing work.

▶ Control systems control the quality of residential construction.

CHECK YOUR FACTS

1. What are the four parts of a system?

2. Give three examples of natural structures.

3. Explain the difference between a static load and a dynamic load.

4. Describe the loads and forces that act on structures.

5. What internal force would be present in a twisted beam?

6. What major force must the foundation of a home withstand?

7. How are beams and columns used in a structure?

8. What information does an architect need before designing a home?

9. Explain the difference between the foundation and the superstructure.

10. Identify the main parts in the system used to build a house.

CRITICAL THINKING

1. Imagine that you are going to build a structure. Discuss the type of materials that you would use for your structure. Explain why you selected them.

2. Name and sketch four different kinds of bridges.

3. How do inspections help ensure the quality of a structure?

4. Explain how a new structure can have both positive and negative impacts on a community.

Power Technologies

CHAPTER 10 *Flight*

CHAPTER 11 *Land and Water Transportation*

CHAPTER 12 *Fluid Power*

What do a rocket, a lawn mower, and a dentist's drill have in common? They are all examples of power technology. They convert and transmit energy to do work.

The chapters in Section 4 describe power technologies. You will learn how airplanes, rockets, and other forms of transportation are able to move. You will learn also about fluid power systems, such as those in the dentist's drill. As you learn about power technologies, think about the many ways they apply to our world.

Technology and Society

A Powerful Technology

Power technology began more than 300,000 years ago, when humans learned to use fire. They converted the energy in wood or other fuel into heat and light. Today every aspect of life is affected by the ways in which humans use energy.

Social Impacts

Power technology affects where we work and live. Tractors and other machines make farming more efficient. Fewer people are needed to grow food. In the United States, only 2.5 percent of the people work on farms. Most people who do work on farms don't live there.

Think of ways city life depends on power technology. What would happen, for example, if there were no electrical power?

Economic Impacts

Power technology has a huge economic impact. For example, the world's factories produce about 50 million motor vehicles (such as cars and trucks) each year. Think of the many jobs in designing, making, selling, and repairing vehicles. The people who earn money at these jobs spend it on housing, clothing, and other items, creating jobs in those industries.

Environmental Impacts

Energy use always creates some waste. For example, car engines create waste heat, as well as air pollution. These waste products pose a health hazard and may contribute to global warming. Yet it is unrealistic to think the world's people will give up cars. What might be some other solutions?

Linking to the COMMUNITY

Find out how electricity is generated in your community. What energy sources are used? How much electricity is generated? Who are the major users? Report your findings to the class.

What Is Energy?

Energy is the ability to do work. Work, in this sense, isn't just what you do to make money. It's any transfer of energy through motion. In a car, energy from gasoline or diesel fuel is transferred through the engine and to the wheels. The car moves; work has been done.

There are several forms of energy. Heat, light, sound, chemical, nuclear, mechanical, and electrical energy are all used by technology. Energy can be converted from one form to another, but it cannot be destroyed under ordinary conditions.

Wherever work is being done, energy is being converted. In the example of the car, chemical energy in the gasoline or diesel fuel is converted to heat energy in the engine, which in turn is converted to mechanical energy to make the car move.

It's important to remember that work involves both force and motion. In a car's engine, for example, the pressure (the

The sun provides solar energy. Indirectly it is also the source of wind and wave energy.

Wind turbines convert the energy of wind to electricity.

Geothermal energy, the heat within the earth, can be used to heat buildings or generate electricity.

ENERGY SOURCES

Fossil fuels--coal, oil, and natural gas--come from plants and animals that died millions of years ago. Because they take millions of years to form, fossil fuels are considered nonrenewable. In other words, the current supplies are all we will ever have.

force) of a burning air-fuel mixture causes a piston to move. If the piston did not move, the energy would simply heat the engine parts. No work would be done.

What is Power?

Power is the rate of work. This can be stated as a mathematical formula:

$$Power = \frac{Work}{Time}$$

From this formula, you can see there are two ways to increase power. You can increase the amount of work or decrease the amount of time.

Power Systems

Power technologies involve systems for moving and controlling power. Power systems can be mechanical, electrical, or fluid. All include a source (input); transmission and control (process); and output (use). In chapters 10-12, you will learn about power systems used for transportation and for tools and equipment.

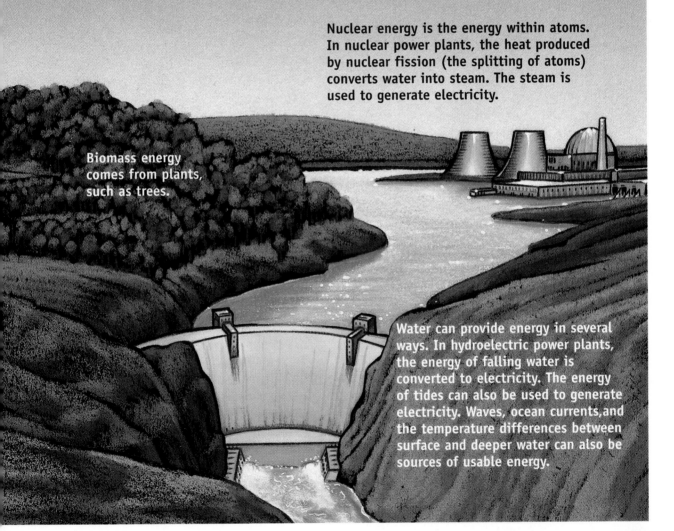

Nuclear energy is the energy within atoms. In nuclear power plants, the heat produced by nuclear fission (the splitting of atoms) converts water into steam. The steam is used to generate electricity.

Biomass energy comes from plants, such as trees.

Water can provide energy in several ways. In hydroelectric power plants, the energy of falling water is converted to electricity. The energy of tides can also be used to generate electricity. Waves, ocean currents, and the temperature differences between surface and deeper water can also be sources of usable energy.

CHAPTER 10

Flight

OBJECTIVES

▶ describe the forces that oppose motion.

▶ identify and discuss the forces that allow an airplane to fly.

▶ identify and discuss the processes used to control flight in an airplane.

KEY TERMS

aerodynamics

aerospace

ailerons

airfoil

Bernoulli effect

drag

fluid friction

force

gravity

inertia

Sarah and David stand motionless on the top edge of the "half-pipe" ramp at Skateboard Park. Their sneakers grip firmly onto the rough surface of their boards. With a quick jerk of their legs, they begin the drop down the ramp. They glide faster and faster. At the bottom of the ramp they're at top speed and begin the climb up the second side of the ramp. They reach the top of the ramp and, with a final burst of energy, they become airborne. For a few seconds, David and Sarah are flying, but in the blink of an eye they return to the ramp and the process begins again. Each time Sarah and David reach the top of the ramp they become airborne—but only for a very short time.

Why are David and Sarah's flights so short? When airborne, why don't they continue to soar like a bird or plane?

Aerospace (air-oh-SPACE) is the study of how things fly. By studying aerospace technology, we can find out the answers to these and other questions about flight.

FORCES

Things move only when a force is applied to them. A **force** is a push or pull that transfers energy to an object. Force can set an object in motion, stop its motion, or change its speed and direction.

On a windy day, you can feel the force of moving air push against your body. The force of a magnet can pull paper clips towards it. The force of gravity pulled David and Sarah down the ramp. Fig. 10-1.

Birds, airplanes, rockets, and all things that fly need to apply force so they can become airborne.

Objects that are very heavy need a greater force to make them move. When heavy objects are moving, it takes a greater force to stop them or change their direction. An object will remain still or will continue to move in the same straight line unless an outside force acts on it. This property of matter is called **inertia** (i-NURR-shuh). It was inertia that pushed

David and Sarah up the second side of the ramp. All things that fly rely on inertia to keep moving.

Newton's Three Laws

Isaac Newton was an English scientist. He lived over four hundred years ago. He proposed three laws of motion. These three laws have helped scientists understand the forces that affect an object.

Newton's *first law of motion* states that a body will remain at rest unless a force acts on it.

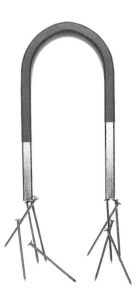

▶ **Fig. 10-1** Forces transfer energy in different ways. What force is pulling down on the skydiver? What force is pulling up on the nails?

Newton's *second law of motion* states that the change of motion is proportional to the force acting on the body.

Newton's *third law of motion* states that for every action there is an equal and opposite reaction.

Each of these laws of motion applies to Sarah and David and their skateboard runs. As you read this chapter, see if you can discover the answer to the following questions:

• What forces caused Sarah and David to begin to move on their skateboards?
• What forces caused them to slow down?

Forces That Oppose Motion

Certain forces oppose or act against objects in motion.

Friction is the force that brings moving objects to rest. Friction occurs when objects come in contact with each other. Friction acts in an opposite direction to motion, causing objects to slow down and then stop. Fig. 10-2.

As soon as David and Sarah began to move up the second side of the ramp they started to lose inertia or energy. They began to slow down and lose motion. The force of friction was working against them.

Gravity is a force that pulls objects towards the center of the Earth. It was gravity that gave Sarah and David the force needed to accelerate down the ramp. The force of gravity also pulled on them as they rose up the second side of the ramp. The pulling force of gravity increases as the mass of an object increases. Gravity is one reason why Sarah's and David's flights

▶ **Fig. 10-2** Friction is a force that opposes the motion of the object. There are different kinds of friction. The landing space shuttle shows rolling friction. Resistance occurs between the wheel and the ramp. Why does it take so long for a parachute to drift to the ground? Air resistance, a fluid friction, slows the fall of the parachute.

were so short. Gravity pulled them back down towards the ramp. It opposed their inertia.

Gravity and friction also pull and slow down airplanes, kites, and rockets. They will slow down any object traveling through the air.

Forces and Flight

Birds, planes, rockets, and kites fly because they generate forces that are greater than opposing forces such as gravity and friction. Fig. 10-3. Sarah and David's flights were short because the forces they generated were quickly overcome by friction and gravity.

Lift

Lift is an upward force used to overcome gravity. The shape of the wings on airplanes and even birds provides the lifting force needed to fly. (This shape is discussed later in this chapter.) When the force of lift is greater than the weight of the plane, the plane will rise into the air.

Thrust

Planes also need a forward force to fly. This forward force is called *thrust*. Thrust helps to provide lift and overcome friction.

What kind of force would Sarah and David have to generate to soar like birds from the top of the ramp?

AERODYNAMICS

A key factor in achieving flight is speed. We have learned that friction is a force that slows down and eventually stops an object in motion.

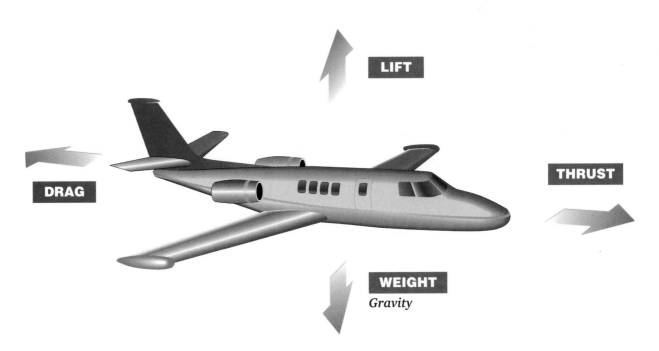

▶ **Fig. 10-3** The combined action of these four forces enables an aircraft to take off, fly, and land. Which of these four forces are opposing forces?

Fluid friction is the cause of the resistance an object meets as it moves through the air. The force of fluid friction is created when particles of air contact the moving object.

Drag is the force of fluid friction on moving objects. Fig. 10-4.

Drag can be reduced by making air move more smoothly over the surface of an object.

Aerodynamics (air-oh-dy-NAM-iks) deals with the forces of air on an object

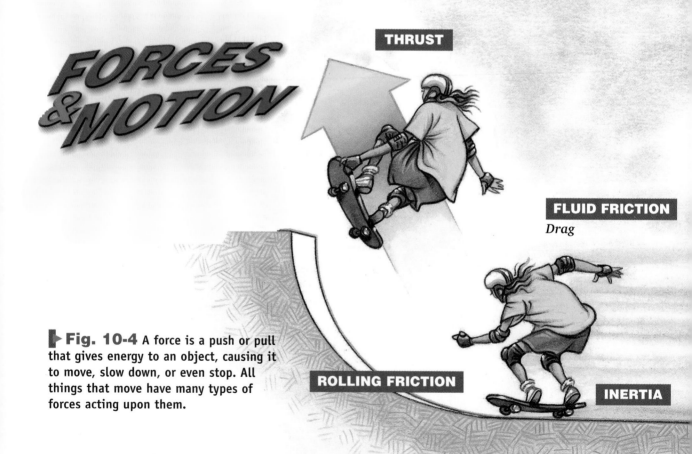

THRUST

FLUID FRICTION
Drag

ROLLING FRICTION

INERTIA

▶ **Fig. 10-4** A force is a push or pull that gives energy to an object, causing it to move, slow down, or even stop. All things that move have many types of forces acting upon them.

moving through it. One goal of aerodynamics is to design objects so that fluid friction is reduced as the objects move through the air. Cars, boats, and even the helmets worn by skateboarders have aerodynamic designs that reduce fluid friction. Aircraft and rockets are designed with pointed noses and rounded, smooth surfaces for the same reasons.

AIRPLANES

The wings of an airplane are used to create lift. Each wing is shaped like a

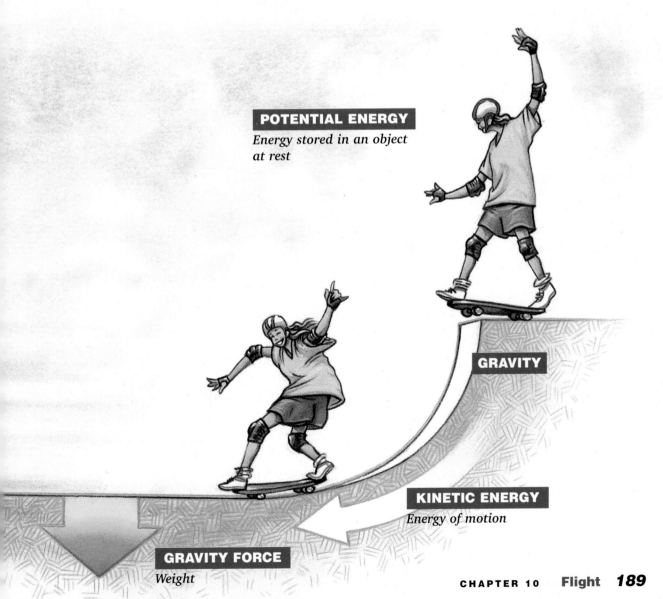

POTENTIAL ENERGY
Energy stored in an object at rest

GRAVITY

KINETIC ENERGY
Energy of motion

GRAVITY FORCE
Weight

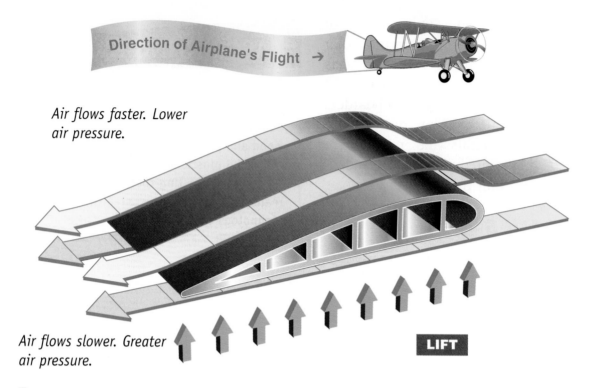

Direction of Airplane's Flight →

Air flows faster. Lower air pressure.

Air flows slower. Greater air pressure.

LIFT

▶ **Fig. 10-5** A wing is an airfoil. As the wing moves through the air, the air divides to pass around the wing. Air passing above the wing moves faster than air passing below the wing. Fast-moving air has lower pressure. This difference in air pressure forces the wing upward. This force is called lift.

Linking to SCIENCE

Bernoulli Effect. The Bernoulli (burr-NEW-lee) effect states that a fast-moving fluid exerts less pressure than a slow-moving fluid. This effect helps airplanes to fly. Because of the airplane wing's shape, air moves faster over the top of the wing than under it. The difference in pressure helps create lift.

Try this activity to see the Bernoulli effect in action. Hold an 8 1/2" x 11" sheet of note paper by its two 8 1/2" ends. The 11" sides will curve downward. Blow across the top of the paper. Observe how the farther side of the paper rises. The fast-moving air above the paper created an area of lower pressure, lifting the sheet.

wedge. It is round in the front, thickest in the middle, and narrow at the rear. A shape such as this is designed to speed up the air passing over the top surface. Such a shape is known as an **airfoil**. The speed of moving air particles affects the amount of pressure that particles place on an object.

Pressure is a force. Airplane wings reduce the air pressure above the wing, allowing the stronger pressure below the wing to lift the aircraft. Fig. 10-5.

An airplane's engine provides the thrust that moves the airfoil through the air. When the force of lift is greater than the weight of the aircraft, the airplane will rise.

Explore

Design and Build an Airfoil

State the Problem
To design and test a model wing section that demonstrates lift when air is moved past it.

Develop Alternative Solutions
Gather diagrams and pictures of various airfoil designs. Take specific note of each shape.

Select the Best Solution
Using graph paper, draw a pattern for an airfoil design you think will provide a great deal of lift.

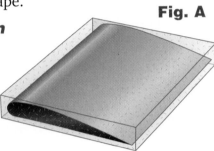

Fig. A

Implement the Solution

1. Trace this pattern onto a Styrofoam® block. Remove material from the block until it takes the shape of your design. See Fig. A.
2. Mount the airfoil model as shown in Fig. B., using wire. Use a block of wood for the base as shown.
3. Set a blow dryer at cool temperature and high speed. Use it to force air past the front edge of the airfoil as shown in Fig. C.

Fig. B

AIRFOIL

Evaluate the Solution

1. Did the airfoil lift? If not, what corrections do you have to make?
2. Experiment with a few additional designs. How does shape affect lift?
3. Sketch the most successful airfoil designs. Study your drawings to see what these designs had in common.

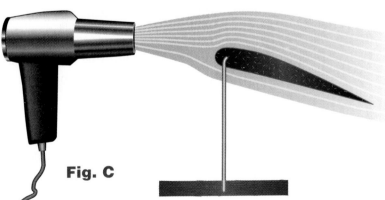

Fig. C

Propellers

Propellers are spinning airfoils. They create low pressure areas in front of the propeller so that high pressure areas behind the aircraft can push it through the sky. The propeller helps to move the wings through the air so they can create lift.

Controlling Airplanes

Pilots control the position of an airplane by adjusting flaps located on the wings and tail section of the aircraft. The flaps act like rudders on a boat. They deflect air and create drag. The increased drag causes the plane to turn.

Wing flaps called **ailerons** (AY-luh-rons) change the shape of the wing, increasing and decreasing the amount of lift the wing creates. Flaps at the front and rear edges of the wing create drag to slow the aircraft's speed during landing and aid in turns. Fig. 10-6.

The rudder, located on the tail section of the airplane, is used to turn the aircraft. Flaps on the tail, called elevators, are used to help the aircraft climb and dive in the air.

Explore

Design and Build a Rocket

State the Problem
To produce a model rocket that will reach the highest possible altitude. The rocket will be powered by the gases produced when water and Alka Seltzer® tablets are combined.

Develop Alternative Solutions
Gather pictures of various rocket and fin designs. Note difference between shapes.

Select the Best Solution
Develop a design for a rocket that you think will be able to attain the greatest altitude. Sketch the design on graph paper.

Implement the Solution
1. The plastic film canister will become the rocket engine. Produce a body tube by wrapping the notebook paper around the film canister.
2. Develop a nose cone.
3. Fasten the nose cone to the top of the body tube.
4. Develop a pattern for the fins. Cut the fins from the notebook paper.

Collect Materials and Equipment

plastic 35-mm film canister. (The canister must have an *internal sealing* lid. These are usually translucent canisters. A canister with an external lid — a lid that *wraps around the canister rim* — will not work.)
cellophane tape
sheet of notebook paper
two Alka Seltzer® tablets
water
graph paper
pencil
ruler
scissors
bucket

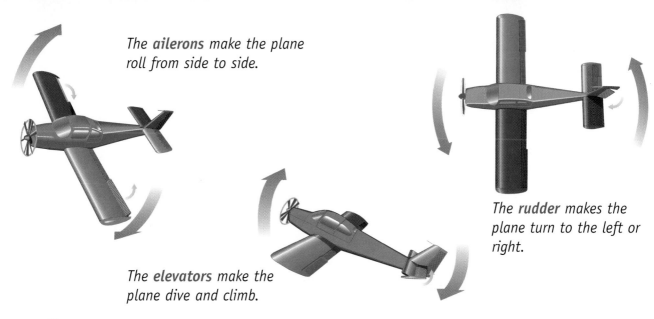

The ailerons make the plane roll from side to side.

The rudder makes the plane turn to the left or right.

The elevators make the plane dive and climb.

▶ **Fig. 10-6** Notice how the parts of an airplane create forces that affect flight.

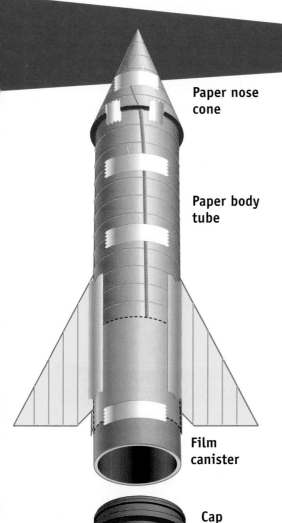

Paper nose cone

Paper body tube

Film canister

Cap

5. Fasten the fins to the rocket.
6. Turn the rocket over. Place about 1/4" of water in the canister. Drop one-quarter of a tablet of Alka Seltzer® into the canister. Quickly cap the canister. Place the rocket in the bucket for launching. Be sure to stand back. Keep classmates away. **NOTE:** All students should be wearing safety glasses.
7. Make note of the height achieved by the rocket.
8. Experiment with the body tube length, mass, and fin size. Test again for the greatest altitude.
9. Conduct four more test flights with your design. Refine the design if needed. Test again. Record the heights reached by each flight.

Evaluate the Solution

1. What effect did the mass of the body tube, fins and nose cone have on the height reached by the rocket?
2. What effect did fin size, shape, and area have on the trajectory of the rocket?
3. Which of Newton's three laws explains why the rocket lifts off?

HELICOPTERS

How is a helicopter different from an airplane? Helicopters can move in any direction. They can even hover, or float, in midair. The rotors, or blades, of the helicopter control its motion. Rotors are shaped like propellers. They are also airfoils, which provide the lift and thrust the aircraft needs.

Controlling Helicopters

Helicopters can have two to eight rotor blades. The flight of the helicopter is controlled by changing the angle, or *pitch*, of the blade. The front edge of the rotor blade can be raised or lowered to vary the amount of lift the blade creates.

As the pitch increases, lift increases and the helicopter moves vertically. When a helicopter hovers, the rotors are pitched so that they produce only enough lift to match the weight of the aircraft. Fig. 10-7.

A helicopter moves forward when the rotor blades are moved forward. The combined pitch and forward position splits the force of lift into a raising force and a thrusting force.

A helicopter that has experienced engine failure can sometimes descend safely. Such a helicopter can *autorotate* to a rough landing. How does autorotation help break the fall of a helicopter?

JET PLANES

Jet planes streak across the sky every day. Have you ever noticed that they have no propeller? How is thrust provided? How do their wings provide lift?

Jet planes use a different process to generate thrust. Jet planes use reaction

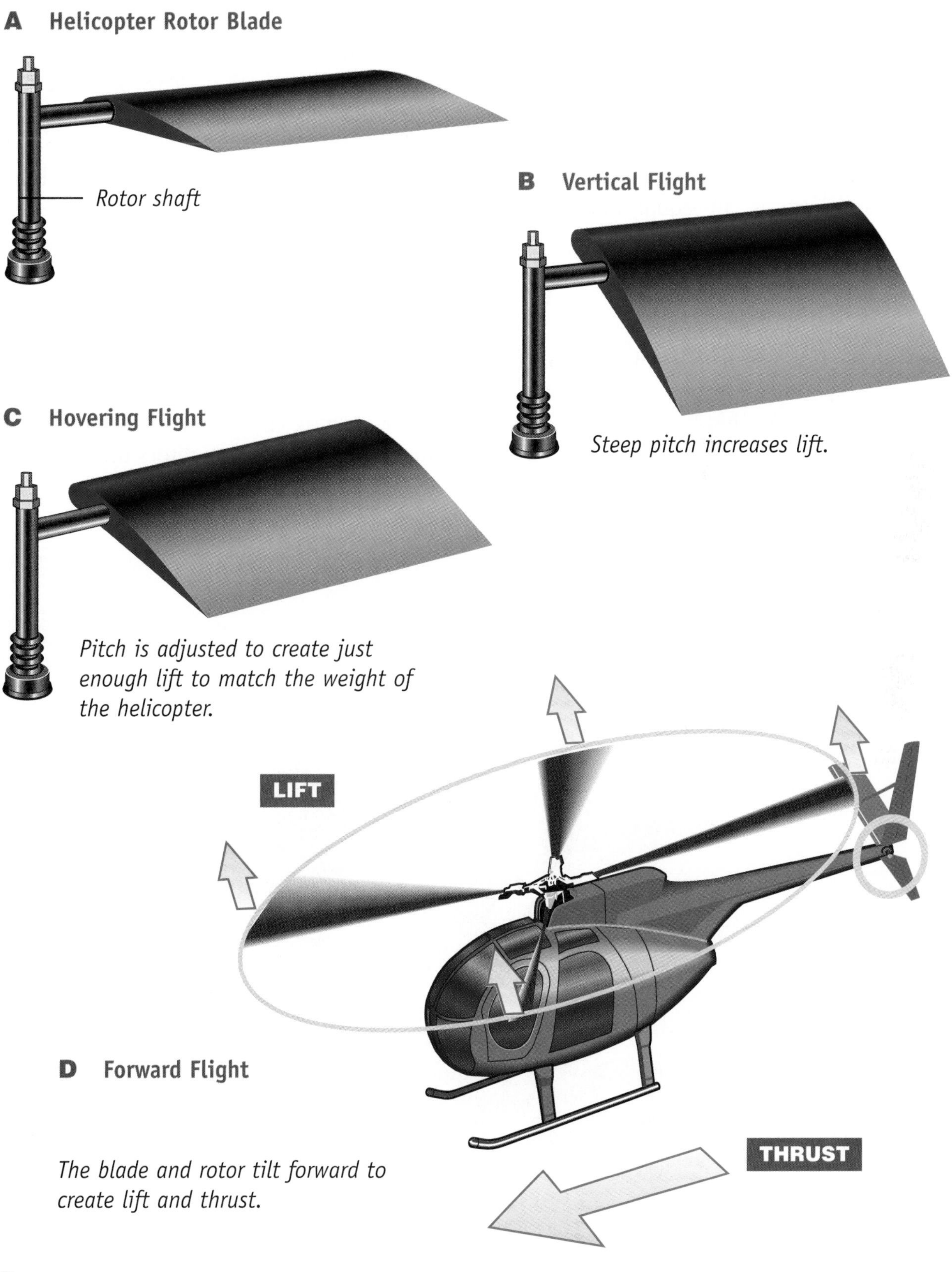

A Helicopter Rotor Blade

Rotor shaft

B Vertical Flight

Steep pitch increases lift.

C Hovering Flight

Pitch is adjusted to create just enough lift to match the weight of the helicopter.

LIFT

D Forward Flight

The blade and rotor tilt forward to create lift and thrust.

THRUST

▶ **Fig. 10-7** How the position of a helicopter rotor blade affects helicopter flight.

engines to provide the force needed to move the wings through the air. Reaction engines work on the principle of action / reaction, which is Newton's third law of motion.

Newton's third law states that for every action there is an equal and opposite reaction. This means that forces come in pairs. Every force must have an equal and opposite force.

Imagine this. You're sitting on a swing. Your feet dangle below you, not touching the ground. You are absolutely motionless. You hold a red brick in each hand. All at once you thrust the bricks forward with all your strength. Which way does the swing move? The swing moves backwards. The force of the bricks moving forward is the action. The opposite, or reaction, force propels the swing backwards.

Have you ever seen firefighters handling a fire hose at a fire scene? If you have seen this, you have probably noticed that several firefighters may be holding the hose itself. Another firefighter may be holding the nozzle of the hose to aim the stream of water on the fire.

The hose is under pressure from the water rushing into it. To compensate for this pressure, the firefighters must keep a very tight grip on the hose. As shown in

▶ **Fig. 10-8** Forces come in pairs. Every force must have an equal and opposite force. Notice the reaction of these firefighters to the thrust of the water within the hose.

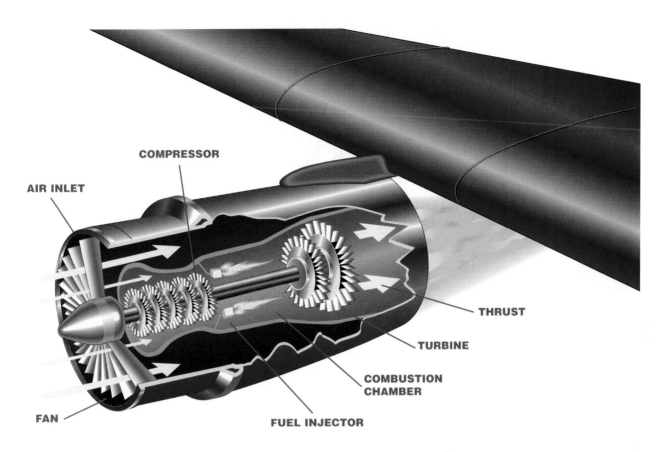

AIR INLET

COMPRESSOR

THRUST

TURBINE

COMBUSTION CHAMBER

FUEL INJECTOR

FAN

▶ **Fig. 10-9** The jet engine shows the forces of action and reaction. When air enters the front of a jet engine, it is compressed and squeezed in the combustion chamber of the engine. The jet moves forward in reaction to the gases rushing out the back of its engine.

Fig. 10-8, the firefighters are struggling to keep the stream of water on target.

Jet engines suck air into the front of the engine. Then they compress it, heat it, and eject it from the back at a very high speed. The air streaming out the back is the action force. The aircraft moving forward is the reaction force. The reaction force moves the wings through the air, creating lift. Fig. 10-9.

Aircraft powered by jet engines are more powerful and efficient than propeller-driven aircraft.

FASCINATING FACTS

North American Aviation built the X-15 to be the fastest plane in history. In 1962, it entered space. On October 3, 1967, the X-15 reached 4,534 mph, flying faster than any other plane. The temperature on the plane surface was 3,000 degrees Fahrenheit, blistering the outside of the plane. Luckily, all the fuel had been exhausted, preventing a major explosion.

Rockets

SOLID FUEL BOOSTERS (rocket engines) contain a solid propellant. A circular or star-shaped channel runs down the center of the fuel, forming the combustion chamber. The fuel burns along this channel. Solid fuel, once ignited, must burn completely.

IGNITER

PROPELLANT

CHANNEL

SWIVELING NOZZLE

LIQUID PROPELLANT TANK

OXIDIZER TANK

FUEL PUMPS

COMBUSTION CHAMBER

SWIVELING NOZZLE

LIQUID FUEL BOOSTERS (rocket engines) usually burn liquid hydrogen and liquid oxygen. These are fed from separate tanks into the combustion chamber. Liquid fuel engines can be turned on and off.

Rockets are the largest objects that fly. Like jets, rockets use the forces of action and reaction. Rockets move forward by pushing out powerful streams of hot gases. These gases are made by burning fuel.

Rockets are unique because they must work in outer space, where there is no oxygen. Rocket engines must carry oxygen with them.

Future space missions will require rockets to travel longer distances. Nuclear rocket engines could replace liquid hydrogen/ liquid oxygen engines. Nuclear engines are more efficient and provide more thrust per pound of fuel. A nuclear reactor superheats liquid hydrogen. The hydrogen does not burn, but passes through the nozzle of the rocket at high velocity, creating thrust. Nuclear engines are not designed for launching rockets into space. They are designed as power plants for shuttle transports to the moon and beyond.

CONTROLLING ROCKETS

The nozzle at the base of the engine can swivel. It can direct the burst of hot gases in different directions. The rocket reacts by changing direction. The swivel of the nozzle is used to steer the rocket. The space shuttle orbiter is steered by small rocket engines.

SPACE SHUTTLE ORBITER

How does the space shuttle orbiter differ from an airplane? First, the shuttle has six powerful rocket engines. Three of these engines are detachable. The remaining three are on the orbiter itself. Firing these engines allows the shuttle to break out of its orbit and return to earth. To assist in reentry, the orbiter also uses two orbital maneuvering system engines.

Solid Rocket Booster

External Tank

Provides fuel for orbiter's main engines. Separated from orbiter just before orbiter enters orbit. Reenters atmosphere and burns up.

Orbiter

Solid Rocket Booster

Provides thrust in liftoff. Separated from orbiter at height of 28 miles (45 km). Parachuted into ocean. Recovered by U.S. Navy ship. Reused.

SPECIFICATIONS

Operational life: 100 flights

Length: 123 ft (37 m)

Wingspan: 78 ft (24 m)

Frame material: mainly aluminum

Exterior covering: heat-reflecting tiles on underside

Crew: Flight crew of 3, plus 4 mission specialists

Weight: 114 tons (103 metric t)

Number of main engines: 3

Total thrust: 470,000 lbs (211,500 kg)

Apply What You've Learned

Design and Build a Propeller

State the Problem
Research, design, build, and test a model air propeller.

Develop Alternative Solutions
Gather diagrams and pictures of various air propeller designs. Take specific note of each shape.

Select the Best Solution
Evaluate the various shapes. Select the shape that you think will be most effective.

Implement the Solution
1. Shape the propeller. Fig A.
2. Construct the deflection gauge. Fig. B.
3. Mount the propeller to the motor stand. Connect the motor to the battery. Fig. C. Place the deflection gauge in front of the motor.
4. Test the effectiveness of the propeller. Measure the amount of air moved by the propeller. Do this by determining the angle the paper swings from vertical when the propeller is blowing air on it. Plot this amount on a graph. Label this entry "Prop 1."
5. Reverse the direction of the motor. How does this affect the output of the propeller?
6. Construct a different propeller. Do this by changing the shape and size of the airfoil. Test its effectiveness.
7. Measure the amount of air moved by the second propeller. Plot this amount on the graph. Label this entry "Prop 2."

> **Collect Materials and Equipment**
>
> wood strip 1" wide, 1/4" thick, and 4" long
> abrasive paper
> wood support for motor
> wood base for motor support
> DC toy motor
> files and other shaping tools
> deflection gauge
> safety motor mount
> DC battery

First, lay out the strip of wood. Locate its center. Create a 3/4" center hub.

Then, cut or file away diagonals on the propeller, leaving the center hub flat.

Finally, round the leading edges. This will help shape the wood strip into an airfoil design. Drill the center hold for the motor shaft.

▶ **Fig. A** Shaping the propeller. The small drawings on the right show a cross section of the propeller. The shaded part shows the shaped propeller.

8. Reverse the direction of the motor. How does this affect the output of the propeller?

9. Compare the output of the first propeller with the output of the second propeller.

Evaluate the Solution

1. How does changing the shape or size of the propeller affect its output?

2. Write a brief report (150-200 words) regarding this activity. Your report should contain no spelling errors. It should be written in complete sentences. In your report, be sure to mention the difference in the amount of air moved. Be sure also to mention what accounted for this difference. Discuss the effects of propeller design differences. Be sure to relate any differences in the amount of air moved to the Bernoulli effect.

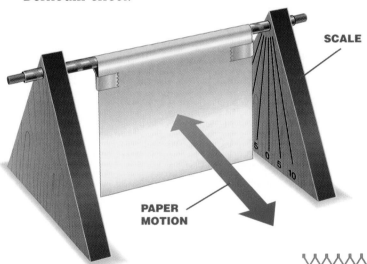

SCALE

PAPER MOTION

▶ **Fig. B** Construct the deflection gauge by assembling a dowel rod, a drinking straw, and a sheet of 8 1/2" x 11" paper as shown. The greater the angle of the paper from vertical, the greater the force of the propeller.

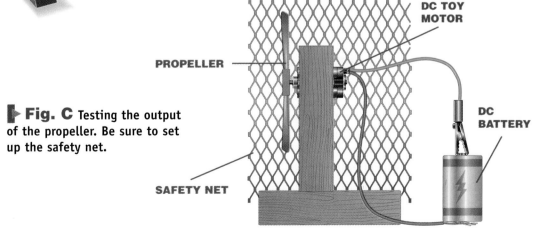

DC TOY MOTOR

PROPELLER

DC BATTERY

▶ **Fig. C** Testing the output of the propeller. Be sure to set up the safety net.

SAFETY NET

CAREERS IN
Aviation

RESERVATION ASSISTANT

Friendly, outgoing person to handle airline reservations in busy airport. Reservations experience preferred, but we will train appropriate candidate. Excellent benefits. Send resume to: Comet Aircraft Transport, Inc., 4560 Grainbelt Avenue, Oak Tree, MO 83932.

AIR FREIGHT COORDINATOR

Manufacturing company with growing international business requires someone to manage and track air shipments. Ideal candidate will have experience in air-freight operations. Must be able to work against tight deadlines. Must be able to work easily with others and quickly resolve problems. Send hand-written resume to Integral International Parts, P.O. Box 190, Little ...

CATERING COORDINATOR

Large regional airline requires person to manage in-flight meals. Responsibilities will include the purchase and inventory of in-flight meals and snacks. Must be willing to relocate to Chicago. Administrative experience in food service helpful. Send resume to InterAir, 2319 W. Smithville Road, Pontiac, IN 37321.

AIR TRAFFIC CONTROLLER

Prior experience at a medium-size airport a must. Ability to communicate clearly, pay attention to detail, and work calmly under pressure. Attractive wage and benefit package to right candidate. To schedule a confidential interview, call Todd Foster, (378) 672-9061.

AIRCRAFT MECHANIC

Graduation from a certified technical school required, plus 3 years on-the-job experience in repair of propeller-driven aircraft. Must be able to work under tight deadlines. Prior references essential. Send resume to: **Ace Aircraft, 2987 Rampart Drive, Topeka KS 45987. No phone calls, please.**

FLIGHT ATTENDANT

Major domestic airline has immediate openings for flight attendants. No experience needed. If you enjoy working with the public, like to travel, and are interested in an exciting career in aviation, this could be the job for you. Must be flexible and at ease with people. Must be willing to relocate. Selected candidates will receive free paid training at our company training institute. Call for an interview: (110) 345-9467. Ask for Anne Smith.

Linking to the WORKPLACE

You can explore a career in aviation through library research or a personal interview. Choose any of the careers highlighted above. Pretend that you have a career goal of obtaining employment in that job. Identify what you would have to do to obtain that job. Then draw up a list of the specific steps you would need to take. Share that list with the class in a brief report. Ask for class feedback on your plan.

Chapter 10 Review

SUMMARY

▶ Several forces affect an object in flight. These forces include friction and gravity, which oppose motion.

▶ Lift and thrust also affect an object in flight. They help the object overcome gravity.

▶ An airplane is able to fly because its wings are airfoils. Their shape reduces the air pressure above the wing. This allows stronger air pressure below the wing to lift the plane.

▶ An airplane is controlled by movable parts of the wing (ailerons) and tail (rudder).

▶ A helicopter is controlled by the angle or pitch of the blade. A helicopter rotor is also an airfoil.

▶ Rockets are powered by hot gases made by burning fuel. The fuel may be solid or liquid.

CHECK YOUR FACTS

1. Sarah's sneakers gripped the rough surface of her skateboard, keeping her from slipping off the board. What is the name of the force at work?

2. Describe two forces that oppose motion.

3. Name two forces that help an airplane overcome gravity.

4. Identify the parts of an aircraft that help control flight. Briefly describe how they work.

5. What force acting upon the wing flaps of an airplane causes the plane to turn or slow down?

CRITICAL THINKING

1. Would it be more difficult to stop a rolling bowling ball or a rolling tennis ball? Explain your answer.

2. Describe how thrust is used to create lift in an airplane.

3. Describe how an air propeller works. Does a water propeller work in the same way?

4. Make a sketch of an airfoil. Describe the forces that act on the airfoil as it moves through the air.

5. Early in this century, some inventors tried to fly by fastening birdlike wings to their arms. They then flapped their arms as hard as they could while running down a steep hill. Why didn't they fly?

Land and Water Transportation

OBJECTIVES

▶ describe various modes of land and water transportation.

▶ discuss power sources used in transportation.

▶ describe the operation of a four-stroke cycle gasoline engine.

KEY TERMS

buoyancy

diesel engine

gasoline piston engine

maglev

transportation

vessel

How did you get to school this morning? Did you walk? Did you take a bus or a subway? Were you driven in a car? Perhaps you used more than one form of transportation. Technology links one form of transportation with another to make transportation more efficient. The high-speed train shown here is an example of the use of tehnology to improve transportation.

MODES OF TRANSPORTATION

Transagement is the process by which people, animals, products, and materials are moved from one place to another. Fig. 11-1. A *mode* is a method of doing something. Many different modes of transportation are used to move people, products, and materials. Modes of transportation can be organized according to the pathways, or "ways," used by transportation systems. Land transportation vehicles travel on highways and railways. Water (marine) vessels use seaways and canals, while planes travel through airways. This chapter will focus on land and water modes of transportation.

Let's trace the path of the CD player from the factory in Japan to your home.

Most likely the completed product moved down a *conveyor belt* in the factory to the shipping department. There, the product was packaged and stacked on pallets. The stacked pallets were loaded by *forklifts* into a large *container*.

The container was mounted on sets of wheels to become the trailer portion of a

▶ **Fig. 11-1** This oceangoing cargo ship carries containers on its deck and below deck. These containers carry a variety of products. When the ship reaches land, the containers will be unloaded at the dock. They will then be carried by land transportation systems such as railroads and trucks.

tractor-trailer (also known as a *semi*). The trailer was pulled by the tractor to a ship docked at a harbor. A large *crane* removed the container from the wheels and placed it onboard a large ship.

The *ship* traveled across the ocean on sea-lanes and docked at a port in the United States. Again, a *crane* removed the container from the ship. This time the container was placed on a railroad flatcar.

A *train* pulled by a diesel-electric locomotive moved the products along railways to a warehouse for storage. *Trucks* carried the product from the warehouse to your neighborhood store. Finally, you might have traveled to the store on your *bicycle* to purchase your CD player and bring it home.

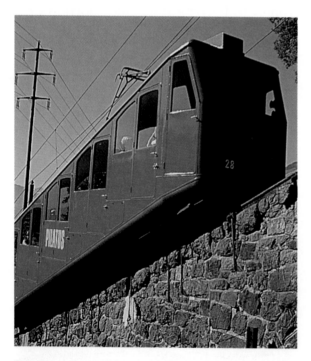

▶ **Fig.11-2** Because they provide transportation in nearly every environment, railroads have tested the design skills of engineers. This electric railroad has been designed to provide transportation to people living in villages in the Swiss Alps.

SYSTEMS IN TRANSPORTATION

A *system* is a group of related parts organized to work together to accomplish a particular goal. Fig. 11-2. The word *system* is often used with other words to describe something. For example, in your community there is a *school system*. Your community also has a *governmental system* to regulate, manage, and enforce its laws.

Let's examine the school system. It consists of teachers, principals, secretaries, and custodians. They all work together to educate you, the student. That is the goal of a school system.

Systems exist in technology. A *technological system* is also a group of related parts working together to achieve a goal.

A system may contain smaller systems within itself. These smaller systems are called *subsystems*. The subsystems work together to accomplish the system's goals. For example, a stereo system consists of speakers, a tuner, an amplifier, a CD player, and perhaps a cassette player. These individual parts, or subsystems are linked to provide music. This is the goal of a stereo system.

Transportation systems can be simple or complicated. Count the vehicles used in the trip described above. This trip used an *intermodal transportation system*. More than one type, or mode, of transportation was used.

Each day people, products, and materials are moved along a system of roads, bridges, tunnels, rails, elevators, conveyors, waterways, and airways. Together these subsystems make up a complex transportation system that transports people and delivers products throughout the world.

Common Elements

All systems have common elements. Transportation systems, whether it's your bicycle or a locomotive, are alike in many ways. How can this be possible? Let's examine the common elements of a transportation.

In all transportation systems, the *input* is energy. Your legs provide the energy to move your bicycle. What provides the energy for a locomotive? Most use diesel or electric motors. Bicycles, trains, planes, cars, and boats all require that energy be input into the system. Energy is needed before the system can achieve its goal of moving people, products, or materials.

The *process* part of a transportation system is the action stage. The *feedback* part of the system checks that the process is being correctly performed. For example, in the case of a locomotive, the engineer would obtain feedback from the gauges that monitored the engine's performance.

The result produced by the process part of a system is called the *output*. If the system is working correctly, the output should be exactly what was expected. In a transportation system, the desired output is the successful transportation of people and products.

LAND TRANSPORTATION

Early land transportation methods were slow and unreliable. At first, people traveled by walking. When animals were domesticated (tamed), they provided the power to move people and materials. Sleds loaded with goods were pulled by horses or oxen. The invention of the wheel changed the ways in which goods were moved. Later improvements in land transportation systems were focused on powering and guiding wheeled vehicles.

Rail Transportation

Powering Rail Systems

Railroading on a wide scale became possible with the invention of the steam locomotive. Fig. 11-3.

▶ **Fig. 11-3** Because they burn wood and coal, steam-powered locomotives release pollutants into the air. Worldwide, the number of steam-powered locomotives is diminishing.

Explore

Design and Build a Model Maglev System

State the Problem
Design and build a model maglev vehicle and guideway.

Develop Alternative Solutions
Gather resources that show the basic design of a maglev system. Prepare several sketches. These should show the guideway and the placement of magnets on the vehicle and the guideway. Using graph paper, draw a pattern of the side and top view of your maglev vehicle.

Select the Best Solution
Choose the design that you think is most effective.

Implement the Solution
1. Construct the guideway. Refer to the drawings and the dimensions assigned to you by your teacher.
2. Fasten the magnets to the guideways. Be sure that the same pole is facing up on each magnet.

Collect Materials and Equipment
Guideway
magnets 1" x 3/4" x 3/16"
pine wood (for base)
acrylic plastic (for sides)
1/4" dowel rod
Body
insulation board
cardboard
foam plastic
Chassis
masonite
magnets (same as above)
standard material processing equipment

GUIDEWAY

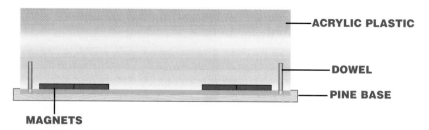

ACRYLIC PLASTIC

DOWEL

PINE BASE

MAGNETS

A *locomotive* is a self-propelled vehicle used to pull or push trains of *rolling stock* (all types of railroad cars). The steam locomotive harnessed the energy of compressed steam to power rail traffic.

Steam locomotives boiled water using coal or wood fires. The pressure of the steam was used to move pistons. These pistons, in turn, caused the wheels to turn. Until the 1940s, steam locomotives pulled most of the trains in the United States.

A steam locomotive required a great deal of maintenance and regular refilling

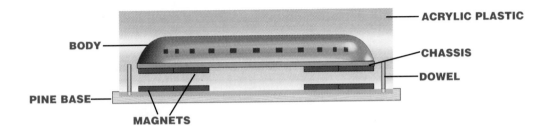

BODY

ACRYLIC PLASTIC

CHASSIS

DOWEL

PINE BASE

MAGNETS

3. Construct the vehicle body.
4. Levitate the chassis by fastening magnets to the bottom. Be sure the same magnetic poles are facing each other. You should feel the force of magnetic repulsion in each magnet.
5. Shape the body of the vehicle.
6. Install the vehicle body onto its chassis.

Evaluate the Solution
1. Does the vehicle levitate on the guideway? If it does not, make changes until the vehicle levitates.
2. Can the vehicle travel in either direction? Explain your answer.
3. Are there design changes that would allow better operation?

with water and fuel. The black smoke spewing from its smokestack fouled the air with soot.

In the 1940s, the diesel-electric locomotive was introduced as an alternative. Diesel-electric locomotives convert fuel oil into power using a diesel engine. (Diesel engines are discussed later in this chapter.) The engine drives a generator that produces electric current. The current turns an electric motor, which powers the wheels.

An all-electric locomotive also uses an electric motor for power. However, it obtains its electricity from overhead power lines or electrified rails. Some

▶ **Fig. 11-4** This high-speed TGV train runs in France. It provides transportation at speeds of 186 mph.

electric locomotives use their powerful motors as generators during coasting periods. This allows them to feed electricity back into the power source.

Rail Transportation Today

The popularity of passenger rail travel over long distances is growing again. A combination of economics and new technology has aided this century-old mode of transportation. The rising cost of air travel has made rail travel economically pleasing to many travelers.

The speed at which passenger trains now move has made rail travel very attractive. The electrically powered Japanese "Bullet Train" and the French-built TGV trains provide passenger service

at speeds of 130-190 mph. Such high speeds have made rail travel competitive with automotive and air transportation. Fig. 11-4.

Linking to MATHEMATICS

Figuring Distance. The Eurostar is the train that travels through the English Channel Tunnel (Chunnel). A trip from London to Brussels, Belgium takes 3 1/2 hours on the Eurostar. If the Eurostar travels at an average speed of 97 mph, what is the distance in miles between London and Brussels using the Eurostar route? Use the formula:

Distance = Rate x Time
▲

Commuter trains transport people in large cities. Chicago, New York, and San Francisco have commuter train systems. Often referred to as *mass transit*, these trains provide transportation within the city and its suburbs. By doing this, they help relieve the congestion of city automobile traffic.

Because real estate is expensive in very large cities, many people living in such cities live in apartments. Garaging a car in a large city is expensive and spaces for on-street parking spaces may be very scarce. For apartment-dwellers, access to mass transit is essential. A good mass transit system provides an efficient and economical mode of transport.

Freight trains carry a large percentage of the materials and products transported in the United States. Freight cars come in many shapes. Boxcars, flatcars, hoppers, and dump cars move solid materials. Tank cars move liquids such as oil, gasoline, and paints. Refrigerated and heated cars carry perishable materials. Fig. 11-5.

Look at the railroad cars shown in Fig. 11-5. Because rail links were important to developing industries, it was important that the cars be designed to meet the customer's needs. Which of the car designs shown here would have been among the first? Which designs would have been among the later designs?

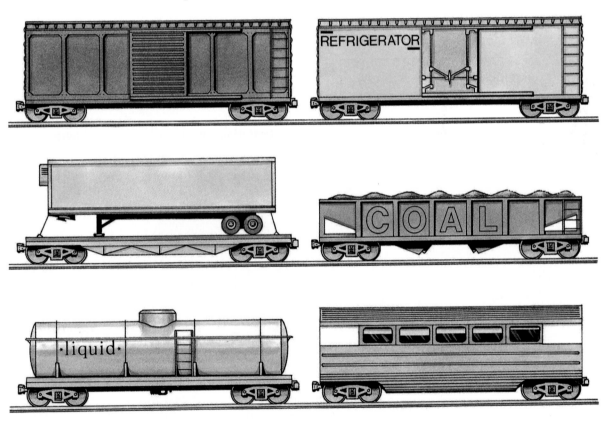

▶ **Fig. 11-5** The design of a railroad car is determined by its use. Because railroad cars are designed to carry people, as well as various types of cargo, different designs are needed. These are six of the more common railroad cars. Would you be able to identify the primary use for each car if some of the cars had not been labeled?

Maglev

Imagine a train without wheels. Imagine a train that glides along an invisible field of energy. Picture a train without an engine that is propelled along a guideway with the same invisible force that keeps it suspended in the air. If you visited Japan or Germany you could see such a remarkable train. The technology that makes this transportation possible is called *magnetic levitation*.

Maglev (short for <u>MAG</u>netically <u>LEV</u>itated) trains are levitated, or floated, above a *guideway* (track) and propelled (moved forward) by magnetic fields. The magnetic fields are created by large electromagnets. Fig. 11-6.

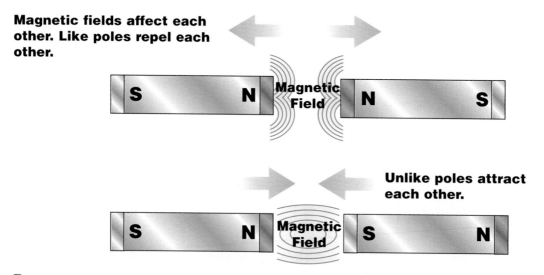

Magnetic fields affect each other. Like poles repel each other.

Unlike poles attract each other.

▶ **Fig. 11-6A** Maglev trains are powered by magnetic levitation. This principle is shown here.

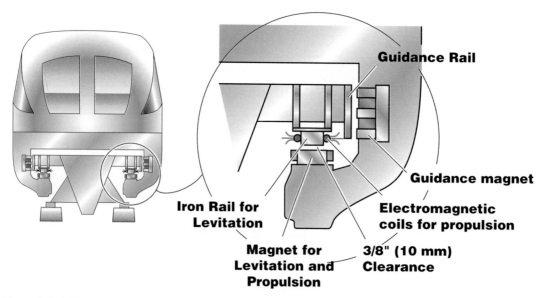

Guidance Rail

Guidance magnet

Iron Rail for Levitation

Electromagnetic coils for propulsion

Magnet for Levitation and Propulsion

3/8" (10 mm) Clearance

▶ **Fig. 11-6B** Note the way in which the lower part of a maglev train car wraps around the rails. Only one rail is shown here. The car wraps around both rails. The magnet in the rails works with the electromagnetic coils in the train car. The resulting levitation and propulsion keeps the train just above the rails and moves it forward.

Have you ever experienced the force generated when you try to push together the like poles of two magnets? Have you felt the force of attraction between two unlike poles? It is this push-and-pull force that levitates the train above the guideway and propels it down the track.

Maglev trains have several advantages over trains commonly used today. One difference is that maglev trains do not touch the guideway. Because of this, problems faced by steel train wheels on steel tracks are avoided. With very little friction to rob power, maglev trains can rapidly accelerate to over 300 mph. Without tracks that creep out of alignment, their ride is smooth and quiet. Less maintenance is required. Snow, ice, and heavy rain have little effect on the operation of maglev trains. Perhaps their greatest advantage is a reduction in pollution. Maglev trains do not burn gasoline or diesel fuel.

There are disadvantages to maglev trains. Maglev guideway construction costs millions of dollars per mile. Because travel routes between large cities usually have many stops along the way, maglev trains would not be able to travel these routes at top speed. Perhaps the greatest criticism is the amount of electricity the electromagnets require to levitate and propel a maglev train. This electricity must come from electric power plants. Most of these burn polluting fossil fuels.

Motor Vehicles

Motor vehicles are the most widely used mode of powered land transportation. Automobiles, vans, buses, and trucks are all considered motor vehicles. We rely on these vehicles to transport people, products, and materials. Fig. 11-7.

▶ **Fig. 11-7** The development of the automobile was a milestone in technology. However, had the automobile been unaffordable, it would not have changed society in the ways that it has. Automobiles were made affordable by the process of mass production.

Explore

Design and Build a Cam Operating System

Engines used in transportation change motion from one direction to another. For example, an automobile engine changes the up-and-down motion of its pistons (linear motion) into rotary motion (circular motion). The direction of motion is changed by using a mechanical device called a *cam*. Cams are arranged along the *camshaft* in the car's engine.

A cam is an offset wheel. Cams are connected to rods. Fig. A. A second rod, or follower, rests on the top of the cam. As the cam turns, the top rod moves up-and-down.

State the Problem

Design and build a cam system that operates a device by changing rotary motion into linear motion. At least one part of the device must move as a result of the cam's action.

Develop Alternative Solutions

Develop three sketches of devices that can be operated by a cam system.

Collect Materials and Equipment
1/4" dowel rod
container for cam operating system (orange juice can or milk carton)
plastic tubing (must fit snugly over the dowel rod)
cardboard
foamcore board
construction paper
brass paper fasteners
pipe cleaner
glue
scissors
pliers

Fig. A

FOLLOWER — LINEAR MOTION

CAM

ROTARY MOTION

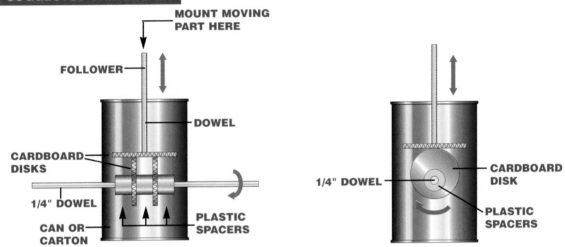

SUGGESTED CAM DESIGN

MOUNT MOVING PART HERE

FOLLOWER

DOWEL

CARDBOARD DISKS

1/4" DOWEL

CAN OR CARTON

PLASTIC SPACERS

1/4" DOWEL

CARDBOARD DISK

PLASTIC SPACERS

Select the Best Solution

Consider the complexity of the design and the amount of time you have to build. Review your sketches. Select the design you feel best meets the specifications and solves the problem.

Implement the Solution

1. Prepare any patterns you might need for the construction of the device.
2. Construct the cam system.
3. Place the cam system within the device.
4. Attach the device to the shaft of the follower.

Evaluate the Solution

1. Test your system. Does the device move in the expected way?
2. Make any needed changes.
3. Test the system again.

The Gasoline Piston Engine

A **gasoline piston engine** is an engine that creates power by burning fuel inside the engine. During this burning, or combustion, the fuel combines with oxygen. It burns to give off heat. The *heat engines* used in motor vehicles control combustion. They change the heat of combustion into mechanical energy in order to do work. Heat engines are classified according to where the combustion takes place.

In *external-combustion engines*, fuel is burned outside the engine. The steam locomotive engine mentioned earlier is an external-combustion engine. Steam is created in a boiler outside the engine and passed into the engine through valves. The expanding steam pushes against a piston inside the engine. The motion of the piston sets the wheels in motion.

When fuel is burned inside the engine, the engine is called an *internal-combustion engine*. Most cars use an internal-combustion engine that burns gasoline for power. Gasoline engines depend on hot expanding gases. These gases provide the mechanical energy needed to turn the wheels of the vehicle.

In both the steam locomotive and internal-combustion engine, power comes from expanding gases within a cylinder. These gases cause a piston to move. A *cylinder* is a closed container. In most engines they are made of steel. A *piston* is a disk or short cylinder that moves up and down within a hollow cylinder. It must fit tightly enough within the cylinder that air cannot escape. Yet it must be able to move up and down within the cylinder.

A gasoline engine operates on a *four-stroke cycle*. In this cycle, four separate operations convert heat energy into mechanical energy. During each stage of the cycle, a piston slides up and down within a cylinder. Fig. 11-8.

During the first stroke, a gasoline and air mixture is pumped into the cylinder. This stroke is called the *intake stroke*.

FOUR-STROKE CYCLE ENGINE

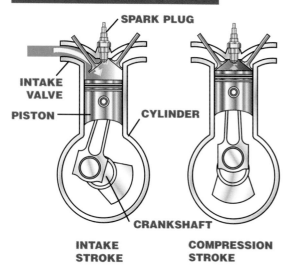
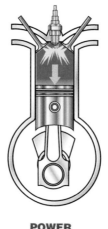
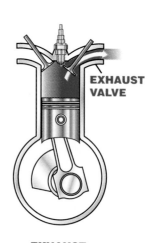

INTAKE
STROKE

COMPRESSION
STROKE

POWER
STROKE

EXHAUST
STROKE

▶ **Fig. 11-8** The four cycles of a gasoline engine. This engine has one cylinder. Note that the valves (at the top) open and close. Note also the camshaft at the bottom. The crankshaft is the first link in a system of drives and gears that transfers the power of the engine to the wheels.

The second stroke compresses the fuel mixture by moving the piston up in the cylinder. A spark plug produces an electric spark that ignites the compressed fuel. This is known as the *compression stroke*.

The burning fuel creates expanding hot gases. An explosion takes place in the cylinder. This explosion forces the piston back down the cylinder. This third stroke is the *power stroke*. A series of shafts and gears connected to the piston transfers the energy from the moving piston to the wheels.

Finally, the piston moves back up to clear the cylinder of unburned gasoline and waste gases. This fourth stroke is called the *exhaust stroke*. The four-stroke cycle then begins again.

Most automobile engines have four, six, or eight cylinders. Each cylinder operates through the four-stroke cycle. The way in which the energy of the pistons is converted into rotary motion is shown in Fig. 11-9.

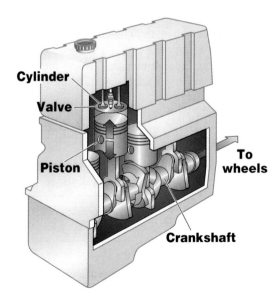

▶ **Fig.11-9** The vertical motion of the pistons is converted into rotary motion through the crankshaft.

Linking to SCIENCE

Scientific Principles in Engine Operation. In an internal-combustion engine, the ignition of the fuel creates a controlled explosion. The burning gases expand rapidly, creating additional power. Fig.11-10.

In a diesel engine, the compressed air becomes hot enough to ignite the diesel fuel sprayed into the cylinder.

A. As the pressure exerted on the gas inside the cylinder increases, the volume of gas decreases. As the volume of gas is pressed into a smaller space, the gas molecules press with greater energy against the cylinder walls and piston.

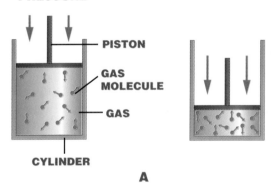

B. As a gas is compressed into a smaller space, the temperature of gas increases.

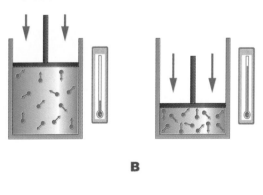

▶ **Fig.11-10** (A) Boyle's law. (B) Charles' law.

The Diesel Engine

A **diesel engine** is an internal-combustion engine that burns fuel oil by using heat produced by compressing air. The diesel engine operates in a different way than the gasoline engine. During the *intake stroke*, the diesel engine injects only air into the cylinder. The *compression stroke* squeezes the air. This causes the air to increase in temperature. It becomes much hotter than the air in a gasoline engine. At this point, diesel fuel is sprayed into the cylinder.

The hot air ignites the fuel. An explosion takes place. The hot, expanding gases drive the cylinder down, beginning the *power stroke*. As in the gasoline engine, the power stroke provides the mechanical energy needed to turn the wheels of the vehicle. Why doesn't a diesel engine require a spark plug?

Both gasoline and diesel engines have advantages and disadvantages. Table 11-A.

On-Site Land Transportation

At times it is necessary to move people and products a short distance. The use of motor vehicles and locomotives is not practical in such cases. *On-site transport systems* are designed for this purpose. For example, malls use elevators and escalators to transport people between floors. Large airports rely on *people movers* to transport passengers to terminals and gates. Moving walkways help travelers move from one part of the terminal to another quickly and easily. Fig. 11-11.

▶ **Fig. 11-11** The vastness of modern airline terminals has suggested clever ways to move travelers using on-site transportation. Key points are linked by "people movers." These moving pathways use the principle of the conveyor belt.

Table 11-A		
Comparing	**Advantages**	**Disadvantages**
Gasoline Engines	Good acceleration Good power	Uses expensive fuel Pollutes the air
Diesel Engines	More powerful than a gasoline engine More fuel-efficient than a gasoline engine	Very noisy Pollutes the air more Provides slow acceleration

WATER TRANSPORTATION

Water transportation, or marine transportation, refers to the systems used to transport people, products, and materials across waterways. For thousands of years people have engineered new ways to increase the capacity, comfort, and speed of the vessels. A **vessel** is a water vehicle that transports people and products.

Types of Vessels

Vessels are designed to perform specific jobs. The design depends upon what the vessel will transport. The three broad classifications of vessels are commercial, recreational, and utility.

▶ **Fig.11-12** Barges provide economical delivery of low-cost products such as coal. In many countries, barges are an important part of the inland water transportation system. What factors would limit the use of barges on rivers? What types of cargo would not be suitable for shipment by barge?

Commercial vessels transport people and products to make a profit. Ocean liners and ferries are commercial ships that transport people. Cargo ships move products. *Barges* carry a variety of solid materials such as coal, iron ore, and even garbage. Fig. 11-12.

Container ships carry large containers packed with products. The containers are lifted on and off the ship by cranes. On land, the containers are placed onto a rail system or transported by motor vehicles.

Freighters are large vessels that carry solid materials and products. Freighters are used to transport large items such as automobiles.

Tankers are vessels that carry cargoes of liquid products such as oil. Fig. 11-13.

Utility vessels do work on water. Tugboats and towboats perform the delicate task of moving ocean liners and large freighters in and out of harbors. Such ships are too big to maneuver

▶ **Fig. 11-13** Tankers are used for the transoceanic shipment of goods.

unaided in shallow water and tight spaces. Commercial fishing boats are also utility vessels. Many of the large North Sea fishing boats not only catch the fish, but also process it onboard the vessel.

Recreational vessels are used for fun. Sailboats, speedboats, and cabin cruisers are just a few kinds of recreational vessels.

Hull Design and Buoyancy

The *hull* of the vessel is the part of the boat that sits in the water. The shape of the hull on a barge is different from the shape of a speedboat hull. Why is this so?

Many factors affect a vessel's hull design. The main design feature of the hull is the shape of the bottom. Hulls that sit deep in the water have a *displacement-style* hull. Displacement-style hulls are very buoyant. This means they float easily. They ride smoothly in the water. These hulls can hold a great amount of weight. Tankers, barges, ocean liners, and freighters have displacement-style hulls.

Boats that require speed usually have a *planing-style* hull. Planing-style hulls are designed to rise out of the water at high speeds. Water moving across a hull creates drag and friction. Friction robs energy from the boat. With the hull partly out of the water, less surface area is in contact with the water. This reduces friction and drag. Speedboats and many other recreational vessels have planing-style hulls. Figs. 11-14.

Have you ever experienced the feeling of being almost weightless while under water? You were experiencing the force of buoyancy. **Buoyancy** (BOY-ann-see) is the upward force a fluid places on an object placed in it. This force is referred to as upthrust. Objects float because of upthrust.

Boats are made of a variety of materials. Wood, steel, and fiberglass are a few of the more common boat-building materials. Solid pieces of these materials would sink in water. Then why do boats float? The answer to this question is explained by Archimedes' principle. The principle states that a buoyant force on an object is equal to the weight of the fluid displaced by the object. An object floats in a fluid when the weight of the fluid it displaces is equal to or greater than its own weight or when the upward push on the object is greater than the object's own weight. Fig. 11-15.

▶ **Fig. 11-14** The purpose of a boat or ship can be determined from its hull type. (A) A displacement-style hull is necessary in utility ships and oceangoing ships that carry heavy cargo. (B) A planing-style hull is used on boats designed for speed. Speedboats use this hull.

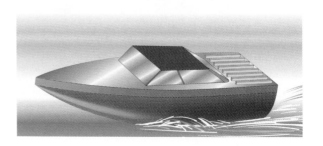

DISPLACEMENT-STYLE HULL

PLANING-STYLE HULL

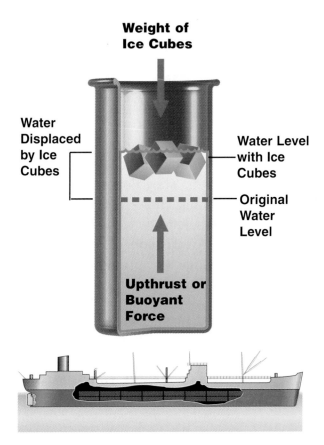

Weight of Ice Cubes

Water Displaced by Ice Cubes

Water Level with Ice Cubes

Original Water Level

Upthrust or Buoyant Force

Powering Vessels

Early vessels relied upon wind power to move them through the water. The size, shape, and arrangements of the sail allowed the wind to push or even pull the vessel through the water. The *lateen sail* allowed ships to move *across* the direction of the wind by taking a zigzag path. Modern sailboats still use the same principles. Fig. 11-16.

The wind is an uncertain source of power. Ships often drifted for days waiting for the wind to fill their sails. In 1807, Robert Fulton built a ship powered by steam. The steam engine powered paddles on the back or sides of the vessel. His invention changed marine transportation in the same way the locomotive changed rail travel.

▶**Fig. 11-15** (A) Ice cubes are like miniature icebergs. Most of an ice cube is underwater. An ice cube floats because of the upthrust. (B) A cargo ship floats for the same reason that an ice cube floats. This cargo ship, shown in a cutaway view, has enough buoyancy to float both itself and the cargo of containers it carries.

▶**Fig. 11-16** Each sail on a sailboat is an airfoil. Sails can even propel a vessel into the wind. The two sails on the sailboat shown here act as large airfoils. The wind travels through the channel between the two sails. This movement of wind creates a powerful suction force, which pulls the vessel through the water. The sailboat follows a zigzag course, using a sailing technique known as tacking. This keeps the sails at the correct angle to the wind. The keel adds stability.

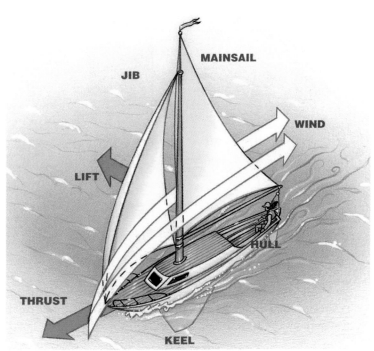

JIB

MAINSAIL

WIND

LIFT

HULL

THRUST

KEEL

By the 1820s, ships with steam engines and iron hulls began to replace wooden sailing ships. By the mid-1850s, *screw propellers* powered by steam engines began to replace paddles. The first diesel-powered vessels appeared after 1911. Modern seagoing vessels use steam-turbine engines and gas-turbine engines to power their massive propellers.

Some commercial ferries (called *hydrofoils*) use hydrofoil technology to develop high speeds. A hydrofoil is like an underwater wing. At certain speeds, its winglike structure creates *lift*, much like an airplane wing. (See Chapter 10.) The hull of the ship rises out of the water, reducing drag. Fig. 11-17.

Some hydrofoils are powered by *waterjets*. Waterjets take water in at the front of the vessel and force it out the rear using large pumps. The high-speed waterjet propels the vessel. Propellers with specially designed shafts are also used to propel hydrofoils.

Air-cushion vehicles are sometimes called Hovercraft®, or ground-effect vehicles. These vehicles work equally well on land or water. They float on a cushion of air. Large fans create a pocket of air under the vehicle. The vehicle rides on this cushion. Air-cushion vehicles can be propelled using air propellers, screw propellers, and waterjets. Fig. 11-18.

IMPACTS

Society has been changed by the development of modern transportation

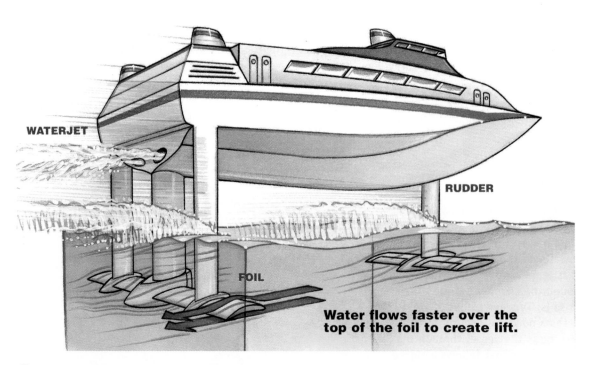

WATERJET

RUDDER

FOIL

Water flows faster over the top of the foil to create lift.

▶ **Fig. 11-17** The hydrofoil takes its name from its use of underwater foils. As shown here, the rapid flow of a fluid (in this example, water) over a foil creates lift. Forward propulsion is made possible by the waterjet.

▶ **Fig. 11-18** Hovercraft provide transportation for short trips over water. A large fleet of Hovercraft are in use in the English Channel, which lies between France and England. These Hovercraft are now competing with the Chunnel, the tunnel beneath the English Channel.

systems. In some societies, workers have always commuted, or traveled to their place of work. Now, though, commuting has become easier. Many are able to travel to their workplace by public transportation systems.

Transportation systems have had adverse effects on the environment. For example, the building of the railroads in the American West transformed that landscape. More recently, the building of the Chunnel required new railroad construction in some of the most beautiful parts of England.

Because most of our transportation systems depend on fossil fuels such as oil, the problem of pollution has increased. Pollutants are carried into the air by vehicle exhaust, as well as the plants that refine and process oil. Ground and water contamination occurs from pollutant particles and from oil that has been improperly disposed of.

THE FUTURE

Electric cars may someday replace gasoline-powered and diesel-powered vehicles that rely on combustion. Electric vehicles use an electric motor to provide energy to turn the wheels. Electrical energy is supplied to the motor from storage batteries in the vehicle. These batteries must be recharged periodically. Electric vehicles are quiet, economical to operate, and nonpolluting. Although the manufacture of electric vehicles is limited, they are available.

In recent times, interest in replacement fuels has increased. One reason is the potential shortage of fuel oil, from which gasoline is made. Also, pollution problems are caused by the exhaust of gasoline and diesel engines. Experimental vehicles have used engines that burn methane or propane gases as fuel. They have demonstrated that a clean and economical fuel alternative to gasoline is possible.

Apply What You've Learned

Design and Build a Catamaran

State the Problem

A catamaran (CAT-uh-ma-ran) is a boat with two hulls. Design and build a wind-powered catamaran. Design the hulls, the sail, and a system to keep the vessel stable in the water. The vessel will sail down a 12' trough. The trough will be made of PVC pipe and filled with water. The vessel will be powered by the wind generated from a 20" box fan.

Develop Alternative Solutions

Gather photos and drawings showing catamaran designs. Study the shape of the hull and the shape of the sails. Prepare sketches of possible hull designs. Remember that the catamaran must easily fit within the trough.

Select the Best Solution

Select the design you feel will be most effective.

Collect Materials and Equipment
high-density construction foam
insulation board
sail material (fabric or plastic)
modeling materials (dowels, balsa strips, etc.)
20" box fan
material processing tools
glue gun, cool melt
trough 8" wide and 12' long

CATAMARAN DESIGN

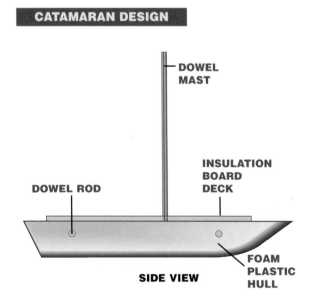

DOWEL MAST

DOWEL ROD

INSULATION BOARD DECK

FOAM PLASTIC HULL

SIDE VIEW

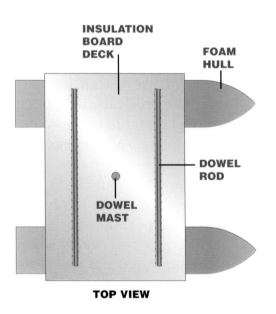

INSULATION BOARD DECK

FOAM HULL

DOWEL ROD

DOWEL MAST

TOP VIEW

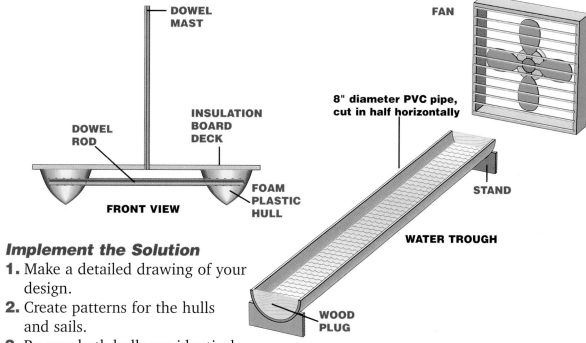

DOWEL MAST

DOWEL ROD

INSULATION BOARD DECK

FRONT VIEW

FOAM PLASTIC HULL

FAN

8" diameter PVC pipe, cut in half horizontally

STAND

WATER TROUGH

WOOD PLUG

Implement the Solution

1. Make a detailed drawing of your design.
2. Create patterns for the hulls and sails.
3. Be sure both hulls are identical.
4. Assemble the hulls.
5. Design a system to support the sail.
6. Install the sail system.
7. Complete the construction of the catamaran.

Evaluate the Solution

Place the catamaran at one end of the water-filled trough.

1. Is the vessel stable? Does it sit level in the water? (Weight might have to be added to the hull to level it in the water or to have it sit deeper in the water.)
2. Test the sail by using the fan as a wind source.
3. If the vessel falls over or sails in an unstable way, a keel may be needed. Research the function of a keel. Prepare keel designs. Add the keel to your catamaran.
4. Make additional modifications to your catamaran until it sails smoothly down the trough.

CAREERS IN
Land and Water Transportation

RAILROAD BRAKER

Entry-level position for high school graduate. On-the-job training provided. Must pass physical exam. Should have good mechanical aptitude and ability to work well with people. May lift up to 50 pounds. Apply in person to S & K Railroad Company, 900 West Stephens Drive, Chicago, IL 60020. No phone calls.

MARINE ENGINEER

Engineer needed to supervise and coordinate activities of crew in operating and maintaining engines and electrical equipment aboard ship. Degree in marine engineering required. Must have previous experience. Excellent salary and benefits package. Submit resume to: North Shore Shipping Company, 2300 East 55th Street, Baltimore, MD 22001.

SAFETY COORDINATOR

Interstate trucking company has immediate opening for a safety coordinator. Oversee safety traffic program and instruct drivers in traffic and safety regulations. Will also investigate accidents and direct transfer of cargo in emergencies. Background in trucking industry needed. Salary and benefits offered. Send resume to: Warner Trucking Lines, 200 South Industrial Drive, Dayton, OH 62402.

TRAVEL AGENT

Travel company specializing in cruises seeks travel agent with exceptional telephone skills, high school diploma, and travel school education. Prefer applicants with computer skills and knowledge of computerized reservation systems. Send resume to: Magic Cruise Specialist, 800 North Second Avenue, Westlake, NJ 24202.

DISPATCHER
****

Trucking company needs dispatcher to relay information and orders that coordinate the movement of vehicles and freight. Entry-level position requires high school diploma and one year of post-secondary training. Knowledge of computer-aided dispatch system a plus. Will provide on-the-job training. Must have good telephone, radio and recordkeeping skills. Apply immediately at: Highway Trucking, 3009 Fulton Parkway, Kansas City, KS.

Linking to the WORKPLACE

When you are exploring careers, it is important to look at what you will like and dislike about the job. Choose one of the transportation jobs listed above. Then list the things that you think you would enjoy about working in that job. After writing down all of the things you would like, make a list of the things that you think you would dislike about the job. Share your list with the class in a brief report. You may find that factors that attracted you to your job are the things that some of your classmates disliked. Fortunately, there are thousands of jobs from which you can choose.

Chapter 11 Review

SUMMARY

▶ Transportation is the process by which people, animals, products, and materials are moved from one place to another.

▶ In all transportation systems, the input is energy.

▶ Maglev trains are levitated, or floated, above a guideway and propelled by magnetic fields.

▶ A gasoline piston engine is an engine that creates power by burning fuel inside the engine.

▶ A diesel engine is an internal-combustion engine that burns fuel oil by using heat produced by compressing air.

▶ Marine transportation refers to the systems used to transport people, products, and materials across waterways.

CHECK YOUR FACTS

1. List three different modes of land transportation.

2. What advantages does a diesel locomotive have over a steam locomotive?

3. What factors have led to an increase in rail travel in the United States?

4. How are most modern locomotives powered?

5. What made the automobile affordable?

6. Describe the operation of the four-stroke cycle gasoline engine.

7. Identify some examples of on-site transportation.

8. Why is a barge designed with a displacement-type hull?

9. List three factors that must be considered in the design of a hull.

10. Identify three propulsion systems used in modern marine vessels.

CRITICAL THINKING

1. Describe how motor vehicles have changed the way people live and work. Include positive and negative impacts of that technology.

2. Recreational boating is a billion-dollar industry in the United States. What impacts has recreational boating had on shore communities?

3. Explain how buoyancy is used to raise and lower a submarine.

CHAPTER 12

Fluid Power

OBJECTIVES

▶ define fluid power.

▶ explain the difference between hydraulic and pneumatic systems.

▶ identify the basic components of fluid power systems.

▶ give examples of how fluid power is used.

▶ discuss the future of fluid power.

KEY TERMS

fluid

fluid power

hydraulic systems

mechanical advantage

Pascal's Principle

pneumatic systems

pressure

How did you get to school this morning? Did you ride a bus or subway? Were you driven in a car? You probably have not spent much time thinking about fluid power. In fact, the term may be unfamiliar to you. However, fluid power helps you in many ways. Cars, buses, and subways use devices that depend on fluid power.

Fluid power is also used in entertainment. For example, fluid power systems are used to give lifelike motion to dinosaur models. Did you know that many amusement park rides depend on fluid power?

Dinamation Dinosaurs ©1997
Dinamation International Corp.

WHAT IS FLUID POWER?

Fluid power is the use of pressurized liquids or gases to move heavy objects and perform many other tasks. A **fluid** is any substance that flows. Liquids and gases are both fluids. When they are not moving, fluids have no power. When they are put under pressure and moved to where they are needed, fluids can perform work. Fig. 12-1.

Fluid power is one of the three basic systems used to transmit and control power: mechanical, electrical, and fluid. *Mechanical power* moves airplanes and other vehicles. *Electrical power* gives us light and operates motors.

The properties (physical characteristics) of air and water allow boats to float and airplanes to fly. Engineers take advantage of the properties of fluids to design power

systems for many different purposes. In this chapter, you will learn about two types of fluid power systems: hydraulic and pneumatic.

FLUID SCIENCE

All objects are made of matter. There are three states of matter: solids, liquids, and gases. The state in which matter exists depends on how tightly its molecules are held together. Solids have molecules that

DIGGER BUCKET RAM

DIPPER RAM

LOADER LIFT RAM

BOOM LIFT RAM

STABILIZER RAM

▶ **Fig. 12-1** Almost every construction job requires the use of earthmoving equipment. Vehicles such as this excavator depend on hydraulic systems. Separate rams (hydraulically-operated pistons) transfer the power of the hydraulic system to the digger bucket, boom lift, stabilizers, and loader.

are strongly linked. The molecules of liquids are loosely held together. Gas molecules can move in all directions.

Because of the space between their molecules, gases are easy to compress. Solids and liquids are not. Solids have a definite shape and occupy a certain amount of space (volume). Liquids have a definite volume and take the shape of the container they are in. Gases do not have a definite volume. They can fill a container of any shape and size. However, as pressure on a gas increases, the volume of the gas decreases. This is known as Boyle's law. Fig. 12-2.

Pressure

Pressure is the force on a unit surface area (such as a square inch). Pressure is essential in all fluid power systems. The formula for calculating pressure is:

$$\text{Pressure} = \frac{\text{Force } (F)}{\text{Area } (A)}$$

The molecules of both liquids and gases (fluids) bump into the walls of their containers. This pushing is pressure.

Blaise Pascal was a French scientist who lived in the 1600s. He found that when force is applied to a confined liquid, the resulting pressure is transmitted unchanged to all parts of the liquid. His discovery became known as **Pascal's Principle.**

FLUID POWER SYSTEM SAFETY

To use fluid power systems safely, follow these rules.
- Always wear safety glasses while operating a pneumatic or hydraulic system.
- Never blow air at yourself or another person.
- Do not place your hand in the path of a moving piston rod.

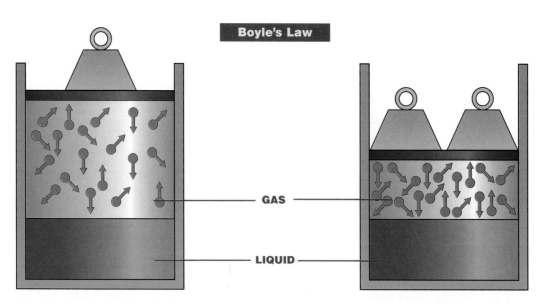

Boyle's Law

GAS

LIQUID

▶ **Fig.12-2** Boyle's law states that as pressure increases, the volume of gas decreases. Note that pressure does not change the volume of liquids under constant temperature.

Explore

Design and Build a Water Squirter

State the Problem

Design and construct a water squirter.

Develop Alternative Solutions

The device should project water to a distance of 3 to 10 feet. Sketch several possible designs as accurately as possible. Design the wood frame to permit comfortable, single-hand gripping, leaving one hand free to operate the piston. One basic design idea is shown here.

Select the Best Solution

Select the design that you think will be most effective.

Implement the Solution

1. Build the unit.
2. Test the completed unit. If no water can be forced out, enlarge the hole in the cork. Keeping the hole as small as possible will permit the maximum range. Too much force may cause the cork to pop out. If this happens, place a pin through both cork and tubing. This will keep the cork in the tube.
3. Demonstrate the water squirter to the class.

Evaluate the Solution

1. Are there any leaks? Can they be stopped?
2. Does the squirter project water accurately?
3. Is the range satisfactory?
4. What could be done to improve the water squirter's performance?

> **Collect Materials and Equipment**
>
> flexible tubing (1/2" to 3/4" diameter)
> plastic syringe units, or parts to fabricate one
> corks, of a size to fit tubing
> wood to use for building frame

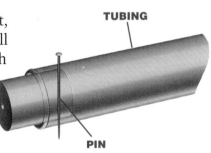

TUBING
CORK
PIN

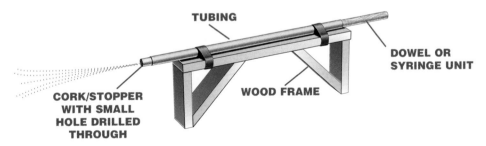

TUBING
DOWEL OR SYRINGE UNIT
WOOD FRAME
CORK/STOPPER WITH SMALL HOLE DRILLED THROUGH

- Disconnect the air-pressure line before working with pneumatic system components.
- Position air and hydraulic lines so that they are not a hazard to yourself or others.

TYPES OF FLUID POWER SYSTEMS

There are two types of fluid power systems. **Hydraulic** (high-DRAW-lick) **systems** are fluid power systems that use oil or another liquid. **Pneumatic** (new-MAT-ick) **systems** are fluid power systems based on the use of air or another gas.

Components

Hydraulic and pneumatic systems are very similar in design. The knowledge you gain by using one system is easy to apply to the other. All fluid power systems have similar basic components, or parts:
- a fluid.
- a compressor or pump.
- a reservoir or receiver.
- control valves.
- actuators.
- flow regulators.
- transmission lines.

Refer to Fig. 12-3 as you read about these components.

For fluid, most hydraulic systems use oil to transmit power. Pneumatic systems typically use air.

The *compressor* or *pump* supplies fluid under pressure to the system. A hydraulic system uses a pump to move oil. Most pumps are motor-driven. A pneumatic system uses a compressor that draws air into a chamber. The air is then compressed or squeezed into a smaller space. The compressor can be operated by hand or motor-driven. A bicycle pump is not really a "pump." It is actually a manual compressor. Most pneumatic systems use motor-driven compressors.

The oil in a hydraulic system is pumped into a *reservoir*, where it is stored. In a pneumatic system, a *receiver* takes the air from the compressor and stores it. Later the air is released as needed.

Control valves open and close passages to direct air or liquid to the proper location in the system. They also regulate the fluid pressure and the rate of flow. Valves are used to control the actions of cylinders. Valves can be controlled manually, electrically, or by air pressure.

Actuators change pressure into mechanical motion. The actuator is usually a cylinder or motor.

Cylinders are made up of the cylinder body, piston, and piston rod. *Single-acting* cylinders are designed so that air pressure is applied to only one side of the piston. When the air pressure is released, a spring returns the piston to its original position.

Double-acting cylinders are designed so that air can be applied to either side of the piston. With this type of cylinder, air pressure can be used to extend and retract the piston.

Cylinders produce linear (straight line) or reciprocating (back and forth) motion. Applying the brakes in a vehicle is linear motion. Fluid motors produce rotary

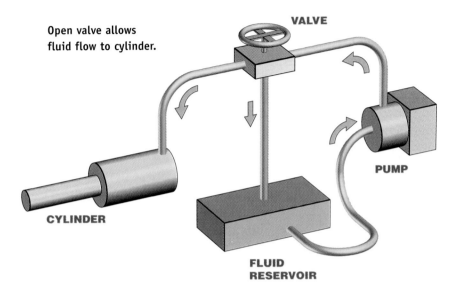

Fig. 12-3A Open valve allows fluid flow to cylinder.

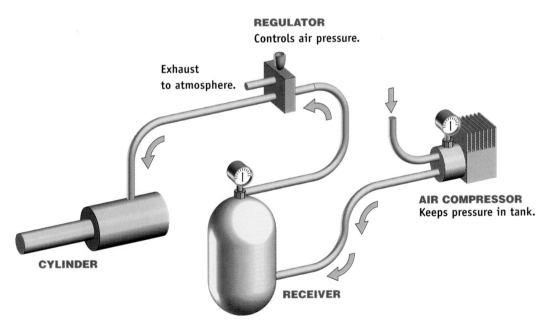

Fig. 12-3B A simple pneumatic system.

Linking to COMMUNICATION

Word Roots. Many words in English are based on word parts from the ancient Greek language. Two important terms in this chapter—hydraulic and pneumatic—use Greek roots. *Hydro* means "water." *Pneuma* means "blast of air." Discuss with your study group how these two root words apply to hydraulic and pneumatic. Use the dictionary to prepare a list of other words using the same root words. Be sure each word in your list refers in some way to water (hydro) or air (pneuma).

▲

(circular or spinning) motion. Rotary motion is produced in a tool such as a drill. Complicated industrial equipment often uses combinations of the different types of motions.

Flow regulators control the speed of piston travel in a cylinder. They do this by restricting the flow of air in one direction. Two flow regulators are needed to control the motion of a double-acting cylinder.

Transmission lines are pipes and hoses. Pressurized fluid is moved through these to the other components in the system.

Hydraulic Systems

Liquids cannot be compressed. Therefore, they can be used to transfer force. Figure 12-4 shows this. The force applied to piston A puts pressure on the fluid. The fluid then exerts the same amount of pressure in all directions (Pascal's Principle). Thus the 10 pounds of force from piston A is transferred to piston B. Note that the two pistons are the same size.

Hydraulic systems can also multiply force. The increase in force gained by using a machine is called **mechanical advantage.** Figure 12-5 shows how pistons of different sizes can produce a

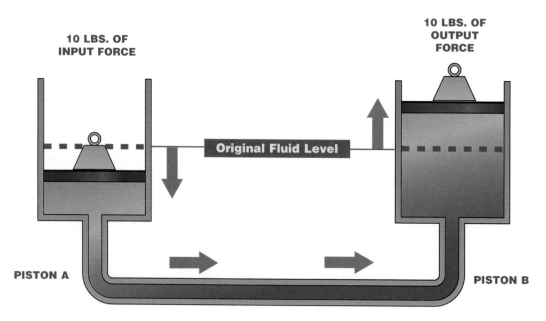

10 LBS. OF INPUT FORCE

10 LBS. OF OUTPUT FORCE

Original Fluid Level

PISTON A

PISTON B

▶ **Fig. 12-4** Piston A and Piston B are the same size. When force is applied to Piston A, the force is transferred by the liquid to Piston B.

mechanical advantage. Remember that pressure is calculated by dividing the force by the area.

$$P = \frac{F}{A}$$

$$P = \frac{50 \text{ pounds}}{5 \text{ square inches}}$$

$$P = 10 \text{ pounds per square inch (psi)}$$

Piston B has an area of 10 square inches. To calculate the force applied to piston B, multiply the pressure times the area as follows:

$$F = P \times A$$

$$F = 10 \text{ psi} \times 10 \text{ square inches}$$

$$F = 100 \text{ pounds}$$

Note that 2 inches of input movement were required to produce 1 inch of output movement. A gain in force results in a loss of distance.

Automobile jacks use a simple hydraulic system. The handle must be moved up and down many times to raise the jack even a few inches.

Hydraulic systems are ideal for use when strength and accuracy are required. This is why they are used on heavy construction equipment such as backhoes and bulldozers.

Pneumatic Systems

Pneumatic systems have some advantages over hydraulic systems. The air they use is usually readily available. If a pneumatic system leaks, there is nothing to clean up. No hazardous materials are

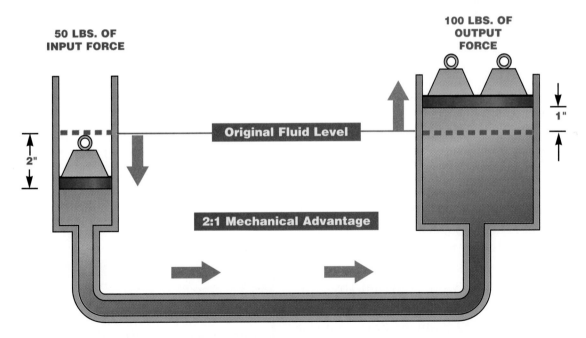

▶ **Fig. 12-5** Force can be multiplied by using pistons and cylinders of different sizes. In the system shown here, 50 lbs. of input force produces 100 lbs. of output force. The dotted line in each cylinder shows the fluid level before the application of input force.

released. This is particularly important in the food processing industry. Pneumatic systems are also useful in locations such as spray-painting booths. Another advantage is that pneumatic systems are fast.

Pneumatic systems have some disadvantages. The energy required to compress air can be expensive. Also, many pneumatic devices are noisy.

FLUID POWER SYSTEM DIAGRAMS

Engineers plan pneumatic systems by drawing *schematic circuit diagrams*. They use symbols to represent components. Figure 12-6 shows a few of these symbols. These symbols were devised by the American National Standards Institute (ANSI). Figure 12-7 shows how some of these components are arranged to create a circuit. Fluid power system diagrams are read from bottom to top.

HOW FLUID POWER IS USED

Fluid power can be applied in a variety of ways. Many small hand tools like drills, wrenches, and sanders are operated by fluid power. So are the large machines used to crush automobiles for recycling. Fluid power is used to lift and move heavy objects. It can also open and close doors. Today most industries use fluid power.

Manufacturing

Fluid power systems are used in almost every manufacturing plant. Hydraulic lifts move materials to where they are needed. Pneumatic devices perform tasks such as painting. Many of the material-moving devices on assembly lines are hydraulic or pneumatic.

Many of the devices used to move materials in factories are powered by hydraulic or pneumatic systems. These systems are also used to power many of the tools used on assembly lines. Pneumatic systems provide an inexpensive source of energy when a lot of power is needed. For example, the tool used to tighten and remove the bolts on an automobile wheel is usually a pneumatic tool.

FASCINATING FACTS

In Paris, a system of pipes carries compressed air to users throughout the city. The system was installed in the 1880s.

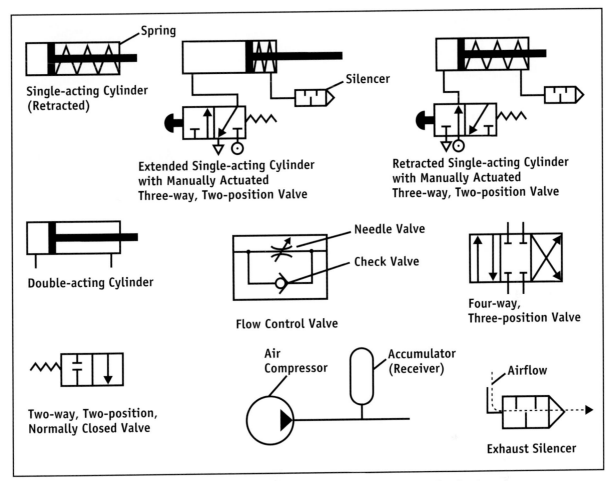

Fig. 12-6 These are some of the symbols used to diagram a pneumatic circuit.

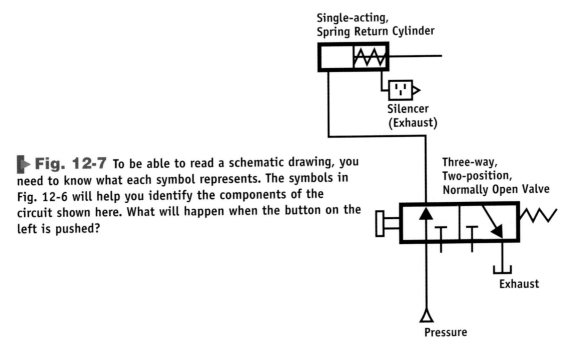

Fig. 12-7 To be able to read a schematic drawing, you need to know what each symbol represents. The symbols in Fig. 12-6 will help you identify the components of the circuit shown here. What will happen when the button on the left is pushed?

Explore

Design and Build an Air Cushion Vehicle

State the Problem

Air cushion vehicles (ACVs) use a pneumatic system to provide lift as they travel. Design and build a model ACV capable of transporting as many pennies as possible.

Develop Alternative Solutions

A simple ACV is shown here.. A model built with this design will work. How can you modify this design to lift and move as many pennies as possible? Sketch several possible designs.

Select the Best Solution

Select the design that you think will be the most effective.

Implement the Solution

1. Make your first ACV using a 4" diameter base. Use the pencil to make a hole in the center of the base.

> **Collect Materials and Equipment**
>
> corrugated cardboard
> compass
> scissors
> pencil
> hot glue gun
> balloons in assorted
>   sizes and shapes
> empty thread spools
> pennies

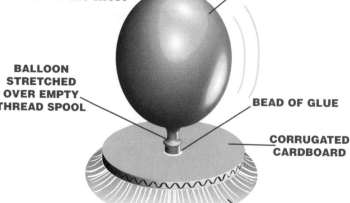

DEFLATING BALLOON

BALLOON STRETCHED OVER EMPTY THREAD SPOOL

BEAD OF GLUE

CORRUGATED CARDBOARD

AIR CUSHION

Transportation

Fluid power systems make automobiles steer easily, ride smoothly, and stop. The equipment used to repair flat tires is hydraulic and pneumatic. Protective air bags work because they are pneumatic.

The controls that pilots use to guide airplanes, helicopters, and even the space shuttle are fluid-powered. Aircraft landing gears are lowered and raised hydraulically. Also, hydraulic shock absorbers built into landing gears make landing safer and more comfortable.

2. Center the thread spool over the hole. Attach it to the base using hot glue. Allow to dry.
3. Inflate the balloon and stretch it over the spool. Pinch the balloon at its base so that the air cannot escape. Place the ACV on the floor and release it.
4. Add several pennies to the top of the ACV and repeat the test.
5. Repeat the test using balloons of various sizes and shapes. You might also use bases of different sizes. Determine how these changes affect lifting power.

Evaluate the Solution
1. Make a table to record the results of each test. Include information about the size and shape of the balloon, the size of the base, and the number of pennies moved.
2. Which combination of balloon and base worked best?
3. What might be done to improve control of the ACV?
4. Describe one use of an ACV in business or industry. How might the design of such an ACV differ from the design of your model?

Construction

In home construction, pneumatic nailers are replacing hammers. Pneumatic framing, roofing, and trim nailers have greatly increased the amount of work a carpenter can complete in a day. The lifting, pushing, and digging mechanisms on heavy construction equipment are hydraulically operated. Fig. 12-8.

Agriculture

Hydraulic devices are used in agriculture on the equipment used for planting and harvesting. Pneumatic power systems are commonly used in factories that wash, can, and package food. Fig. 12-9.

Health Care

Fluid power devices are also important in health care. Your dentist uses a high-speed pneumatic drill. The dental chair is raised and lowered by a hydraulic system.

Medical personnel can now use pneumatic devices to give injections. These devices force vaccines through the skin without piercing it.

Pneumatic artificial kidney machines are used by thousands of people every day. These machines are controlled by a computer.

IMPACTS

People have used fluid power for thousands of years. Windmills have been used to pump water. Waterwheels were used to turn millstones to grind grain. Many of the devices that make our life easier depend on fluid power. Today, fluid power systems perform important tasks in manufacturing, transportation, construction, agriculture, and health care.

Some have said that certain uses of fluid power have had a negative impact on our environment. As examples, they point to the building of dams on rivers. Dams are built for several reasons. For example, they are built to help prevent or control flooding. Dams are also built to generate electric power. The fluid power of the water through the dam powers generators.

▶ **Fig. 12-9** Pneumatic systems are used in all phases of agriculture and food processing. Many of the machines used in food canning are powered by pneumatic systems.

Some have been critical of such dam building. They argue that the damming of rivers has impacted the reproduction cycle of fish such as the salmon. They also point out that such construction alters the landscape. They point out that the advantages gained by building a dam do not outweigh the negative impacts.

THE FUTURE

The fluid power industry will continue to develop new uses for pneumatic and hydraulic control systems. In the future, individual components are expected to be smaller, lighter, and less expensive.

Engineers will continue to develop new uses for fluid power in manufacturing and other industries. Microprocessors will combine fluid and computer control. These devices will play an important role in automated systems.

Linking to MATHEMATICS

Figuring Area. Apple cider is juice that has been pressed from apples. A large commercial cider press uses hydraulic power to press the juice from the ground-up apples. Assume that a cylinder with a diameter of 4" is used. Figure the area in square inches of the 4" diameter cylinder. Use the formula $A = \pi r^2$. *A* stands for Area; π has a value of 3.14; *r* stands for radius.

Apply What You've Learned

Design and Build a Gameboard

State the Problem

Design and build a game that uses pneumatic power.

Develop Alternative Solutions

A player should be able to move the marble from hole to hole using the pneumatic syringe hookups (possibly one in each corner). The panel may be held down with springs or rubber bands.

The basic setup is a flat panel (perhaps 8" x 10") with several holes. Each hole must be smaller than the marble that will be used in the game. Sketch several possible designs. Note that smaller holes will allow the marble to move from hole to hole more easily. It might look similar to the gameshown here.

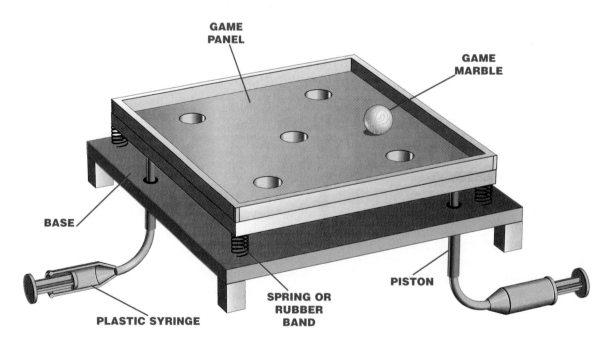

GAME PANEL

GAME MARBLE

BASE

PLASTIC SYRINGE

SPRING OR RUBBER BAND

PISTON

Select the Best Solution

Select the design that you think will be most effective.

Implement the Solution

1. Prepare the game panel.
2. Cut and assemble the parts for the base.
3. Mount the game panel on the base.
4. Mount the syringes. It is often a good idea to mount the syringes in a secure holder to allow one-hand operation.
5. Present your finished game to the class. Explain what problems were encountered and how they were solved.

Evaluate the Solution

1. Does the game work as intended?
2. Is the game reasonably easy to operate?
3. Is the game made solidly enough to be used again and again?
4. Changing the angle of a horizontal panel in small stages through the use of pneumatic power (such as supplied by plastic syringes) is a difficult task. Making small, gradual motions takes practice and skill to achieve. How could this game be redesigned so that it would require less skill to play?
5. How do you think a game such as this could be made using other sources of power? Why might it be convenient to control motion using fluid power?

CAREERS IN
Fluid Power

WASTE WATER PLANT OPERATOR

Industrial wastewater treatment plant needs operator with mechanical aptitude and competence in basic math. Two-year degree and certification required. Will perform tests, keep records and do repairs and maintenance. Send resume to: Burlington Industries, 903 West Town Street, Athens, NE 63205.

NUCLEAR ENGINEER

Engineer needed to monitor nuclear tests and examine operations of facility utilizing radioactive material. Direct operations and maintenance activities. Degree in nuclear engineering required. Competitive salary and benefits package. Submit resume to: Tennessee Nuclear Plant, 3355 Kingston Parkway, Knoxville, TN 32204.

PETROLEUM RESEARCH ENGINEER

Oil refining company seeks petroleum research engineer for designing new petroleum blends and troubleshooting. Bachelor's degree in chemistry or chemical engineering required. Internship or summer job experience helpful. Work in lab setting. Computer skills a must. Send resume to: Clark Oil Company, Human Resources Director, 1800 King Drive, Oklahoma City, OK 77320.

PRECISION INSTRUMENT TECHNICIAN

Technician needed to inspect and repair mechanical systems used at power generating plant. Involved in all facets from regular inspection and preventive maintenance, to actual adjustment or replacement of faulty parts. Two-year training required. Knowledge of hydraulics, digital electronics and electricity required. To apply, submit resume to: Personnel Department, Blue Water Power, 1415 Lakeside Drive, Alcoa, TN 33956.

HYDRAULIC REPAIRER

Manufacturing company needs hydraulic repairer to inspect, maintain and repair robotics equipment. Knowledge of hydraulics and mechanical ability required. Computer skills helpful. Apply in person to: North Start Industries, 900 Stafford Parkway, Kewanee, IL 64302.

AUTOMOBILE TECHNICIAN

Expanding auto repair shop needs technicians with own tools. ASE Certification preferred. Two years of formal training required with knowledge of electronics. Numerous benefits with high hourly pay. For more information, call Joe at (212) 473-4500.

Linking to the WORKPLACE

The job ads listed above describe openings that an employer has written. They tell you what's needed for a person to get that job—the job requirements. Read the jobs listed above and write down all the different requirements you find. Even if you are not sure what a requirement means, write it down anyway. Were some of the listed skills needed in more than one job? What are the requirements for the job you are most interested in at this time?

Chapter 12 Review

SUMMARY

▶ Fluid power is the use of pressurized liquids and gases to perform work.

▶ There are two types of fluid power systems. Hydraulic systems use oil or another liquid. Pneumatic systems use air or another gas.

▶ All fluid power systems have similar basic components: a fluid, a compressor or pump, a reservoir or receiver, control valves, actuators, and transmission lines.

▶ Hydraulic systems are used when strength and accuracy are required. Pneumatic systems are well suited for food processing and spray-painting situations.

▶ Fluid power systems are used in most industries.

▶ Industry will continue to develop new uses for fluid power control systems.

CHECK YOUR FACTS

1. Define fluid power.

2. Describe the two main types of fluid power systems and explain the difference between them.

3. Identify the basic components of a fluid power system.

4. What part of a fluid power system changes pressure into mechanical motion?

5. What is the proper name of the drawing used to represent a fluid power system? How is the drawing read?

6. Name at least four industries that use fluid power systems. Give one example of how fluid power is used in each industry.

CRITICAL THINKING

1. Explain some of the important differences between the two types of fluid power systems.

2. Describe how the molecules of solids, liquids, and gases differ.

3. Use an example to explain how a hydraulic system can multiply force.

4. Plan a simple pneumatic system using the proper symbol for each component.

5. Do research to find out how fluid power devices are used in your community and nearby areas.

6. Do research to find out how fluid power might be used in the future.

Bio-Related Technologies

CHAPTER 13 *Health Technologies*

CHAPTER 14 *Environmental Technologies*

The prefix "bio" means life. What words include the prefix "bio"? You may be thinking of *biology* and *biodegradable*. These words have a common theme. Each of them relates to living organisms. Biology is the study of living things. A biodegradable material is one that can be broken down by living organisms, such as bacteria.

The chapters in this section discuss the ways we use bio-related technologies. For example, we use health technologies to improve the quality of life. We use environmental technologies to develop new methods of agriculture. Each of these technologies offers real and immediate benefits.

Cereal Technology

We use technology to make the grasses we eat, called cereal grains, both chewable and tasty. By grinding the seeds of wheat, corn, rice, oats, and rye very fine we obtain flour.

Cereal History Rolls Onward

In the late 1800s, a health craze brought forth "granula." This was whole wheat coarsely ground and soaked and then passed between rollers. The rollers formed the wheat dough into sheets. These sheets were baked, broken up, and baked again to produce a wholesome and crunchy snack food.

Borrowing this idea, the Kellogg brothers used rollers to flatten wheat dough and toast it into flakes. Cornflakes, oat flakes, and bran flakes soon followed.

In the 1940s cereal technologists embraced the spray gun. At first, this was used to replace vitamins broken down in cooking processes. Soon this technology was used to spray on sugar and produce the frosted flake.

Cereal Shapes

Cereal makers also perfected a method of pressurized cooking. When the cooker was opened, the still-hot moisture inside the kernels caused them to increase ten times in size. Toasting these exploded grains made them snap, crackle, and pop when milk was added.

Makers of spaghetti have known that a dough of finely ground flour can be forced through a die to form various shapes. This process is known as *extrusion*. The first breakfast cereal made by this process was Cheerioats, later known as Cheerios.

Linking to the COMMUNITY

At your grocery store, make a list of cereal brands. Catalog them according to the technology used to make them. Were they rolled, popped, or extruded? Was more than one process used?

CHAPTER 13

Health Technologies

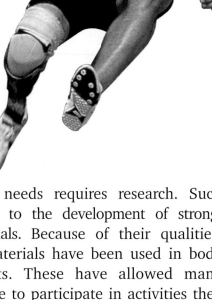

OBJECTIVES

▶ describe five technological activities associated with health technologies.

▶ define biomechanical engineering.

▶ discuss the impacts of technology on diagnosis in medicine.

▶ identify future trends in biorelated technology.

▶ define biomechanical engineering.

▶ identify the items a human factors engineer must consider.

KEY TERMS

biomechanical engineering

diagnosis

ergonomics

genetic engineering

human factors engineering

immunization

physical enhancements

prosthesis

Can you imagine a world in which diseases could not be cured? Advances in health technologies have extended the human life span.

They have also expanded the range of activity for many people with health concerns. The use of technology to meet human needs requires research. Such research has led to the development of strong, lightweight materials. Because of their qualities, some of these materials have been used in body replacement parts. These have allowed many people to continue to participate in activities they have always enjoyed.

PHYSICAL ENHANCEMENTS

Human factors engineering is a design process. It is used in product design. It gives special attention to the strengths and limitations of the human body. One activity within human factors engineering is the design and construction of physical enhancements. **Physical enhancements** are replacement body parts.

Biomechanical engineering is the use of engineering principles and design procedures to solve medical problems. This includes the development of physical enhancements such as artificial implants and artificial limbs. Fig. 13-1.

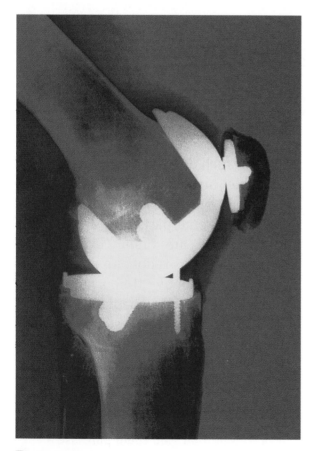

► **Fig. 13-1** Many body parts can be made artificially. The technology of making artificial body parts is called *bionics*.

A **prosthesis** (prahs-THEE-sis) is a people-made device used to replace human body parts. The plural of prosthesis is *prostheses* (prahs-THEE-sees). They are used to replace limbs, teeth, or even heart valves. Biorelated technology is one family of technology. It is responsible for these developments. Today, it is common for people to receive a prosthetic device after a body part has been injured due to accident, disease, or birth defects. Prostheses replace joints, tissues, organs, and organ systems. Such prostheses are designed by biomechanical engineers.

HUMAN TECHNOLOGY RESOURCES

Information

The information explosion of the last ten years has set the stage for today's developments in health technologies. *Information resources* continue to fuel new discoveries by scientists, engineers, technicians, and doctors. Health workers must keep up with new information on biology, chemistry, material science, and engineering processes. Fig. 13-2.

Fig. 13-2 An *angiogram* is an X-ray photograph of the heart. This angiogram shows an artificial valve in the *aorta* (a-OR-tah), which is the main artery carrying blood from the heart.

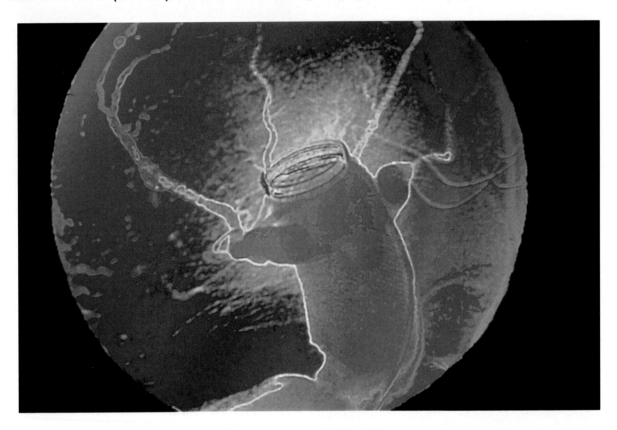

Materials

Physical enhancements could not be made without new *material resources*. The human body is a harsh environment for foreign (outside) materials. Chemicals that make up our tissues and organs often react negatively with materials placed inside the body. Some materials are toxic (poisonous) to our natural systems. Others are rejected by the body's natural defense system, its *immune system*.

Titanium (tie-TANE-e-um) is a metal that is commonly accepted by the human immune system. Titanium is strong and lightweight. It is often used in prostheses. Hip and knee joints are made from titanium.

Plastics such as woven acrylics are used as flexible artificial arteries and blood vessels. Silicone interacts well with living tissues. Electronic parts that are inserted into the body are often covered in a silicone wrap. This is done to protect their circuits from body fluids. Electronic devices are small and adaptable. Because of this, they are widely used as physical enhancement aids.

FASCINATING FACTS

For $25 million, you could purchase a complete set of the different artificial body parts now made. There are nearly 200 different ones.

Explore

Design and Build a Human Joint Replacement

State the Problem

Design and build a model of a prosthesis for a joint found in the human body. You might consider the hip joint, elbow joint, knee joint, or finger joint.

Biomechanical engineers are responsible for designing physical enhancements or body replacement parts. Their job requires a good understanding of human anatomy and how our natural systems work. They must also have a background in materials science and material processing techniques.

In this activity you will have the opportunity to explore some of the skills needed to be a biomechanical engineer.

Develop Alternative Solutions

Examine reference material on the human skeletal system. Select a body joint for which you wish to design an artificial replacement. Research details on the joint. What type of joint is it? How does it move? What muscles control it? How does it attach to other parts of the body? What size is the joint? Select modeling materials. Develop drawings for three models of the same joint.

Select the Best Solution

Select the design that you feel meets the criteria for the artificial joint you want to design and build.

Implement the Solution

Following your design, build a working model of the prosthesis.

Evaluate the Solution

1. Compare your design to the way the natural joint operates.
2. Measure the freedom of movement and the ease of movement.
3. Measure the degree of movement.
4. How will your design fit into the body?
5. Modify the design, if needed.

Collect Materials and Equipment
wood
metal
wire
modeling clay
plastic
cardboard
Styrofoam™
standard material processing equipment
books and videos on the skeletal system
charts on the skeletal system and human joints
plastic models of the hand, foot, etc.

Linking to COMMUNICATION

Uses for Titanium. Using search engines on the Internet in your school's media center, identify uses for titanium. In a brief essay, discuss those characteristics of titanium that make it suitable for each of these uses.

▲

HEALTH CARE TECHNOLOGY

Do you have asthma or know someone with asthma? An old "cure" for this breathing illness was to swallow a handful of spider webs rolled into a tight ball. Have you ever had a headache? In earlier times you might have tied a flour sack around your head or smeared crushed onions on your forehead. Folk remedies (cures) like these were once the basis for health care.

The diagnosis and treatment of illness has changed a great deal over the last hundred years. Here again, information is the reason for these changes. Through research, we have learned how body systems work. Modern health care technology focuses on three areas: prevention of disease, diagnosis of illness, and treatment. Fig. 13-3.

Prevention of Disease

Have you heard the statement, "An ounce of prevention is worth a pound of cure?" You can be immunized against many diseases. *Vaccines* are substances placed in the body to artificially increase its immunity to various diseases. The design and production of vaccines are health technology activities. The phrase "to immunize" means "to protect." **Immunization** (IM-mew-ni-ZA-shun) is

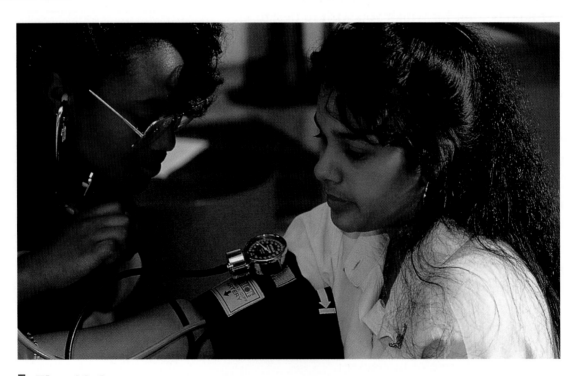

▶ **Fig. 13-3** Blood pressure readings are a valuable tool in medical diagnosis.

an action that protects the body against a disease.

Tetanus (TET-uh-nus) bacteria can enter your body when your skin is punctured by a rusty nail. The bacteria travel through your body in your bloodstream. Tetanus was once the leading cause of death among farmers in the United States. Now the tetanus vaccine is injected into humans to protect them from the tetanus bacteria.

Injecting medicine through the skin can be painful. Scientists are working on alternatives to needle injection. Stick-on patches can be used with some substances. Methods involving electrical pulses and sound are being tested.

Diagnosis of Illness

Diagnosis is the process of examining a patient and studying the symptoms to find out what illness the patient has. Like engineers, doctors use many resources to solve medical problems. Machines developed by bioengineers can monitor and measure many of the body's natural functions. Doctors can measure your temperature, blood pressure, pulse rate, breathing rate, blood chemicals, and your body's electrical waves. Blood tests help doctors check your health. Fig. 13-4.

▶ **Fig. 13-4** Blood analysis provides key information on health.

An *electrocardiograph* (e-LEC-tro-car-dee-oh-graph), or EKG, records the electrical currents of the heart. An *electroencephalogram* (e-LEC-tro-en-sef-oh-low-gram), or EEG, records the electrical activity in the brain. Many tools of medical technology can be used to allow technicians to "see" into the body. The *CAT scan*, *MRI*, and *ultrasound* are used in diagnosis. Fig. 13-5.

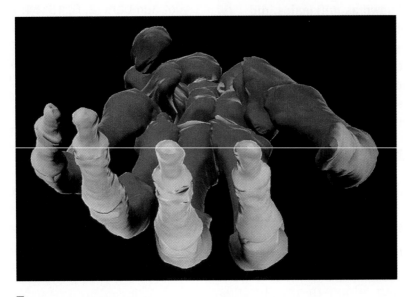

▶ **Fig. 13-5A** A CAT scan is made by a rotating X-ray machine that uses a computer to enhance the image.

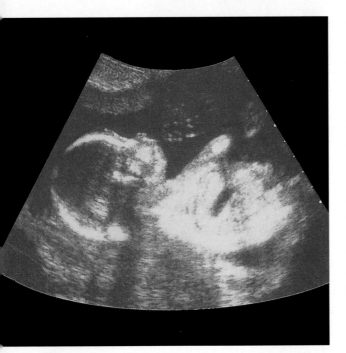

▶ **Fig. 13-5B** Ultrasound uses sound waves to generate an image. Can you recognize the image?

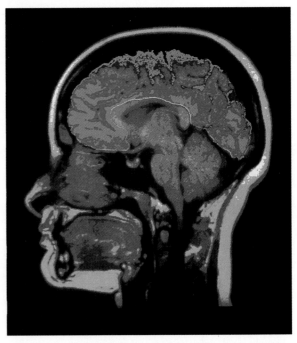

▶ **Fig. 13-5C** Magnetic resonance imaging (MRI) uses magnetic energy to create an image of body tissue.

Treatment

After an illness is diagnosed, treatment usually follows. Treatment may take the form of drugs. Health technologies are responsible for the design and production of these drugs. Treatment might also include surgery. Today's operating room is filled with technological marvels. Some assist doctors in monitoring and regulating the patient's natural systems. Others maintain the clean, healthy environment in which surgery must take place.

ERGONOMICS

Why is a dentist's chair different? It has been designed to support your body in a comfortable way. A dentist's chair is designed ergonomically.

Ergonomics (erg-oh-NOM-icks) is the study of designing equipment and devices that fit the human body, its movement, and its thinking patterns. Ergonomic designs help people work more efficiently. Human factors engineers must consider the anatomy (structural makeup) and psychology (mind and behavior) of human beings when they design. Fig. 13-6.

Suppose you are a human factors engineer. You are asked to design a workstation for your classroom. What factors would you consider?

You might start with the design of the chair. You would want your chair to have an adjustable height. This would allow your feet to comfortably touch the floor. A lumbar support would brace your lower back. The seat would be contoured to your thighs and lower body.

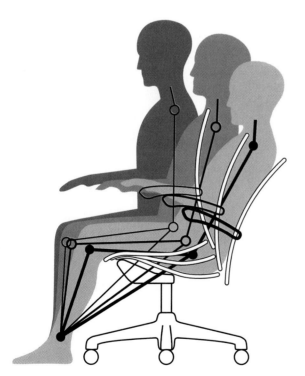

▶ **Fig. 13-6** The ergonomic design of this chair is based on human needs. It provides comfortable support. Note the contoured seat and back.

Explore

Design and Build an Assisted-Living Product

State the Problem

Of all the products manufactured, some are for shelter or protection. Some are for amusement or entertainment. Some are for transportation. One product category is called products for assisted living. These items make it easier for a physically challenged person to perform tasks. These include holding eating utensils, opening doors, and fastening buttons.

Simulate a physical challenge (such as limited finger dexterity). Then design a device that will help you compensate for that disability.

Develop Alternative Solutions

You may be wondering why some people need help with simple, everyday tasks. Can you think of a reason? Do you know anyone who might need such assistance? What alternatives would such a person have if devices to assist in everyday living were not available? If you can answer these questions, you will have come a long way in understanding how some people cope with tasks that do not pose a problem to most of us.

<table>
<tr><th>Collect Materials and Equipment</th></tr>
<tr><td>Your instructor will supply building materials (perhaps wood, cardboard, cutting tools, and more). You can use these to make a model of your design. Materials availability and specific instructions will be announced in class.</td></tr>
</table>

The desk height would have to be adjustable to match the distance from your shoulder to your elbow. This would allow you to keyboard comfortably. The position of the computer monitor would be important to avoid eye fatigue and neck strain. Your new workstation would make you more productive.

Environment Design

Human factors engineers also design environments as well as individual items. For example, they design living areas and working areas. When an environment is created, human anatomy and psychology must be considered. Lighting, noise,

Alone or in small groups, decide which physical limitation you will imitate. Decide how this can be done. For example, you might simulate a finger disability by taping one or more fingers to a stiff piece of wire. Decide how a person with such a physical limitation could successfully accomplish a common task, such as making a telephone call. Sketch what your device would look like. You may need to prepare several sketches.

Select the Best Solution
Select the design that you think will be the most effective.

Implement the Solution
1. Make the device, following your design. A working model is the most effective way to demonstrate how the device will actually work.
2. Present your project to the class. Include your original concept. Explain how you solved the problem of physical limitation.
3. Demonstrate your model.

Evaluate the Solution
1. Was the selected disability clearly described?
2. Was the selected disability effectively simulated?
3. Did the assisted-living product work well?
4. Was your presentation well organized?

temperature, and air quality must also be considered. What changes would you make to your classroom to improve its ergonomic design? What factors would you have to consider in the design of an operating room? How would these factors differ from the design of the welding area of an assembly line?

In 1990, then-President George Bush signed into law the "Americans with Disabilities Act" (ADA). This federal law sets requirements for the design of public places (environments) that do not restrict persons who are physically challenged from using that facility. Schools, government buildings, restaurants, and other places of business must consider the needs of persons with disabilities. Access ramps, wide hallways, wide doors, lever door openers, and grasp bars are examples of ergonomic features. All of these have been designed for persons with disabilities.

IMPACTS

Health technologies have improved the quality of our lives. People today live longer and are healthier than at any other time in history. Health technologies provide society with products, processes, and procedures. These meet many of our physical and psychological needs.

Quality of Life

Artificial limbs have dramatically improved the quality of life for persons who are physically challenged. Myoelectric prosthetics allow people to live independently and continue to contribute and enjoy life activities. (The prefix "myo" refers to muscles.) Health technologies are responsible for these and other developments that focus on the human body and other living organisms. Fig. 13-7.

▶ **Fig. 13-7** Burnt skin from a burn patient is placed in a culture. It is then reattached to the patient. This procedure is preferred to a skin transplant from a donor because it avoids the possibility of rejection of the donor's skin.

Linking to SCIENCE

Hearing a Pin Drop. The human ability to hear is remarkable, even without a bionic device. You and your partner can measure and compare hearing ability. You'll need a pin, a ruler, measuring tape, paper, and a pencil. You'll also need a quiet place where you can sit at a table.

Your partner should stand by the table with his or her back to you. You should hold the ruler vertical, with one end resting on the tabletop. Drop the pin from the 4-inch (10-cm) mark onto the table. If your partner hears it, he or she should raise his or her hand. Have your partner walk away two steps. Then drop the pin again from the same height. If your partner hears it, he or she should raise his or her hand. Continue the experiment. After each pin drop, your partner should walk away two steps. Your partner will finally reach a point where he or she can no longer hear the pin drop. Measure the distance to this point using the measuring tape.

Repeat the process, changing places with your partner. Whose hearing is better? Compare your results with those of your classmates. Compute the average distance at which the pin drops could not be heard. ▲

Questionable Developments

At times, the impacts of technology can be negative and unpredictable. Many years ago, a drug taken by pregnant women caused severe birth defects in children.

Some health technologies have forced society to make very personal decisions about the quality of life. For example, machines can now keep the body alive even though the brain is dead. What questions regarding their use are raised in your mind?

THE FUTURE

Scientists are working on devices that will assist sight and hearing. Scientists are continuing to work on developing new drugs. Some diseases can no longer be cured by antibiotics that were formerly effective. Fig. 13-8.

Enhanced Senses

Biomechanical engineers are now working to develop artificial limbs that will both sense temperature changes and

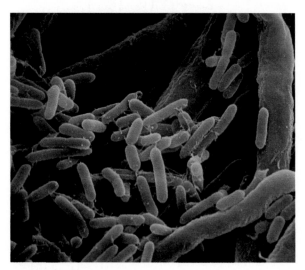

▶ **Fig. 13-8** A *micrograph* is a reproduction of an image formed by a microscope. This micrograph shows bacteria that are resistant to penicillin.

control the pressure the limb can exert. Adding superstrength to an artificial limb may be closer than we think.

Bioengineers are working to develop artificial eyesight. They have built a system based on a tiny microchip that fits into the eye. The microchip processes information sent to it by a small camera mounted on a pair of eyeglasses. The microchip is connected to nerve cells in the eye that are connected to an optic nerve. The nerve carries the signals from the microchip to the brain and the person "sees."

Prototype artificial hearing is already available. A small microphone and amplifier clipped onto a belt sends sound to a microprocessor (computer chip). The processor changes the sound energy into electrical energy. The electrical energy travels on a cable that goes through the skull and connects to the nerves in the ear.

Genetic Engineering

Genetic engineering is the process of changing the genetic materials (genes) that make up living organisms. Genetic engineering is changing the way doctors fight diseases caused by genetic defects.

Cloning (CLOH-ning) usually refers to the process of creating genetically identical organisms. Genetic material from one individual can be used to grow a new individual, or clone, that is identical in every respect. This technology may eventually lead to the development of animals that can supply organs for use in humans. The organs would be produced in cloned animals. The animals would have been genetically engineered to reduce the chances for organ rejection after transplanting.

Apply What You've Learned

Design and Build a New Computer Keyboard

Collect Materials and Equipment

Paper layout of a blank computer keyboard (letters not printed on keys)
Computer keyboard

State the Problem

How fast can you keyboard, or type? Did you know that the design of the computer keyboard can be traced back to the development of the first typewriter? Why do you suppose the letters on the keyboard are placed in such hard-to-reach positions? For example, "e" and "a" are common letters. However, they are positioned near the fingers that you have the least control of—the little finger and the ring finger. The typewriter designers did this on purpose. Many people could type faster than the letters could strike the paper. The typewriter would jam. The solution was to slow down the typist. This was done by placing the most commonly used letters in hard-to-reach spots.

As typewriters were improved, speedy typists were no longer a problem. However, the keyboard design didn't change. People were accustomed to using the traditional keyboard layout. They didn't want to learn a new pattern. Today, if you're willing to relearn the keyboard layout, you can use more efficient keyboards.

In this activity you will take on the role of a human factors engineer. Design and test a new keyboard configuration that is more efficient and increases the typing speed.

Develop Alternative Solutions

1. Examine the human hand. Determine the strongest and most flexible fingers.
2. Determine the most comfortable position for keyboarding.
3. Examine the traditional keyboard.
4. Using the paper keyboard layout, experiment with the placement of letters on the keys. Be sure to make your own keyboard layout. Don't write on the keyboard layout on the opposite page.

Select the Best Solution

Experiment with the designs you have developed. Time yourself while simulating typing a short paragraph. Select the most efficient arrangement of keys.

Implement the Solution

Follow your new keyboard design. Tape the letters of the alphabet as well as the other characters used in word processing to a computer keyboard.

Evaluate the Solution

Practice keyboarding on your new keyboard. Compare the speed and ease of typing with typing on a traditional keyboard. Make modifications if necessary.

CAREERS IN
Health Technologies

HOME HEALTH AIDE

Home health-care provider needs aides to provide personal care and home management services to patients in their homes. Will help patients bathe, exercise, and dress. Will check vital signs to evaluate condition of patients. Must have valid driver's license. Outgoing personality and good communication skills needed. On-the-job training for high school graduate. Courses in health and knowledge of nutrition helpful. Call Carol Kublick to schedule an interview: (217) 528-3703.

DENTAL ASSISTANT

Team of professionals in search of dental assistant with outstanding technical and verbal skills to provide quality care to our patients. Must be graduate of an approved dental-assisting program. Must have ability to learn on the job. Challenging full-time position with benefits package. Call (912) 458-8897.

BIOTECHNOLOGY RESEARCHER

Small research company seeks self-motivated biotechnology researcher to conduct studies of chemicals and therapeutic effect. Will track data and conduct testing. Entry level requires bachelor's degree in biochemistry, molecular biology, chemistry, or chemical engineering. Excellent salary and working environment. Submit resume to: Miller Research Group, 978 Highland Avenue, Raleigh-Durham, NC 43098.

BIOMEDICAL EQUIPMENT TECHNICIANS

Technician needed to work in hospital setting installing and testing new equipment. Will operate, service, and teach others how to use specialized machines. Associate degree required. Work rotating shifts and be on call for emergencies. Knowledge of electrical and electronics fundamentals necessary. Send resume to: Hospital Supply Center, Manager, 234 North Royal Street, Chicago, IL 60032.

DENTAL HYGIENIST

Enthusiastic and well-organized hygienist needed. Experience in sterilization procedures and X-ray certification preferred. Excellent opportunity with growing practice. Mayfield Heights area. Call (812) 785-3365.

Linking to the WORKPLACE

In many careers in health technologies, workers wear a uniform in the work setting. List five jobs that are not in health technologies that require workers to wear uniforms. What would you like about wearing a uniform to work? What would you dislike?

Chapter 13 Review

SUMMARY

▶ Human factors engineering is the process of designing for people.

▶ Biomechanical engineering refers to the use of engineering principles and design procedures to solve medical problems.

▶ Prostheses are people-made devices used to replace human body parts.

▶ The immune system is the body's natural defense system.

▶ Ergonomics is the study of designing equipment and devices that fit the human body, its movement, and its thinking patterns.

CHECK YOUR FACTS

1. Describe at least five activities associated with health technologies.

2. Define the term *biomechanical engineering*.

3. Define the term *prosthesis*.

4. Identify uses of technology in diagnosis in medicine.

5. Identify some positive and negative impacts of the health technologies.

6. What is genetic engineering?

7. Define the term *ergonomics*.

8. What must a human factors engineer consider in preparing a design?

9. List the "human factors" that must be considered in the design of an apartment for a person in a wheelchair.

CRITICAL THINKING

1. Health technologies have made organ transplantation possible. There are more people who need organs than there are organs available for transplant. What criteria would you use to rank people on the waiting list for an organ transplant?

2. The "Americans with Disabilities Act" requires some private businesses to alter their business environment. They need to do this to meet the needs of persons who are physically challenged. Changes are sometimes costly. How would you respond to business owners who say this is not fair?

3. What new developments may occur in health technologies in the future? How will these affect society? (Try to identify developments not mentioned in this text. Use your imagination.)

Environmental Technologies

OBJECTIVES

▶ identify the five activities in traditional soil farming.

▶ identify the advantages of controlled environment agriculture (CEA).

▶ identify two products created by bioprocessing.

▶ give one example of a bioprocessing technique used to process materials.

▶ discuss the impacts of bio-related technologies.

KEY TERMS

bioprocessing

bio-related technology

chromosome

controlled environment agriculture (CEA)

genetic engineering

genetics

heredity

hydroponics

photosynthesis

Imagine . . .

McIntosh, Delicious, and Granny Smith apples—all growing on the same tree. An apple tree supplier advertises a tree that produces three different varieties of apples.

Do you know what root knot *nematodes* (NEM-uh-toads) are? Root knot nematodes are worms. They attack the root structure of tomato plants. If you buy tomato seeds from a certain seed catalog, chances are your tomatoes won't suffer from them.

What's the connection between tomato seeds and apple trees? Both have been the target of bio-related technology research. In this chapter we'll discuss how technology has been used to change plants.

AGRICULTURE AND ENVIRONMENTAL TECHNOLOGIES

Have you ever visited a farm? Farming, or agriculture, is a technological system. It produces plants and animals for food, clothing, and other products. Farms are like outdoor food factories.

Farms have been improved by the use of mechanization and bio-related technologies. A **bio-related technology** is a technology that improves the techniques used to nurture life. Bio-related technologies are also called *environmental technologies*.

Bio-related technologies help the farmer produce greater quantities of products with higher quality. Mechanization, genetic engineering, grafting, and controlled environment agriculture (CEA) have changed agriculture as much as the invention of the plow.

Mechanization

The use of machines to make work easier is called *mechanization* (MEK-uhn-eye-zay-shun). Mechanization is not a bio-related technology. However, it is responsible for some of the greatest increases in agricultural production. Farmers now grow plants and animals with more efficiency than ever before.

Growing crops involves five activities: clearing the soil, tilling the soil, planting the seeds, cultivating the crops, and harvesting the crops. Fig. 14-1. Long ago

▶ **Fig. 14-1** These are the five main steps in traditional farming. Today, many farmers practice conservation tillage, in which crops are grown with minimum disturbance of soil.

CLEARING	TILLING	PLANTING	CULTIVATING	HARVESTING

Clearing the land gives the farmer a flat surface to plant.

Tilling loosens the soil so that the plants' roots can take hold.

Seeds are dropped into holes and covered with soil.

Cultivating is the process of caring for the plant.

Harvesting removes the grown plants or products for processing.

these tasks were done by hand. When people learned to train animals for work, the farmer's job was made easier. The plow was one of the earliest tools used to mechanize agriculture. Plows break up the soil so seeds can be planted. Today, *cultivators* may be used to break up the soil. They also drop the seeds and cover the seeds with earth. All of this is done in one operation.

Harvesting is the process used to remove the grown plants or products for processing. Combines are huge machines. They can cut a path of wheat thirty feet wide, separate the grain, and package the stalks in one operation. Combines have greatly increased the amount of products a farmer can bring to market.

Agricultural mechanization is not limited to crop production. Labor-saving machines have also changed how animal products are produced. Dairy farming is a branch of agriculture. It is concerned with producing milk and milk products. Mechanization has made dairy farming a year-round activity. Today, dairy cows are milked every twelve hours. In the past, dairy farming was limited to the spring and summer. That was when the pastures were green with food.

The glass tubes of vacuum milking machines carry the milk to milk houses for processing. Feed conveyors, refrigeration, and sterilizers have lightened the tasks. Dairy farm mechanization has increased milk production. Formerly, one cow could supply milk products for every four persons in the United States. Today one cow can supply milk products for every six people in the population.

Plant Growth

Technology can influence plant growth. We can study this influence. First we must understand the natural processes of plant growth.

Like people, plants need food and water. Plants produce food through a series of chemical reactions. The energy that drives these reactions comes from sunlight. **Photosynthesis** is the process plants use to convert sunlight to energy. Using photosynthesis, plants use light to change water and carbon dioxide into sugar and oxygen. *Chlorophyll* (CLORE-oh-fill) is the green substance in the plant. It traps light for the process of photosynthesis. Fig. 14-2.

Sugar produced by the plant is used as food. Plants also receive food from nutrients in the soil. Nutrients are nourishing substances. Water in the soil helps carry the nutrients into the plant.

Some of the oxygen produced is used by the plant. Some of the oxygen is passed from the plant through its leaves. The oxygen is used by other organisms, including ourselves, for respiration (breathing).

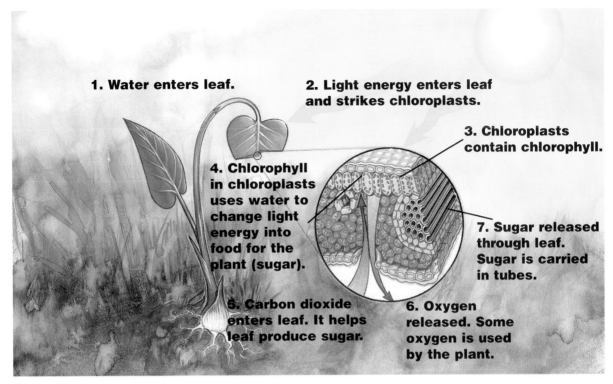

1. Water enters leaf.

2. Light energy enters leaf and strikes chloroplasts.

3. Chloroplasts contain chlorophyll.

4. Chlorophyll in chloroplasts uses water to change light energy into food for the plant (sugar).

7. Sugar released through leaf. Sugar is carried in tubes.

5. Carbon dioxide enters leaf. It helps leaf produce sugar.

6. Oxygen released. Some oxygen is used by the plant.

▶ **Fig. 14-2** The green parts of plants make food during a process called photosynthesis. As shown here, sunlight shining on the plant leaf strikes the chloroplasts. Chlorophyll in the chloroplasts uses water to change this light energy into food. Carbon dioxide also enters the leaf to help produce sugar. Tubes carry the sugar throughout the plant. Some of the oxygen created by the process is used by the plant.

Genetics

Why do some people have blue eyes and others have brown? Why are some kernels of corn white and others yellow? The answers to these questions lie in genetics. **Genetics** (jen-ET-icks) is the science that studies the laws of heredity.

Heredity is the study of the passing on of certain traits from parents to offspring. A *trait* is a physical characteristic. For example, heredity determines the traits that are passed from parents to offspring (children). Your eye color is a trait passed on from your parents. Plants and animals also inherit their traits from their parents.

Linking to SCIENCE

Root Pressure. Plants get nutrients from the soil in the water they take up. In plants the water flows *up*, against the force of gravity.

You can see how this occurs. Roll a paper towel lengthwise to form a tight tube. Stand it on end in a glass containing an inch or so of water. You will see that the water is absorbed and rises upward into the paper.

Eventually, the water will evaporate from the soggy surface. More water will move upward. More evaporation will occur. The same process occurs in plants. Evaporation of water from plant leaves generates a constant pull of water up into the plant. This pull is called *root pressure.*

Explore

Design and Build a Hydroponic Growing System

State the Problem

Hydroponic growing systems vary in the aggregate used, plant support system, nutrient feeding system, plants and container design. Design and build a hydroponic growing system in which you can grow flowers or food.

Develop Alternative Solutions

Select the crop to grow. Gather diagrams and pictures of various hydroponic systems. Research the requirements (light, pH, temperature, nutrients) for the plants you selected to grow. Research feeding designs. Select a feeding method and aggregate, if necessary. Research container designs. Using graph paper, draw a pattern for a hydroponic growing system that you think would serve your purpose.

Select the Best Solution

Select the design that you think will be the most effective.

Implement the Solution

1. Assemble the growing unit. Follow the design you have chosen.
2. Develop a schedule and monitoring system to record the progress of your plants.

Evaluate the Solution

1. Did the hydroponic growing system produce healthy crops?
2. If healthy crops were not produced, what was the state of the crops?
3. What changes might be made to correct the problems?
4. Make these changes. Check to see what effect they have on the crops.

Collect Materials and Equipment

Containers
two one-liter plastic
  bottles
food cans
plastic food containers
clay flower pots
plastic pipe
Aggregate
perlite
vermiculite
sand
marbles
gravel
aquarium gravel
Seeds or Plants
lettuce
beans
tomatoes
herbs
marigolds
rye grass
alfalfa sprouts
Nutrients
Purchase commercially
  prepared solutions. Be
  sure your teacher or
  parent helps in mixing
  the appropriate
  proportions.
General Building Supplies
pine wood
sheet metal
plastic tubing
fish tank aerator
acetate (for greenhouse
  covering)
Equipment
material processing
  equipment

Remember the nematodes? The seed company researched until they developed a tomato plant that had a natural resistance to the worms.

Pollination (pah-la-NAY-shun) is the fertilizing of a flower. Pollination depends on the transfer of pollen. *Cross-pollination* is the transfer of the pollen from the male organ (stamen) of one plant to the female organ (pistil) of another plant. The seed company was able to develop varieties of tomatoes that were resistant to nematodes. The genetic information from the original resistant plant was transferred to the other plants. This transfer was made through the pollen. Fig. 14-3.

▶ **Fig. 14-3** The stamens produce pollen that contains sperm cells. The pistils produce eggs. During pollination, fertilized eggs develop into seeds. Some plants can self-pollinate. Pollen can also be transferred from plant to plant in a process called cross-pollination. Scientists may cross-pollinate plants to create plants with special traits.

Linking to *MATHEMATICS*

Apples and Seeds. New varieties of apples are developed through cross-pollination at blossom time. The Jonagold, a cross between a Jonathan and a Golden Delicious, is such an apple. The seeds are grown to fruiting size. The apples are then tasted and evaluated.

If you cut an apple crosswise, you will find five *carpels* (seed cavities in the core). Each carpel contains an average of two seeds. A semi-dwarf apple tree can produce 3 1/2 bushels of apples. There are 42 pounds to the bushel and 3 medium-size apples to the pound. How many apples and how many seeds are available from this tree? This is the first generation.

Assume that 90 percent of the first generation seeds sprout. At five years each tree produces 3 1/2 bushels. How many seeds are available to plant?

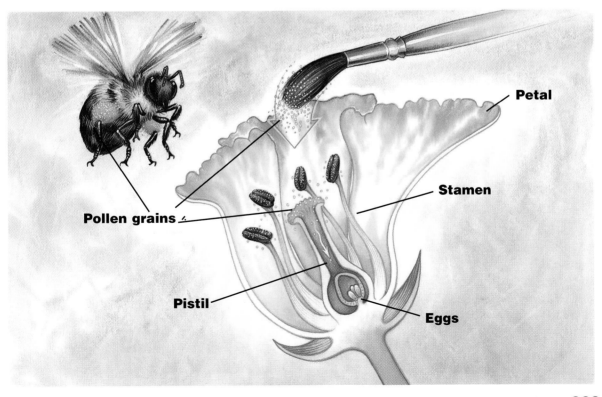

Pollen grains

Petal

Stamen

Pistil

Eggs

Crop Defenses

Farmers in the United States produce twice as much food today as they did fifty years ago from the same amount of land. They are able to do this because they practice intensive agriculture. This includes using:

- plant varieties that have been genetically improved for higher yields.
- pest control to reduce crop losses due to animals, insects, weeds, bacteria, fungi, and viruses.

Today farmers are using integrated pest management. This approach combines pesticides, biological control, and the planting of genetically improved crops. All of these strategies are used to control pests with minimum damage to the environment.

PEST CONTROL THROUGH GENE-SPLICING

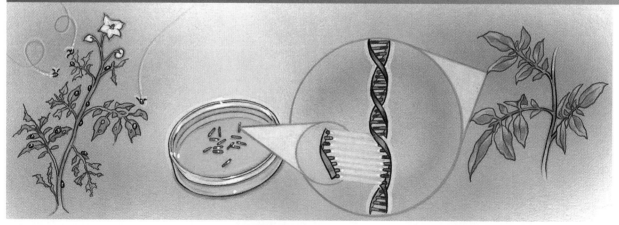

Some crops can be genetically altered to help fight off insect infestation.

The bacterium *Bacillus thuringiensis* (Bt) produces a natural insecticide.

The insecticide gene is snipped from Bt DNA. It is then mixed with plant DNA, which has been snipped at a compatible site.

When planting transgenic (trans-JEN-ick) crops, farmers are required to plant a "refuge" field of natural crops. The crops in the refuge field help to ensure that a few susceptible pests survive. This delays the rise of resistant pests and the need for new pesticides.

PESTICIDES

Natural pesticides are chemicals made by plants to fight off pests. Synthetic pesticides, which are manufactured, expand on this tactic. Limiting factors: Pests can develop resistance. Long-lasting synthetic pesticides accumulate in the food chain.

BIOLOGICAL CONTROL

Importing predators of targeted pests has sometimes proved successful. Limiting factor: Imported predators may overrun non-pest species.

GENETICALLY IMPROVED CROPS

A transgenic crop is a plant that results from gene-splicing. Cross-pollination within species has also improved many crops. Using these methods, disease-resistance has been transferred from one variety to another. Limiting factors: A loss of genetic diversity might increase the risk of massive infestation by a new pest. There is also concern that genes from transgenic crops might transfer to wild counterparts, creating superweeds.

The spliced DNA fragments are joined. When inserted into plant cells, this new DNA will produce a plant that manufactures its own Bt insecticide.

The genetically altered transgenic plant is not attractive to insects.

Genetic Engineering

A *gene* (JEAN) is a unit for an inherited trait, or characteristic (like blue eyes). **Genetic engineering** is the introduction of genes from one cell to another cell. The genes for a desired trait are moved from one plant or animal to another. Genes are located on chromosomes.

A **chromosome** is a tiny particle that carries the genes that pass on inherited characteristics. It is located in the nucleus of a cell. As cells multiply, they get instructions from the genetic information in the genes. The genes are arranged on chromosomes much like beads on a string. These coded instructions produce different traits. These traits include hair color, eye color, and body size.

Through genetic engineering, scientists have improved food production. For example, frost can destroy a potato crop. Bacteria that occur naturally on the leaves of the potato plant freeze at 30° F. The ice crystals that form destroy the plant cells. Scientists have genetically engineered the bacteria. The bacteria cell walls now contain a protein that blocks ice crystal formation. These genetically altered bacteria are then sprayed on the potato plants. They have removed the gene that forms the bacteria. The resulting potato plant won't freeze until the temperature is 23° F. Fig. 14-4.

Selective Breeding

The use of genetics is not limited to plants. Animals can also be genetically designed. *Selective breeding* is a technique used to combine the traits of one animal with those of another animal. It is used to produce offspring with desired

characteristics. Dairy farmers often breed cows that produce milk with a low fat content with healthy, lean bulls. The offspring are strong and healthy. They also produce milk with a low fat content.

Vegetative Propagation

Some plants can be reproduced without seeds or pollination. This is possible through vegetative propagation. *Vegetative propagation* (prop-uh-GAY-shun) is the process of creating plants by using the cuttings of roots, stems, or leaves.

Remember the apple tree with three varieties of apples growing on it? Through a process called *grafting,* growers graft (attach) branches from different varieties of apple trees onto an established apple tree. The grafted branch will grow, supported by the original plant. Fig. 14-5.

Using this technique, growers can change the variety of apples in an entire orchard without planting new seedlings. This saves time and money. Vegetative propagation grows plants more quickly and reliably than planting seeds.

How would the grafting process be used to grow seedless oranges or grapes?

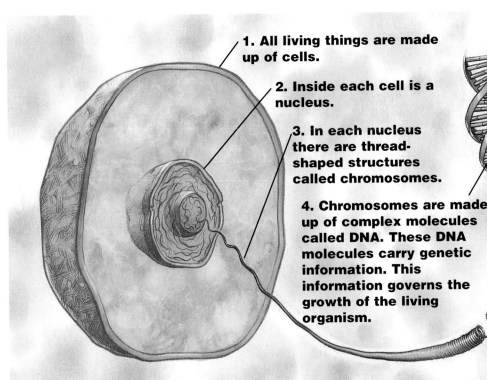

1. **All living things are made up of cells.**

2. **Inside each cell is a nucleus.**

3. **In each nucleus there are thread-shaped structures called chromosomes.**

4. **Chromosomes are made up of complex molecules called DNA. These DNA molecules carry genetic information. This information governs the growth of the living organism.**

▶ **Fig. 14-4** In genetic engineering scientists use enzymes, bacteria, and viruses to isolate genes or genetic traits. They then place these in new cells to change the characteristics of the organism. DNA that controls a specific trait can be isolated and removed. The isolated DNA can then be attached to different chromosomes. These chromosomes will now carry the new trait.

Scion

Rootstock

▶ **Fig. 14-5** Vegetative propagation is a bio-related technology that creates new plants. This is done by attaching roots, stems, and leaves from one plant to another plant of the same species.

Controlled Environment Agriculture (CEA)

Did you ever sing "Old McDonald Had a Farm?" Developments in bio-related technology just might change *where* we grow plants. How about "Old McDonald Had a Food Factory"?

Every plant has its own growing needs. **Controlled environment agriculture (CEA)** recreates these conditions. It produces the perfect growing environment so plants can thrive. Humidity, temperature, lighting, watering, and feeding are five of the conditions controlled through this technology. The *pH level* of the growing environment can also be controlled. The pH level is the level of acidity and alkalinity. Table 14-A.

GROWING ENVIRONMENTS FOR VEGETABLES

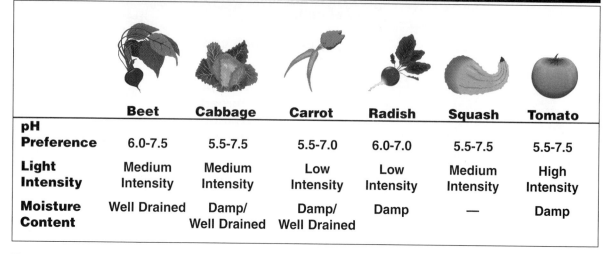

	Beet	Cabbage	Carrot	Radish	Squash	Tomato
pH Preference	6.0-7.5	5.5-7.5	5.5-7.0	6.0-7.0	5.5-7.5	5.5-7.5
Light Intensity	Medium Intensity	Medium Intensity	Low Intensity	Low Intensity	Medium Intensity	High Intensity
Moisture Content	Well Drained	Damp/ Well Drained	Damp/ Well Drained	Damp	—	Damp

▶ **Table 14-A** Controlled Environment Agriculture (CEA) creates the perfect conditions for plants to thrive.

A greenhouse or building is the factory. Often computers are used to control temperature, humidity, lighting, pH level, and plant feeding schedules. Plants may even rest on a conveyor. This conveyor moves slowly through the factory as the plants grow. Artificial lights provide the energy for photosynthesis. Misters keep the humidity at perfect levels. When the plants reach the end of the line, they are mature. Flowering plants, herbs, and food products can be grown in this way. They are packaged and shipped to market from the conveyor. Fig. 14-6.

Linking to COMMUNICATION

Acronyms. An *acronym* is a type of abbreviation. It is a term formed by using the first letter of each word in the term. For instance, "UNICEF" stands for "United Nations International Children's Emergency Fund." Using a dictionary, look up each of the following acronyms: laser, radar, sonar, scuba, and NASA. Write the words whose first letters are used to create the term.

▲

▶ **Fig. 14-6** Controlled environment agriculture allows the growing of food crops in barren landscapes.

Explore

Design and Build a Plant Watering Device

State the Problem

Design and build a device that will automatically deposit a certain amount of water into a house plant container. An improved model would allow adjustment of the quantity of water being deposited.

Collect Materials and Equipment

small plastic (or other leakproof) containers
tubing
funnels
bottles
straws

Develop Alternative Solutions

To design an effective watering device, you will need to ask several questions. Can a plant store water in preparation for a dry spell? What happens if a plant gets too little moisture? What happens if it gets too much moisture? Do plants need warm or cool water? Can a plant receive two days' worth of water at one time, or must the water be evenly distributed over the time period involved?

Implement the Solutions

1. After answering these questions, consider possible solutions. If you are working in a group, brainstorming would be a good start.
2. Make sketches of possible solutions.
3. Construct the device.
4. Test the device. Can it be improved? Does it work well enough as is?
5. Present your project to the class.

Evaluate the Solution

1. Does the device work as intended?
2. Is there undesirable leakage?
3. Can the device be adjusted to regulate the flow of water?
4. Is the device dependable enough to work for an extended period of time?

Hydroponics

Does a plant need soil to grow? It doesn't need soil in a hydroponics system. **Hydroponics** is the process of growing plants in a soilless environment. The word *hydroponics* comes from two Greek words—*hydro* meaning "water" and *ponos* meaning "work."

Hydroponics is a form of CEA. Plants are grown in a controlled environment that supplies the light, humidity, food, and water needed for rapid growth.

Plants do not need soil to grow. They do need the nutrients (food) in the soil and the support that soil gives. Nutrients are chemicals that plants need to grow. In hydroponics systems, a water/nutrient solution is fed to the plants. This is done by flooding, spraying, and pumping the chemical nutrients past the roots, stems, and leaves.

The plant may be supported in a variety of materials. Some plants are hung in mid-air with their roots exposed.

Water culture systems support the plants using wood fiber, gravel, rock wool, or rice hulls. The roots of the plants are submerged in the nutrient solution. Air is bubbled through the solution so the plant roots can get oxygen. Fig. 14-7A.

An *aggregate* (AG-greg-et) is a material such as vermiculite, sand, or perlite. Vermiculite (ver-MICK-u-light) is a mineral substance that is very water absorbent. Perlite (PEARL-ite) is a lightweight volcanic glass. *Aggregate systems* support and grow the plant in an aggregate.

The nutrient solution is flooded into the aggregate. When the aggregate is moist, the nutrient solution is allowed to drain out. The plant's roots feed off the nutrient-soaked aggregate. Fig. 14-7B.

Aeroponic (air-o-PON-ic) *systems* suspend the plant. This allows the exposed roots to be misted with the nutrient solution. Fig. 14-7C.

Advantages of Hydroponics

Is fertile soil in short supply on Earth? Why else should farmers consider growing crops without soil? There are many advantages to soilless agriculture.

Some areas in the world have poor soil. Such soil does not contain the nutrients or drainage plants need to thrive. Hydroponics offers an alternative to traditional agriculture for farmers in these areas. Imagine a barren desert dotted with hydroponics greenhouses growing vegetables. Such greenhouses can be found in the Middle East.

Hydroponics can also be used to grow food on a smaller scale. Metropolitan areas usually have little room for traditional agriculture. Hydroponics systems can be used to grow food in a courtyard or an alleyway.

FASCINATING FACTS

Superweeds? Several crops are being engineered to resist herbicides. In some cases, crops can give this genetically engineered trait to certain kinds of nearby weeds. Super-weeds are possible. However, such weeds must be a relative of the genetically engineered plant.

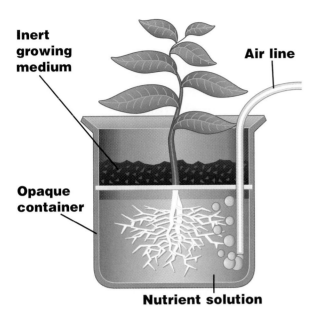

Inert growing medium

Air line

Opaque container

Nutrient solution

▶ **Fig. 14-7A** Water culture systems support the plants using wood fiber, gravel, rock wool, or rice hulls. The roots of the plants are submerged in the nutrient solution. Air is bubbled through the solution so the plant roots can get oxygen.

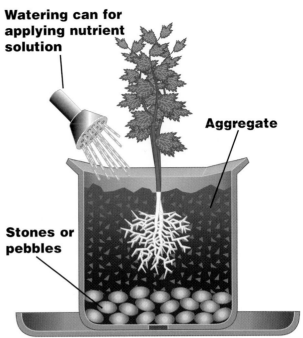

Watering can for applying nutrient solution

Aggregate

Stones or pebbles

▶ **Fig. 14-7B** Aggregate systems support and grow the plants in a material (aggregate) such as vermiculite, sand, or perlite. Vermiculite is a mineral substance that is very water absorbent. Perlite is a lightweight volcanic glass. The nutrient solution is flooded into the aggregate. When the aggregate is moist, the nutrient solution is allowed to drain out. The plant's roots feed off the nutrient-soaked aggregate.

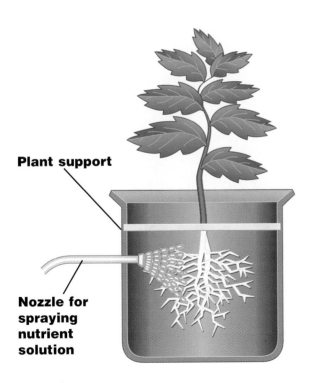

Plant support

Nozzle for spraying nutrient solution

▶ **Fig. 14-7C** Aeroponic systems suspend the plant. This allows the exposed roots to be misted with the nutrient solution.

Compared with traditional soil farming, hydroponics has clear advantages.
- More food can be harvested per square foot of growing area.
- A sufficient supply of nutrients is always available.
- A sufficient supply of water is always available.
- The pH level can be maintained.
- Plenty of oxygen can reach the roots.
- The roots remain clean; entire plant can be harvested.
- Replanting is quick.
- Crops can be quickly changed.
- Nutrient solution can be reused.
- Nutrients are not wasted.
- The system can be automated.
- Pests are more easily controlled.
- The system is environmentally friendly.

Disadvantages of Hydroponics

Hydroponics also has disadvantages.
- Disease and insects can spread rapidly.
- The pH of nutrient solution can change quickly.
- New nutrient solution is needed at all times to make up for evaporation.
- Start-up expense is high.
- High degree of expertise is needed to manage the system.

Aquaculture

Controlled environment agriculture has been adapted to raise fish and aquatic plants. *Aquaculture* is the raising of fish and food plants in water. The water environment is carefully controlled. Aquaculture enables the farmer to raise fish and food plants, such as rice, in

perfect conditions. This increases productivity.

Most aquaculture farms in the United States raise catfish in warm climate ponds. A pond the size of an acre of land can produce over 5,000 pounds of fish per year.

Raceway aquaculture farms fish in long troughs, or raceways. Fresh water is pumped in at one end of the raceway. Dirty water is cleaned, recycled, and recirculated through the system at the other end. Temperature, food, and oxygen content can be closely monitored. If necessary, they can be adjusted. Fig. 14-8.

▶ **Fig. 14-8** Dams on rivers prevent salmon from swimming upstream to spawn, or reproduce. These salmon, raised in a salmon hatchery, will be placed back in the river, above the dam.

BIOPROCESSING

A technological process consists of a series of steps. These steps are used to change materials from one form to another. Try to imagine all the processes needed to manufacture an automobile. Materials need to be separated, joined, and formed.

Bioprocessing also processes materials. **Bioprocessing** is a bio-related technology that uses living microorganisms or parts of organisms to change materials from one form to another. Microorganisms are living creatures too small to be seen by the unaided eye.

What products are created through bioprocessing? You probably eat one each day—bread. In breadmaking, living yeast cells are added to the dough. The cells digest the sugar and starch in the dough. They also release carbon dioxide. The carbon dioxide forms pockets of gas in the dough. This causes the dough to rise.

Without yeast, bread would be a thick mass. Cheese, yogurt, sour cream, vinegar, and sauerkraut are also made using bioprocessing microorganisms.

Bioprocessing and the Environment

Do you want to clean up oil spills? Do you want to filter pollutants from rivers? Do you want to clean pesticides from farm fields? Bioprocessing techniques can help.

Algae are microorganisms. Algae are being used as toxic vacuum cleaners to clean up all kinds of messes. For example, mats made from fermenting grass clippings are seeded with blue-green algae. These mats are then placed in

ponds. There they form a slimy cover on the surface. The algae eats the pollutants in the pond—as well as the grass mats. The pollutants are released as carbon dioxide. Such mats have been used to digest acid runoff from coal mines, uranium, and manganese.

Microorganisms are also used in sewage treatment plants to digest human waste. *Digestion* is a process that changes the sludge from human waste into methane gas, carbon dioxide, and a fertilizer-type material. The process uses an acid-producing bacteria. Fig. 14-9.

IMPACTS

Research in bio-related technology has had a great impact on agriculture. In the 1800s seventy percent of all Americans were involved in farming. The farms were needed to provide food for a growing nation. Today, less than three percent of

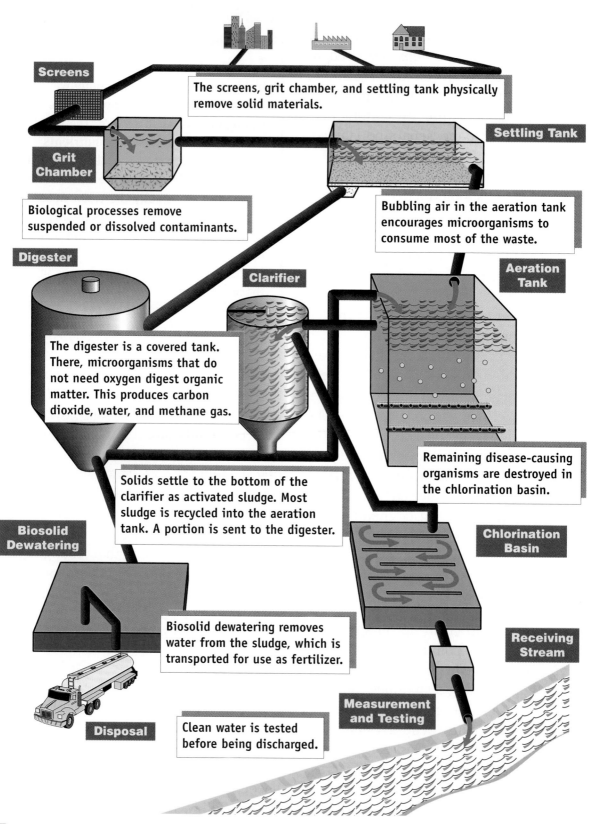

Screens

The screens, grit chamber, and settling tank physically remove solid materials.

Grit Chamber

Settling Tank

Biological processes remove suspended or dissolved contaminants.

Bubbling air in the aeration tank encourages microorganisms to consume most of the waste.

Digester

Clarifier

Aeration Tank

The digester is a covered tank. There, microorganisms that do not need oxygen digest organic matter. This produces carbon dioxide, water, and methane gas.

Remaining disease-causing organisms are destroyed in the chlorination basin.

Biosolid Dewatering

Solids settle to the bottom of the clarifier as activated sludge. Most sludge is recycled into the aeration tank. A portion is sent to the digester.

Chlorination Basin

Biosolid dewatering removes water from the sludge, which is transported for use as fertilizer.

Receiving Stream

Measurement and Testing

Disposal

Clean water is tested before being discharged.

▶ **Fig 14-9** The wastewater treatment process.

the population of the United States are farmers. However, farmers in the United States today do more than feed only the people in the United States. They also export large quantities of food.

Bio-related technologies have changed farming. They have allowed the farmer to produce more food per acre than at any other time in history. This same superproductivity takes place on livestock farms. Genetic engineering and breeding techniques have produced larger, healthier, and leaner cows, pigs, and poultry.

Unfortunately, this superproductivity has changed the environment. Chemicals used to improve plant growth and kill harmful insects have found their way into groundwater. This has polluted once-clean drinking water.

Farmland once rich in topsoil is now covered with dust. It has been robbed of its nutrients. This is due to poor land management and over-planting. Many of the particles in polluted city air were blown in from farmlands with poor soil quality. Fig. 14-10.

THE FUTURE

What about the future? Giant space farms will one day orbit Earth. Astronaut farmers will tend the crops. Soilless farms may dot desert landscapes. Desert biospheres will provide perfect growing conditions. A *biosphere* is an enclosed unit that contains everything needed for life. In these biospheres, climate, light, nutrients, and air will be controlled. Imagine farm factories that use computers to monitor the growing environment. Robots will be used to tend the crops.

Gene farming may become routine. Gene farming involves placing human genes in animals to produce useful drugs. For example, goats are used to produce a certain drug. The human gene is placed into a fertilized goat egg. This egg is then placed into the female goat to develop. The newborn goat may now carry the characteristics of the drug in its milk.

▶ **Fig 14-10** Blowing topsoil has contributed to air pollution and also to the silting of rivers.

Design and Build a Casein Glue Manufacturing System

State the Problem

Bioprocessing is a bio-related technology that uses living organisms or parts of living organisms to process or change materials. Casein glue is a very strong glue made from milk. It is made using bioprocessing techniques. Design a bioprocessing system to manufacture casein glue. Test the strength of the glue on a variety of materials.

Develop Alternative Solutions

Casein is a protein found naturally in cow's milk. To remove the casein from the milk, bacteria is allowed to form lactic acid. This acid curdles the milk. The curds are then separated from the liquid. Casein glue is contained in

Collect Materials and Equipment
Note: To prepare smaller batches of glue, reduce the ingredients proportionally.
Glue
1 cup hot water
1/3 cup nonfat dry milk
3 Tbsp. vinegar
1 Tbsp. cold water
1/2 Tbsp. baking soda
Materials
softwoods
hardwoods
fabrics
paper
acrylic
ceramic
Equipment
1 piece of cheesecloth
measuring tools
paper cups
mixing sticks
clamps

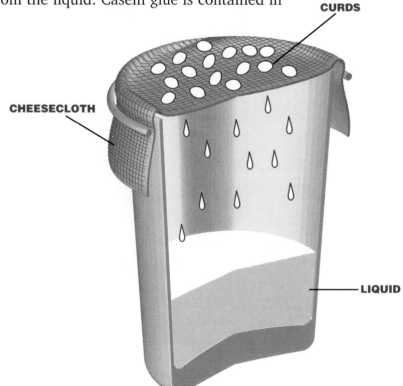

CURDS

CHEESECLOTH

LIQUID

the curds of curdled milk. Bioprocessing is the technique used to obtain the curds. It is bioprocessing because a living organism, the bacteria, was used to process the materials.

Casein glue depends on spoiled milk. Assume that you do not want to wait for the milk to spoil naturally. What can you do to spoil the milk? One solution is to use vinegar to help spoil the milk. By doing this, you are *modeling* the bioprocessing technique.

You will need to record the results in a table. Identify the items that should be listed in the table.

Implement the Solution

1. Mix the hot water with the dry milk.
2. Add the vinegar. Stir gently for three minutes. Allow the milk to curdle.
3. Using the cheesecloth, filter the curds from the liquid. Place the curds in a paper cup. The curds contain the casein.
4. Add the cold water and baking soda to the curds. This will neutralize the vinegar. Stir to a smooth, creamy consistency.
5. Glue together different combinations of test materials. Clamp them or place weights on them. Allow them to set overnight.
6. Test the strength of the bonds.
7. Prepare a table to record the data.
8. Enter the data in the table.

Evaluate the Solution

1. With which family of materials does the casein glue make the strongest bonds?
2. If the material were not strongly joined, what was the cause?
3. What changes were made to correct the problems?

CAREERS IN
Environmental Technologies

GEOLOGISTS

Government environmental protection agency seeks geologist to serve as project leader. Will work with industry to ensure compliance with government regulations. Bachelor's degree required; master's preferred. Should have working knowledge of laboratory and field techniques. Submit resume to: State EPA, 2500 North Stephens Parkway, Columbus, OH 64380.

BIOLOGICAL SCIENTIST

Research opportunity for biological scientist to conduct research for chemical manufacturing company. Work in laboratory setting. Master's degree required with outstanding analytical skills a must. Computer experience helpful. Excellent salary and compensation package. Please send resume to: Ackerman Laboratories, 1902 North Circle Drive, Denver, CO 77202.

PARK RANGER

Ranger needed at state park to enforce regulations, register visitors, provide information, and patrol area. Good communication skills, mechanical aptitude, and knowledge of first aid needed. Must be a high school graduate. Apply to: Dept. of Conservation, 3300 Monroe Street, Jacksonville, MS 42020

FISH AND WILDLIFE TECHNICIAN

Technician needed to collect and record data on wildlife distribution and develop habitat improvement programs. Need knowledge of biological sciences and mechanical skills to perform general maintenance. Two-year degree preferred. Apply to: Harper City Metroparks, 3300 Lakeview Drive, Harper, WV 44066.

LANDSCAPE ARCHITECT

Contractor needs landscape architect with knowledge of surveying, landscape design, and construction. Skills in CAD needed. Degree in landscape architecture required. Send resume to: Able Contractors, Inc., P.O. Box 2350, Kennesaw, GA 30020.

Linking to the WORKPLACE

Interest in the environment has led to improvement and expansion of the number of parks across the United States. Parks are operated by local, state, and federal governments. Think about your favorite park to visit. Write down all of the things that you enjoy about the park. What jobs do you think contributed to making the park enjoyable for you?

Chapter 14 Review

SUMMARY

▶ A bio-related technology is a technology that improves the techniques used to nurture life.

▶ Genetics is the science that studies the laws of heredity.

▶ Controlled environment agriculture creates a perfect growing environment for plants.

▶ Bioprocessing is a bio-related technology that uses living microorganisms or parts of organisms to change materials from one form to another.

▶ Hydroponics is the process of growing plants in a soilless environment.

CHECK YOUR FACTS

1. Explain the process of photosynthesis.

2. List the five main activities in traditional soil farming.

3. Explain how selective breeding has made dairy farming more productive.

4. How has genetic engineering increased farmers' productivity.

5. Describe the process used to grow three types of apples on the same tree.

6. What are the advantages of controlled environment agriculture (CEA)?

7. How does hydroponics farming differ from traditional farming?

8. Identify two products created by bioprocessing.

9. Give an example of how bioprocessing can be used to process materials.

10. Identify two impacts of bio-related technology.

CRITICAL THINKING

1. Many food products are irradiated with radioactive waves to reduce the growth of bacteria and reduce food spoilage. The Food and Drug Administration (FDA) has approved this process. It says it is not harmful to humans. What is your opinion? Be prepared to support your opinion with facts.

2. A pond near your home has been polluted by chemical runoff from a local farm. Researchers want to add a chemical-munching microorganism to the pond to clean it up. Would you support this action? Be prepared to support your opinion with facts.

SECTION 6

Control Technologies

CHAPTER 15 *Electricity and Electronics*

CHAPTER 16 *Computer Control Systems*

CHAPTER 17 *Robotics*

Have you ever heard people talk about keeping something "under control"? The word *control* means "to regulate" or "to direct."

Control technology can be used for an act as simple as flipping a light switch. It can also be used to guide a robotic arm through a delicate operation. We use control technology to adjust the temperature in a room or to change the channels on the television. Control technology is used at home, in the workplace, and at school.

The chapters in this section will help you understand how control technologies are used.

The Future Is Fuzzy Logic

There are instruments that can give you precise information about the speed of oncoming cars. However, you don't need such instruments to cross the street. A rough, or fuzzy, estimate of traffic conditions will help you cross.

The same principle works in a fuzzy controller. This is a control system relying on something called fuzzy logic. In some applications, input is complex or quirky. Consider, for example, image-steadying camcorders, anti-lock braking systems, and subway train controls. In each of these, fuzzy controllers can provide smoother performance.

Fuzzy Logic: The World Between 0 and 1

When a computer evaluates input from a sensor—for example, checking if "the door is open"—the answer is given as false (0) or true (1). There are no values in between. When a human evaluates the same condition, he or she might say the door is partly open. Fuzzy logic mimics this aspect of human processing. Because they can handle such input, fuzzy controllers are very good at managing tasks such as steering and balancing.

A fuzzy controller is effective because of its programming. It uses rule statements for ranges of information rather than many lines of computer code. Because rule statements are short, fuzzy controllers run faster.

Fuzzy Manufacturing

For manufacturers, fuzzy control has proved itself not only in product quality, but in product development. New products can be made ready for market faster and at less cost. Manufacturing processes also benefit, especially in CNC applications. Machine settings can be combined to increase productivity.

Fuzzy logic takes some getting used to. But in some situations, an exact answer is not the most useful answer.

Linking to the COMMUNITY

The chart shown here has two overlapping triangles. This chart relates to temperature. Using three overlapping triangles, prepare a similar chart relating to height. This new chart should show whether a person is considered to be tall, of medium height, or short.

CHAPTER 15

Electricity and Electronics

OBJECTIVES

▶ describe the relationships among voltage, current, and resistance.

▶ explain the basic organization of the two main types of circuits.

▶ describe the operation and uses of diodes and transistors.

▶ explain the operation of an electronic or electrical device in terms of input, process, and output.

KEY TERMS

battery

circuit

current

electricity

electromagnetism

generator

Ohm's law

potential difference

resistance

transistor

Imagine that you lived 200 years ago. You would have been living in a world without electricity. Consider the ways in which your life would have been different.

Many people might argue that harnessing electrical energy has been the most significant technological event of all time. Our technologies use thousands of devices that rely on electricity and electronics.

This chapter will help you understand the nature of electricity as an energy source. It will also help you see how we use electricity and electronics to control technology.

WHAT IS ELECTRICITY?

The study of electricity can be difficult. One reason why is that we usually can't see electricity. We know it's there because we can see the results of its work. We feel the heat from a blow dryer and watch the paddles on the ceiling fan turn—but what *is* electricity?

Electricity is the flow of electrons through a pathway that conducts electricity—a wire, for example. *Electrons*, as will be explained, are tiny charged particles of an atom. Atoms are the building blocks of all things. The universe and everything in it is composed of atoms.

Composition of Matter

Matter is anything that occupies space and has mass. Matter is made up of atoms. Matter may take the form of a solid, a liquid, or a gas. It may be an element, a compound, or a mixture.

An *element* is the purest form of matter. An element is made up of atoms. An *atom* is the smallest particle of an element that retains the properties of that element. For example, salt can be broken down into sodium and chlorine. Elements are so pure that they cannot be broken down any further. There are about 105 known elements. Every substance on Earth is made of one or more of these elements. Fig. 15-1.

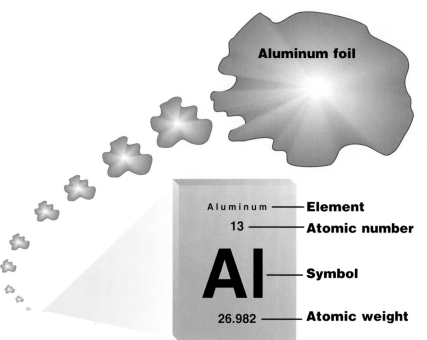

▶ **Fig. 15-1** The smallest part of an element that still behaves like that element is an atom. Assume that you were able to cut a piece of aluminum foil into smaller and smaller pieces. The smallest particle that still behaved like aluminum would be a single aluminum atom.

Aluminum foil

Aluminum —— **Element**

13 —— **Atomic number**

Al —— **Symbol**

26.982 —— **Atomic weight**

There are over 100 elements. Each one has a unique atomic structure. Every substance on Earth is made up of one or more of these elements.

Many times, two or more different atoms combine to form molecules. *Molecules* combine to form compounds. *Compounds* are matter composed of chemically combined atoms of two or more elements. *Mixtures* are also a combination of two or more elements. However, these elements are not chemically combined. They retain their own properties. The *periodic table of elements* lists known elements.

Atomic Structure

An atom has two parts: a center portion, or *nucleus*, and a cloud of *electrons* that surrounds the nucleus. Tightly packed within the nucleus are particles called *protons* (PRO-tahns) and *neutrons* (NEW-trahns). Fig. 15-2.

The number of neutrons in an atom is always equal to or greater than the number of protons (except for hydrogen). Atoms of the same element may vary in the number of neutrons. Such atoms are called *isotopes* (I-so-topes).

▶ **Fig. 15-2** A nucleus in the center of the atom consists of closely packed proton and neutron particles. A cloud of electrons surrounds the nucleus.

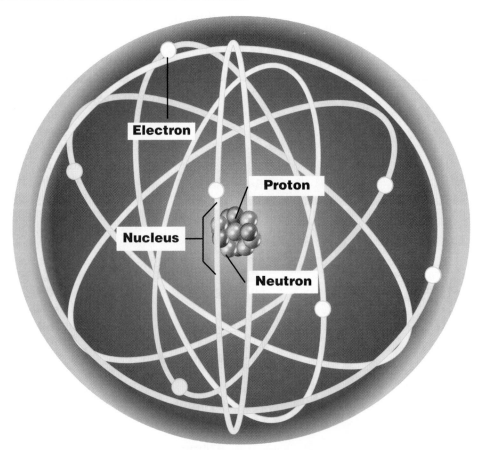

A neutral atom has one electron for each proton in the nucleus. An atom can lose or gain electrons, however. This process, which upsets the balance of neutrality, is called *ionization* (I-on-uh-zay-shun). It is very important to the flow of electricity.

Energy Levels and Electrons

Electrons move about at various *energy levels* at different distances from the nucleus. The area in which the electrons move is known as the *electron cloud*. The energy levels of the atom are often referred to as *shells*. The innermost electrons are at the lowest energy level. Those of the outermost shell are at the highest energy level. The outermost shell is called the *valence* (VAY-lence) *shell*.

Electrons in the outermost subshell of the valence shell are called *valence electrons*. Valence electrons are more loosely bound to the atom than inner electrons. Since valence electrons have more energy and are more loosely bound, they can escape from the atom when enough external energy is absorbed. The escaped valence electron is called a *free electron*.

Atomic Charges

Protons and electrons in atoms contain tiny amounts of electrical energy, or *charges*. Protons have a positive charge, while electrons have a negative charge. Neutrons are neutral. They have no electrical charge.

When ionization occurs, the neutral atom loses or gains an electron and becomes an *ion* (I-on). When a neutral atom loses an electron, the atom becomes a *positive ion* because it has more protons than electrons. When a neutral atom gains an electron, the atom becomes negatively charged, or becomes a *negative ion*.

Similar to the poles on a magnet, like charged particles tend to repel each other. Unlike charged particles attract each other. This phenomenon is known as the *law of charges*. Fig. 15-3. Atoms seek the most stable level possible. It is the interaction between charged particles that causes the flow of electrons that we call electricity.

▶ **Fig. 15-3** When charged particles come close to each other, a force is produced. This force can be either a force of attraction or repulsion.

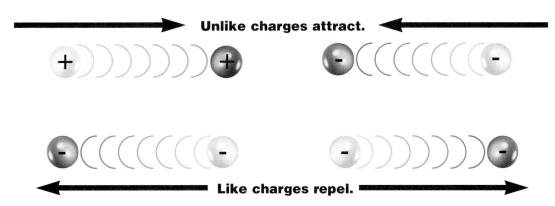

Unlike charges attract.

Like charges repel.

Explore

Design and Build an LED Warning System

State the Problem

Design a new system or model an existing system that uses a flashing light as an indicator light. For this activity, we will use a special type of diode junction that emits light. These are called light emitting diodes, or LEDs. The light must be mechanically activated. For example, it might be activated by pulling a lever or turning a crank.

Develop Alternative Solutions

Look around your school, your home, and your community for devices that use flashing LEDs as indicators or warnings. See if you can identify a need for such a device in your home, school, or community. Sketch designs that will develop a few possible ideas. Draw a schematic diagram showing how the electrical circuit of your system will be assembled.

Select the Best Solution

Make a detailed drawing of the solution you would like to model. Be sure your solution stays within your teacher's guidelines.

Implement the Solution

1. Select building materials to model your idea.
2. Construct the model.
3. Install the electrical components.
4. Test the system.
5. Make needed changes.

Evaluate the Solution

1. Does your system work as planned?
2. What changes can be made to improve the performance of your device?

Collect Materials and Equipment

flashing LEDs
9-volt battery
22-gauge stranded wire
switches
cardboard
foamcore board
Styrofoam®
wood
metal
electrical tape
paint
plastic
solder
soldering pencil
wire strippers
standard material
   processing tools and
   machines

Voltage, Current, and Resistance

A charge has stored energy with the *potential* to do work. For example, the potential energy at the negative terminal of a battery differs from that at the positive terminal. In a complete electric path, this difference in potential causes a charge to move through the circuit. This **potential difference**, as it is called, is the *force* that causes electrons to flow.

A potential difference is also referred to as a *voltage*, or *electromotive force* (*emf*). Voltage is an excess of electrons stored in one location and waiting to move.

Some materials are made of atoms that do not have a strong hold on valence electrons. These materials are called *conductors*. Metals are generally good conductors. Copper, aluminum, silver, and gold, for example, are excellent conductors. Fig. 15-4.

When voltage is applied to a conductor, the excess electrons of the negative end of the wire travel to the positive end, or the end with too few electrons.

The flow of electrons in a wire or other conductor is known as **current**. Current continues until the charges at both ends of the conductor are equal. Figure 15-5 shows how current flows through a conductor.

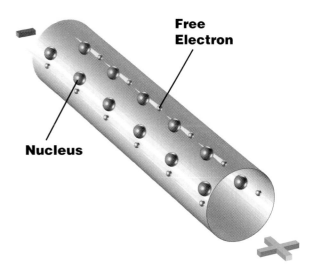

Fig. 15-4 Metals are good conductors of electricity. Materials that conduct electricity have electrons that are not held tightly in place. These electrons are free to move.

In contrast to conductors, some materials are made of atoms that have a tight hold on electrons and a few free electrons. Those materials that resist the flow of voltage are called *insulators*. Plastic and ceramic materials are good insulators. However, if enough voltage is applied, electrons can be forced from atom to atom through the insulator. Insulators have different strengths. The strength of an insulator represents its resistance.

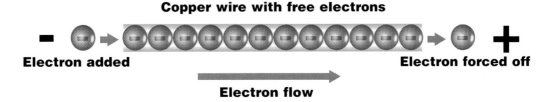

Fig. 15-5 The way in which current flows through a conductor.

Resistance is the opposition to the flow of electrons. Different substances have different resistances. Conductors have low resistance. Insulators have very high resistance.

SOURCES OF ELECTRICITY

As mentioned, for electrons to flow, there must be a potential difference, or voltage. The source that supplies this voltage is called a *voltage source*. It can be produced from a variety of different primary energy sources. These primary sources take energy in one form and convert it to electrical energy. Two voltage sources—batteries and generators—are discussed here.

Cells and Batteries

Have you ever wondered how a battery works? A **battery** is a device that converts chemical energy into electrical energy. A battery generates electrical energy (voltage) with a chemical reaction.

The batteries used in flashlights are also called "cells." A *cell* is a device made of two different conducting materials in a conducting solution. The conductors are called *electrodes* (e-LEC-trodes). The conducting solution is called an *electrolyte* (e-LEC-tro-light).

Dry cells have a paste electrolyte of powdered chemicals. A flashlight battery is a common dry cell. *Wet cells* contain a liquid electrolyte. A car battery is a wet cell. Fig. 15-6.

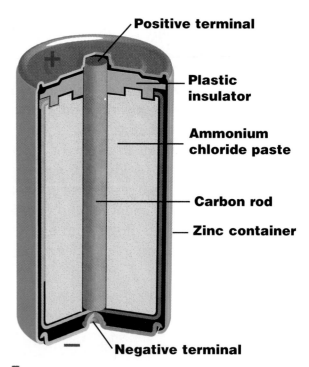

Positive terminal

Plastic insulator

Ammonium chloride paste

Carbon rod

Zinc container

Negative terminal

▶ **Fig. 15-6** Batteries change chemical energy into electrical energy. Voltage is created through chemical reaction. In a carbon zinc cell, electrons travel from the zinc to the carbon rod.

A chemical reaction occurs in the cell between the two electrodes. The electrolyte positively charges one electrode and negatively charges the other. In this way, the cell produces voltage.

When connected to a conductor, cells and batteries produce a current that flows in only one direction. This current is called *direct current (dc)*.

Generators

Mechanical energy is the energy of motion. A **generator** is a device that changes mechanical energy into electrical energy. A generator uses *electromagnetic induction* to force electrons from their atoms. Early scientists found that they could produce an electric current by moving a wire through a magnetic field. When a wire cuts across the invisible lines of force of the magnetic field, voltage is induced in the wire. In addition, if the wire forms a complete circuit, a current is induced as well. Fig. 15-7.

Generators vary in type and in construction. A simple generator is made up of a coil of wire wrapped around a metal core and placed between the poles of a magnet. The wire coil and core assembly is called an *armature* (ARM-uh-chur). This can rotate. As the armature rotates, the coil cuts across the magnetic field. Consequently, voltage is induced in the coil.

Each half-turn, the two connections at the output of the generator change *polarity*. First one end is positive and the other end is negative, then vice versa. The current by such a voltage changes

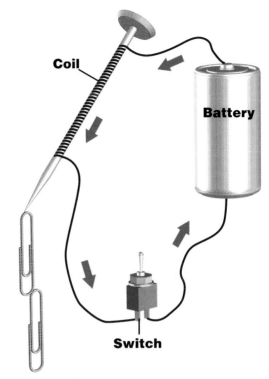

▶ Fig. 15-7 An electromagnet can be made by winding fine insulated copper wire around an iron nail and connecting this device to a battery. An electromagnet can produce a stronger magnetic field than the field produced by a permanent magnet. An electromagnetic can also be turned on and off.

direction each time the polarity changes. This electricity is known as *alternating current (ac)*.

Most of the electricity we use is alternating current generated in power plants that use large generators. Mechanical energy is used to turn the rotating parts of these huge machines. This energy comes from *turbines*. These are bladed wheels that turn when struck by the force of steam or moving water. As the shaft turns, the armature rotates. This generates electrical energy, or voltage.

Electromagnets

If electricity can be induced by cutting across magnetic lines of force, can magnetism be induced from electricity? The answer is yes. An electric current flowing through a wire creates a magnetic field around the wire. The relationship between electricity and magnetism is called **electromagnetism**.

Electromagnets are powerful magnets created by wrapping wire around an iron core and then passing electric current through the wire. The magnetic field can be controlled by turning the current on and off. The strength of the magnetic field can be increased in two ways. It can be increased by increasing the number of coils wrapped around the core and by increasing the current flowing through the coil.

RELATIONSHIPS AMONG VOLTAGE, CURRENT, AND RESISTANCE

You have learned that the force used to move electrons is called voltage. Voltage is a measurable quantity. The unit of measure for voltage is the *volt*. The symbol for volts is *V*, though *E* is sometimes used.

You have also learned that current is the term used to describe the movement of electrons in a wire. Current is measured by counting how many electrons move past a certain point within a wire each second. The symbol for current is *I*. The unit of measure for current is the *ampere*, or *amp*. The symbol for amperes is A.

Finally, you have learned that resistance is the opposition to the flow of electrons. The symbol for resistance is *R*. The unit of measure for resistance is the *ohm*. The symbol for the ohm is the Greek letter Ω (omega).

Ohm's Law

When voltage is applied to a conductor, the current that moves through the conductor is directly proportional to the applied voltage. This relationship is known as **Ohm's law**. This law is basic to all studies of electricity. Table 15-A. Ohm's law is stated mathematically by the formula:

$$\text{Current} = \text{Voltage}/\text{Resistance}$$
$$\text{or}$$
$$I = E/R$$

Table 15-A. Voltage, Current, and Resistance

Electrical energy depends on three factors: voltage, current, and resistance. The relationship of these three factors determines the amount of electrical energy produced.

Physical Quantity	Symbol	Measure of	Unit of Measure
Voltage	$\underline{V}$ or $\underline{E}$	Force	Volts (V)
Current	$\underline{I}$	Electron Flow	Amperes (A)
Resistance	$\underline{R}$	Opposition to Flow	Ohms (Ω)

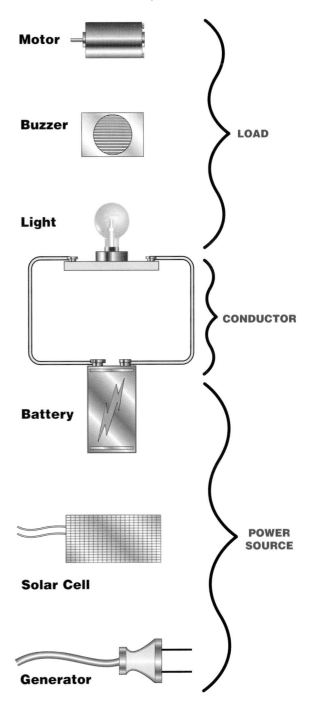

ELECTRICAL CIRCUITS

In electricity, a **circuit** (SIR-cut) is the pathway through which electrons travel. A *simple circuit* consists of a power source, a conductor, and a load. Fig. 15-8. A circuit that uses direct current as a power source might use a battery or a *photovoltaic* (foe-tow-vole-TAY-ik) cell. This is also known as a solar cell. The source of voltage for alternating current is usually a generator at the power plant.

Conductors provide a low-resistance path from the source to the load. Typically, copper or aluminum wires serve as conductors.

The *load* is the device that uses the electric energy. The load could be a motor, a buzzer, or a light. The load converts the electrical energy in the circuit into heat, light, mechanical, or other energy forms.

Main Types of Circuits

The main types of circuits are series circuits and parallel circuits.

Series Circuits

When the components are connected one after another, the circuit is called a *series circuit*. In a series circuit, there is

▶ **Fig. 15-8** An electrical circuit is the pathway that electricity follows. All circuits have a conductor, a load, and a power source.

Motor

Buzzer

Light

Battery

Solar Cell

Generator

LOAD

CONDUCTOR

POWER SOURCE

only one pathway for electrons to follow. Fig. 15-9. A break in any part of the circuit stops all the electrons from flowing. This produces an open circuit. Electrons can flow only in a closed circuit.

Two rules apply to series circuits:
• The current is the same at all points.
• The total resistance of the circuit is equal to the sum of the individual resistance values.

Parallel Circuits

In a *parallel circuit* the components are arranged in separate branches. Fig. 15-10. This arrangement provides multiple pathways in which electrons can flow. A break in one branch of the circuit does not prevent electrons from flowing in the other branches. Two rules apply to parallel circuits:
• All branches are of equal voltage.
• Total current is equal to the sum of the branch currents.

Complex Circuits

A *complex circuit* uses switches or other electronic parts to control the flow of electrons from the source to the load. Switches, resistors, and circuit-protection devices are used in complex circuits.

Switches are used to open and close the circuit. Different types of switches open and close the circuit in different ways.

Resistors are devices that apply resistance to the flow of electrons in the circuit.

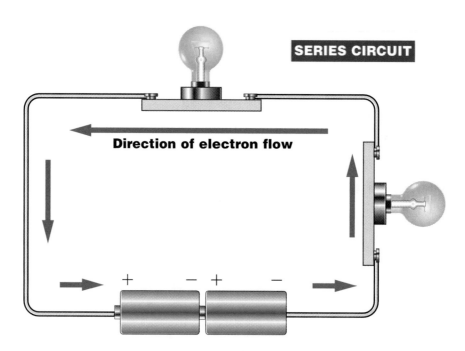

▶ **Fig. 15-9** A series circuit supplies only one path for the flow of electrons. If that path is broken at any point, the flow of electrons will stop.

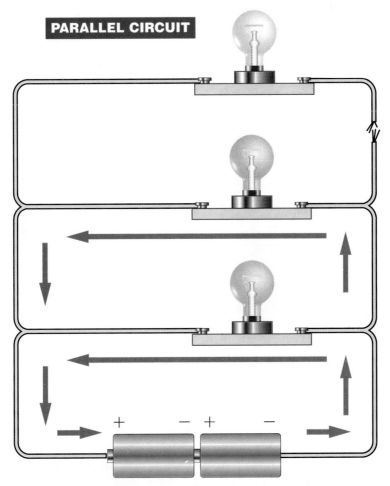

PARALLEL CIRCUIT

▶ **Fig. 15-10** A parallel circuit has many paths along which electrons can flow. A break in one path will not stop the flow of electrons in the other paths.

Potentiometers (puh-TEN-she-om-mutters) and *Rheostats* (REE-uh-stats) are adjustable resistors that can set a range of desired resistances. Potentiometers are used to vary voltage. Rheostats are used to vary current. A dimmer switch on a lamp is an adjustable resistor. The volume control on a radio is another example.

A complex circuit should also contain a circuit-protection device, such as a fuse or a circuit breaker. These devices protect the circuit from large amounts of current. Fig. 15-10.

A *fuse* will open a circuit when its rated current is exceeded. The fuse contains an element, a thin strip of metal, that melts at a specified current. When too much current runs through the fuse, the element will melt. The circuit will then open. Fig. 15-11A.

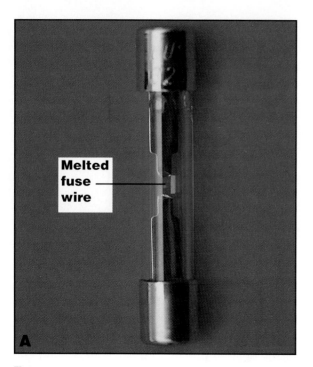

Melted
fuse
wire

A

B

▶ **Fig. 15-11** Fuses(A) and circuit breakers (B) prevent the damage that can be caused by excessive current.

A *circuit breaker* controls a circuit in the same way as a fuse. However, it uses a process that does not require the melting of the fuse element. One advantage of a circuit breaker is that it can be reset and reused after the problem in the circuit has been fixed. Fig. 15-11B.

WHAT IS ELECTRONICS?

Electronics is the study of precisely controlling the flow of electrons. Fig. 15-12. Electronics is the outgrowth of our knowledge and development of electrical energy. Electricity and electronics are not the same.

There are key differences between electricity and electronics. Remember that electronics controls the flow of electrons. Electricity controls voltage levels and the flow of current.

▶ **Fig. 15-12** In electronics, the flow of electricity is controlled by various devices. For example, resistors are placed in a circuit to reduce the flow of electricity. Fuses guard against damage. Transistors control the flow of current.

Semiconductors

The precise control of electrons began with the invention of the vacuum tube. In 1948, however, the Bell Telephone Company invented the transistor. A **transistor** is a semiconductor device that amplifies and acts as an electronic switch. The transistor is tiny and uses very little energy. Transistors rapidly replaced vacuum tubes in electronics. Scientists began using transistors to solve problems in mathematics. That led to the development of today's computers.

Semiconductors are materials with properties that fall between insulators and conductors. They conduct better than insulators, but not as well as conductors. Silicon and germanium are common semiconductor elements. A process called *doping* adds impurities to the pure semiconductor element. This increases its conductivity. Arsenic and gallium are two elements commonly added as impurities.

Some impurities add extra electrons that are free to move about the crystal. This produces an *n-type semiconductor*. Other impurities take away electrons. This produces a *p-type semiconductor*. Unfilled *valence* (VAY-lence) shells result, leaving positive sites in the semiconductor crystal. These positive sites are called *holes*.

The holes have a strong attraction for electrons. Valence electrons from nearby atoms can "fall" into the hole, leaving another hole where it had been. Movement of electrons from one hole to the next makes the holes appear to move in the opposite direction. Conduction through the p-type semiconductor is by holes. Conduction through the n-type semiconductor is by electrons.

Junction Diodes

When a semiconductor crystal is half n-type and half p-type, a *pn junction* is formed between the two regions. Figure 15-13 shows a pn junction device known as a *junction diode*. The n-region has many conduction electrons. The p-region has many holes.

Because of this junction, electrons can flow through the semiconductor in only one direction, from n-type to p-type.

The primary use of a junction diode is to *rectify* current, or to change ac into dc. Computers or televisions would not operate practically with batteries. Therefore, diodes are wired into their circuits to rectify the current. In this way, such devices can be supplied with direct current from a source of alternating current.

Uses and Types of Transistors

Transistors have two main uses. They are used to boost or amplify electronic signals. They also act as electronic

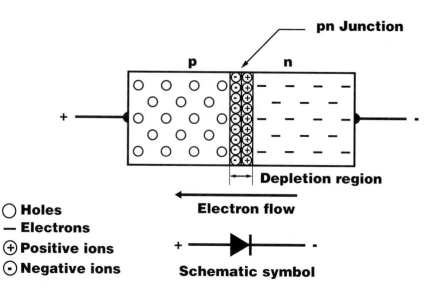

Fig. 15-13 Block diagram of a junction diode. Its schematic symbol is shown below. Electrons can flow in only one direction, from the n-region to the p-region.

pn Junction

p   n

Depletion region

Electron flow

○ **Holes**
— **Electrons**
⊕ **Positive ions**
⊙ **Negative ions**

Schematic symbol

switches. For example, they are used to amplify signals in microphones, radios, televisions, and hearing aids. They are used as switches in computers.

There are two basic types of transistors: npn and pnp. Both have three layers of semiconductor materials: a thin center section, called a *base*, and two thicker outer layers. One layer is called an *emitter*. The other layer is called a *collector*. Wire leads are joined to each of these layers. Fig. 15-14.

The pn junction is basic to the operation of transistors as well as diodes. A *forward-bias voltage* (negative of source connected to the n-region of an npn, positive of source connected to the p-region) is applied to the base-to-emitter leads. At the same time, a *reverse-bias* voltage is applied across the collector-to-emitter

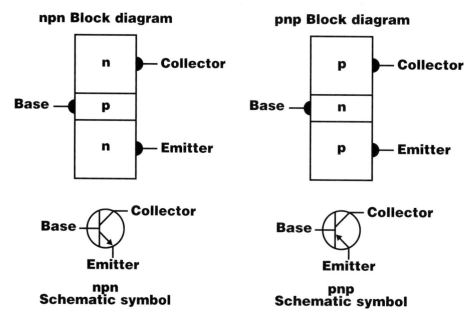

npn Block diagram

n — **Collector**
Base — p
n — **Emitter**

npn
Schematic symbol

Base — Collector
Emitter

pnp Block diagram

p — **Collector**
Base — n
p — **Emitter**

pnp
Schematic symbol

Base — Collector
Emitter

Fig. 15-14 Block diagrams and schematic symbols for npn and pnp bipolar junction transistors.

leads. Current then flows across the forward-biased, emitter-base junction into the base region. The flow of current carriers into the emitter allows current to flow from collector to emitter.

Increasing the base current slightly by increasing the forward-bias voltage causes a much larger collector-to-emitter current. This principle makes the transistor useful as an amplifier. A small input current can control a much larger output current.

In computers, transistors are used to create *logic gates*. Logic gates open and close like electronic switches. They control the flow of electrons at lightning speeds. Computers operate on the *binary number system*. This means that they use 1s and 0s to represent numbers and characters. In the "on" state, the transistor represents the digit 1. The "off" state represents the digit 0. Computers process information by adding, subtracting, multiplying, and dividing the binary 1s and 0s.

ELECTRONIC SYSTEMS

Electronic systems are devices that work together to control, monitor, and measure changes. To work, electronic systems must have *inputs, processes*, and *outputs*.

The input can be a sensor that converts a command into an electrical signal. The input to your oven is your setting of the oven at a certain temperature.

The process is the action that accomplishes a result. For example, in a streetlight control circuit, a photoresistor energizes a relay that switches on the streetlight.

The output is the result of the process. For example, when the battery in your smoke detector is low, it may begin to flash.

IMPACTS

Systems powered by electronics and electricity impact all areas of modern life. Such systems have provided many of the tools we need to earn a living. They have also increased our options for entertainment and travel. The positive impacts of electricity have far outweighed any negative impacts. Fig. 15-15.

THE FUTURE

Electrical engineers face the challenge of increasing the power of a device while limiting its size. This pattern has been present in the development of the radio, the television, the computer, and the cellular phone. Can you think of others?

▶ **Fig.15-15** The full range of invention is shown by the electrical and electronic devices used in a large city.

Apply What You've Learned

Design and Build a Continuity Tester

State the Problem
Design and construct a continuity tester.

Develop Alternative Solutions
The conductivity of a material depends on its atomic structure, or how its atoms are arranged. Materials that are conductive have electrons that are easily moved from their atom by an electromotive force. Materials that act as insulators contain electrons that are held tightly in place. This activity will allow you to test various materials to determine if they are conductors or insulators. The word continuity means "continuous." If a material has continuity, there is a continuous flow of electrons through the circuit made up of the battery, wires, LED, and the material being tested. Refer to the list of materials and equipment. The items needed to construct the continuity tester are shown in Fig. A. Prepare sketches showing how you would construct the continuity tester. Show also how it would be connected to the sample material.

Implement the Solution
1. Construct the continuity tester. Use electrical tape to temporarily hold the wires and components together.
2. Make a table listing the materials you will be testing.
3. Touch the sample materials with the test leads. When the LED lights up, you know that the material is a conductor.
4. After each material is tested, fill in the test results on the table. Did the material have the characteristics of a conductor or insulator?

Collect Materials and Equipment
9V battery
14V LED
22-gauge stranded wire, 18" long
electrical tape
continuity tester
Materials to test
wood
plastic
aluminum
ceramic cup
fabric
sugar and water solution
salt and water solution
steel
copper pennies
glass cup
foam insulation

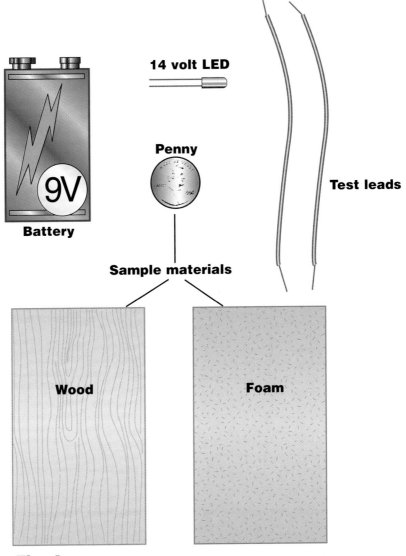

14 volt LED

Battery

Penny

Test leads

Sample materials

Wood

Foam

Fig. A

Select the Best Solution
Select the design that you think will be most effective.

Evaluate the Solution
1. Did the continuity tester work as expected?
2. Think of other uses for a continuity tester. Check with your teacher
before testing your ideas.

CAREERS IN
Electronics

ELECTRICAL AND ELECTRONIC ENGINEER

Growing developer of aerospace, automotive and military-embedded control products is in need of engineers to perform digital and analog electronic circuit design and real-time software programming. Bachelor's degree in electrical engineering required. Please send resume to: P.O. Box 361665, Monroe, MI 65389, or E-mail at ecc@ecc.com.

ELECTRONICS TECHNICIAN

We are a premier printing company expanding our Akron facility and seeking an electronics technician. Responsible for assisting in the design, fabrication, troubleshooting and maintenance of our electrical/electronic manufacturing equipment. Computer programming and problem-solving skills necessary. Interested candidates should forward a resume to: Human Resources, Sealtight Company, 1978 Main Akron Road, Akron, OH 65230.

ELECTRONIC DRAFTER

Electronics manufacturing firm needs drafter to prepare detailed drawings and blueprints for electronic equipment. Draw wiring diagrams, circuit board assembly diagrams, schematics, and layout drawings. Send resume and list of references to: Technical Electronics, Inc., P.O. Box 7022, Chicago, IL 60032.

COMPUTER SERVICE TECHNICIAN

The ideal candidate will have the ability to test, troubleshoot, and resolve hardware problems on personal computers. Effective customer skills needed. Prefer one or two years of formal training in electronics with knowledge of computers. Will work without supervision. Submit your resume to: General Credit Insurance Company, 3201 Beachwood Parkway, Orlando, FL 33238.

COMPUTER SALES REPRESENTATIVE

Computer retail company has grown to include corporate sales division. Challenging opportunity for self-motivated individual to solicit new business while managing existing accounts. Prefer one year of sales experience and PC proficiency. Must have strong organizational, customer service and communication skills. We offer competitive compensation package. Send or fax resume to: American Computer Corporation, 24295 Beechwood Avenue, Cleveland, OH 62240. Fax: (614) 888-3200.

Linking to the WORKPLACE

Advances in technology and manufacturing have impacted our lives. List five electronic products developed in the past ten years that have affected the workplace. How did these products change the ways in which jobs were performed?

Chapter 15 Review

SUMMARY

▶ A battery creates voltage by means of a chemical reaction. A generator creates voltage by means of electromagnetic induction.

▶ When a voltage is applied to a conductor, the current that moves through the conductor is directly proportional to the applied voltage. This relationship is known as Ohm's law.

▶ In a series circuit, components are connected one after another. In a parallel circuit, components are arranged in separate branches.

▶ A diode allows current to flow through a circuit in only one direction.

▶ A transistor controls a large output current with a small input current or voltage.

CHECK YOUR FACTS

1. Draw and label the parts of an atom. Identify the charges of particle.

2. What force causes electrons to flow?

3. Why are the electrons in the outer shell of the atom more likely to flow than electrons in the inner shells?

4. Describe two common sources of electromotive force.

5. Explain how Ohm's law describes the relationship among voltage, current, and resistance.

6. How did the invention of transistors affect the development of electronics?

7. Identify and describe the organization of the two main types of circuits.

8. Describe the operation and uses of diodes.

9. What design feature do all electronic systems have in common?

10. Describe the operation and uses of transistors.

CRITICAL THINKING

1. Describe how a toaster works in terms of input, process, and output.

2. Superconductors are materials that have very low resistance to electron flow. To achieve this, these materials are subjected to very low temperatures. What effect do you think low temperatures might have on electron flow and resistance? Explain your answer.

3. Describe how you would design an electronic fan that turns on when the temperature reaches 88°F.

4. Describe how doping changes the resistance of a conductor.

Computer Control Systems

OBJECTIVES

▶ describe how computers are used to design products.

▶ explain the process of computer numerical control.

▶ identify manufacturing systems that depend upon computers.

KEY TERMS

CAD/CAM

Cartesian coordinates

computer-aided design (CAD)

computer-aided manufacturing (CAM)

computer numerical control (CNC)

flexible manufacturing system (FMS)

simulation software

Imagine that you have been asked to design and manufacture a product. You will need to present a plan for preparing the product drawings. You will also need to explain how the product will be manufactured. Product design and product manufacturing involve problem solving.

The problem-solving process will help you select the most efficient way to prepare a design. It is also useful in deciding on a manufacturing method. This chapter explains how the computer can be used to control the processes of designing and manufacturing a product.

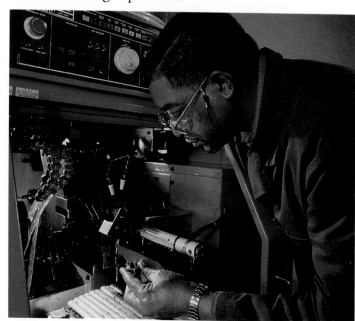

COMPUTERS AND PRODUCT DESIGN

Designers and engineers often say that the hardest part of the job is putting an idea on paper. Usually they begin by making freehand sketches. However, for the manufacture of a product, sketches are not enough. Detailed and accurate drawings called *working drawings* are needed to manufacture a product.

In the past, working drawings were hand-drawn on paper. Tools such as T squares, triangles, compasses, and templates were used. These traditional tools are no longer needed. Products can now be designed on a computer. **Computer-aided design (CAD)** software is software that provides tools for drawing and dimensioning a product.

Using CAD, engineers and designers can draw any line or shape on the computer screen. Fig. 16-1. The designer inputs information into the computer using a mouse, keyboard, and light pen or digitizer pad. The software assembles the information into a detailed drawing. A *plotter*, instructed by the computer, produces the drawing on paper.

CAD programs automatically insert dimensions (sizes) for the object being drawn. CAD programs also contain a library of symbols. Suppose a part is made from wood. The designer can choose from the symbol library the shading that indicates wood. The design can then have the computer shade that part on the drawing.

CAD software enables a designer to produce three-dimensional as well as two-dimensional drawings. If the designer wishes to view the object from a different

angle, the CAD program can turn the object on the screen.

Any image created using CAD software can be saved on the computer's hard drive or on a floppy disk. The image can be called back with a keystroke.

▶ **Fig. 16-1** Using CAD software, this designer prepares detailed drawings that can be quickly altered if necessary.

The greatest benefit of CAD is that it enables the designer to make changes in a drawing quickly. Lines can be added or deleted in seconds. Shapes and sizes are easily changed without having to redraw the entire image. Drawings that once required days to prepare now take only hours.

Computer Modeling

CAD programs are made more powerful by combining them with other engineering programs. Designers and engineers can now test design ideas on a computer using *simulation software*. **Simulation software** is software that can test a product design by simulating the environment in which the product must be able to work. Fig. 16-2.

Engineers will often build prototypes to test design concepts. This can be very expensive. Computer software can accurately test designs. They can perform complex experiments evaluating the performance of the design. For example, aerospace engineers may want to test the lift provided by a new wing design. (Lift is explained in Chapter 10.) Wind tunnel experiments using prototypes can be costly. Computer programs can simulate airflow over the wing and calculate the lift created.

▶ **Fig. 16-2** Through computer modeling, engineers can simulate airflow over a spacecraft. Without computer modeling, such a test would need to be conducted in a wind tunnel.

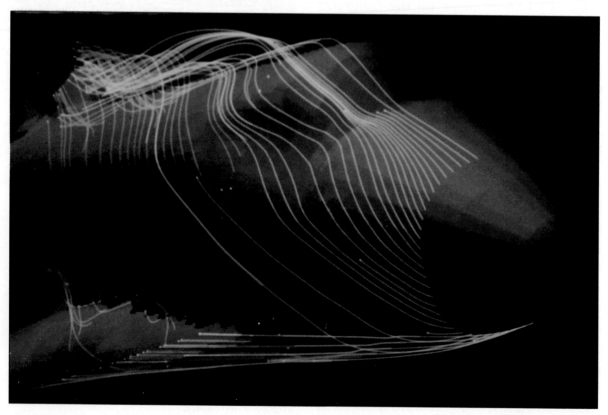

Linking to SCIENCE

Demonstrating Lift. Why does an airplane fly? The difference in air pressure on the top and bottom of the wing provides the lift needed to get an aircraft off the ground. You can demonstrate lift by making a simple *airfoil* (section of an airplane wing). Cut a strip of paper to about 4 inches by 10 inches. Fold it in half. Tape the ends together, forming a loop. Stick a pencil inside the loop of paper and hang the loop from the fold. Hold the pencil horizontal and blow against the fold. The airstream moving across the top reduces the air pressure there. The pressure under the airfoil is then greater than the pressure above and pushes it up.

▲

Artists using animation software can transform CAD drawings into images that look incredibly real. Animation software can "bring an object to life" on the screen.

Imagine viewing a new design concept for a bicycle on a computer monitor. As the bicycle moves about the screen, you can see the gear mechanism shift. You can also watch the shock absorbers absorb the bumps in the road.

COMPUTER NUMERICAL CONTROL

Computer numerical control (CNC) is a control system that uses computers to control the operation of machines during production. The computers do not change the ways in which the machines process materials. They only control the movements of the machine tool.

Machine tools are used to perform separation processes. Separation processes remove material from a workpiece called the *stock*. Fig. 16-3. The person who operates a machine tool is called a *machinist*. Machining is much like

▶ **Fig. 16-3** This CNC cutter can cut parts from sheet steel very precisely.

Explore

Design and Build a Product Concept

State the Problem

Use a computer and a software paint program to design a keychain.

Develop Alternative Solutions

Develop criteria, including specifications and constraints for the keychain. Consider a design for a theme keychain. Theme keychains include monogram keychains, sports keychains, and musical instrument keychains. Using a computer and appropriate software, develop a few design ideas for a keychain. Try to develop designs for a keychain you think might sell in your school.

Select the Best Solution

Ask your classmates to select the best design. Ask them to consider product appeal, ease of manufacture, and availability of materials.

Implement the Solution

1. Bring the paint or drawing program you will be using onto the computer screen.
2. Prepare your design concept.
3. Draw front, top, and side views of the keychain.
4. Save your drawings on a disk.
5. Print your design ideas.

Evaluate the Solution

1. Does the design meet the criteria set up by your teacher and classmates?
2. Is the product interesting enough to create a demand for sales in your school?

carving. The machinist uses the machine to remove material until the desired part is shaped. Many materials can be machined. Wood and metal are the most common.

Machining is a key process in manufacturing. Automobile engine parts, aircraft parts, and even the cranks on your bicycle are machined. Machining is also the process used to create parts for other machines. Examine a vacuum cleaner, power tool, or home appliance. In each of these, you will find many machine parts.

Manufacturers use a variety of machines to shape products. The two most common machine tools are the milling machine and lathe.

A *milling machine*, or mill, spins a cutting tool known as an *end mill*. The spinning end mill is stationary while the stock is moved around it. By moving the stock along the X, Y, and Z axes, a skilled machinist can locate the point on the stock where the cut must be made. You will read more about this locating process later in this chapter.

A *lathe* is a machine tool used to make round and cylindrical parts. The lathe spins the stock and moves a stationary cutter in and around the material. The cutting tool on a lathe moves in only two directions, left and right, in and out. Baseball bats, spindles, and pistons are examples of products made on a lathe. Fig. 16-4.

A lathe can round the corners of square or rectangular stock. A milling machine cannot round corners. Often, both the lathe and the milling machines must be used to manufacture a part.

In the past, skilled machinists controlled the movement of the machine's cutters through the stock. The development of

▶ **Fig. 16-4** Computer numerical control can be used to direct the movements of a lathe. Here, a grinding wheel is being machined. Computer numerical control can help ensure that each grinding wheel has the same dimensions.

machine automation slowly changed this. The operation of machine tools became automatic with the use of punched tapes. The process was improved with the introduction of computers. Modern computer numerical control machines are capable of self-regulation and adjustment.

To gain a better understanding of computer numerical control, imagine a CNC machine that is designed to brush your teeth. The machine needs a tool holder to grasp the handle of the toothbrush. The tool holder must be able to move in any direction. Your head would have to be locked into position so it couldn't move. The machine needs a method to find your teeth and place the brush at the correct starting position. The toothbrush must move in up-and-down strokes as well as in back-and-forth strokes. The brushing machine must also know when the operation is completed. Most importantly, it must be able to repeat the identical process over and over.

Compare the toothbrush example with a machine programmed to drill a hole in a block of steel. The machine has a tool holder to grasp the drill. The block of steel is clamped in position so it will not move during processing. The location of the hole is identified, and the tool is properly positioned. The operation is completed by instructing the machine to push the turning drill into the steel.

Instructions for drilling the hole are stored in a computer. The computer-controlled machine can then repeat the same series of operations automatically, over and over. This is the basic concept behind all computer numerical control machines.

CNC Data Input

Numerical control systems operate machines by giving them a series of instructions. The instructions are coded. The codes are numbers, symbols, or computer language. The codes are used to position the machine tool or the workpiece (material) without human assistance. These coded instructions can be recorded and stored in a variety of ways.

Linking to MATHEMATICS

Hole Depth. A CNC drill is programmed to drill 589 parts. Each part is to be drilled with five holes. Of these holes, two are to be 1/2" deep. Each of the remaining three holes is to be 3/8" deep. If you were to add the depth of all the holes in all of the parts, what would be the total in inches (rounded to the nearest inch)? ▲

With early numerical control machines, instructions or data were input into the machine tool through punched cards or tapes. The location of a hole punched on the tape represented a command to the machine tool. A machine programmer would first construct a *flowchart* of the operations needed to perform a task.

The instructions representing these tasks would be coded and punched into the tape. The tape would then be fed through a "reader." This device would translate the codes on the tape into instructions for the machine tool. The movements of the machine tool would be controlled by the coded instructions on the punched tape.

Today, specialized computer languages have replaced the coded holes punched in tapes and cards. Modern numerical control uses computers to input instructions into machines. The instructions can be stored in the computer and fed directly into the machine. A floppy disk can also be used to transfer the instructions to the machine. All of these changes have allowed more precise machine operations.

Simple operations can be programmed directly into machines. *Manual data input* (MDI), takes place directly on the manufacturing floor. Using a keypad, machine operators can directly instruct the machine tool to perform an operation. Fig. 16-5.

Cartesian Coordinates

How does a CNC program tell a machine tool where to drill a hole and how deep to drill it? How does the program explain to a machine that you want it to cut a slot one inch from the right edge of a block of metal? The answer is simple. The program must give the machine the locations where the cuts will take place.

Locations are specified by sets of letters and numbers. The letter-number combinations that locate points are called **Cartesian** (car-TEE-shun) **coordinates**. The letters identify imaginary lines running through an object. Each line is called an *axis*. The numbers locate points along an axis. A CNC machine runs its tool along the axis lines and performs operations at the points indicated on the axis lines.

CNC machines require only three axes to locate any point on a workpiece. The axes are identified by the letters X, Y, and Z. Fig. 16-6. The X axis runs from left to right

▶ **Fig. 16-5** Keypads give the machinist direct control over the machine.

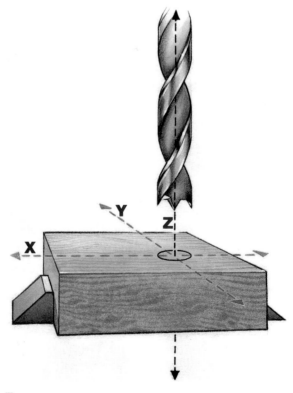

▶ **Fig. 16-6** A CNC drill is guided along three different axes.

Explore

Design and Build a Flowchart

State the Problem
Use a computer to design a flowchart for the manufacture of keychains.

Develop Alternative Solutions
A flowchart is used to show how a product moves down the assembly line, the work that needs to be done, and who will perform each job.

The flowchart should include each production stage. Figure A shows a sample flowchart.

Prepare a few flowcharts that will clearly explain the production sequence for the keychain.

Select the Best Solution
Share your flowcharts with the class. Walk through each step to see how efficiently the materials will move through production. Select the flowchart that is the most efficient.

Implement the Solution
1. Bring up the selected flowchart onto the computer screen.
2. Make any needed changes.
3. Print and save the revised flowchart.

Evaluate the Solution
1. Will materials move efficiently through the assembly line during manufacture?
2. Will adjustments in the flow of processes have to be made during manufacturing?

Collect Materials and Equipment
drawings of the keychain prototype of the keychain computer system

across the workpiece. The Y axis runs up and down, at right angles to the X axis. Both the X axis and the Y axis are parallel to the floor. The X axis runs the length of the workpiece. The Y axis represents the width of the workpiece. The Z axis represents height. It runs vertically, perpendicular to the X and Y axes.

PROGRAM FLOWCHART FOR _____				
STEP NO.	STAGE	DESCRIPTION	LOCATION	WORKER
1	◯	Trace the pattern onto the metal	Bench #1	Worker #1
2	◯	Center punch the areas on the metal to be drilled	Bench #1	Worker #1
3	↓	Move to Bench #2		Worker #1
4	◯	Cut out shape	Bench #2	Worker #2
5	↓	Move to drill press		Worker #2
6	◯	Drill hole in metal	Drill press	Worker #3
7	↓	Move to Bench #3		Worker #3
8	☐	Inspect hole and shape	Bench #3	Worker #4

Fig. A. A sample flowchart.

Points (numbers) identifying locations can be placed along any axis. The location of a point is identified by its coordinates—the letter of the axis with the number of the point on that axis. When a computer is programmed to control a machine tool, the programmer identifies the coordinates where the tool is to do its work.

Part Design

The manufacture of a part on a CNC machine requires planning. Communication is important. An engineer or industrial designer will study the needs the new part must fulfill. This person will then develop designs and determine the specifications for the part. All parts, both simple and complex, need detailed drawings. These must describe the size and shape of the product. *Drafters* confer with the engineers to develop these drawings.

CNC programmers study the drawings to determine the best method for manufacturing the part. They write the program that will direct the machine tool to do the work. The machinist selects the stock, loads the program, and operates the machine. These processes create the product.

Codes

CNC machines require the input of a great deal of information, or data, to manufacture a part. Knowing only the coordinates for a particular cut is not enough data. *Codes* must also be included in the program. Codes tell the machine how fast to move the material or the cutting tool. Codes tell the machine to move the material or tool in a straight line or along an arch. Some codes turn the machine on. Other codes turn the machine off.

A line of machine program code might look like this:

G01X.7Z.8F5

Let's decode this data.

G01 = Cut in a straight line.

X.7Z.8 = To coordinate location X7 and Z8.

F5 = The speed of the cut will be 5 inches per minute.

G CODES:	M CODES:	F CODES:
G00 - Move quickly to a specific coordinate without cutting. G01 - Cut in a straight line to a specific coordinate.	M02 - The program is over. M03 - Turns on the spindle that turns the stock or cutter.	"F" codes determine the feed rate or how fast the material or tool is fed during cutting.

COMPUTER-AIDED MANUFACTURING (CAM) SYSTEMS

Using computers to manufacture a product is called **computer-aided manufacturing (CAM)**. Numerical control, just described, is a type of CAM.

CAD/CAM

CAD/CAM is a combination of two computer control systems: computer-aided design and computer-aided manufacturing. A CAD/CAM system allows engineers to design a product using CAD software. They can then send the design information directly to the computer that controls the machinery that produces the product. CAD/CAM is an example of computers "talking" to each other. Design features such as size and shape are sent from the CAD software to the CAM software.

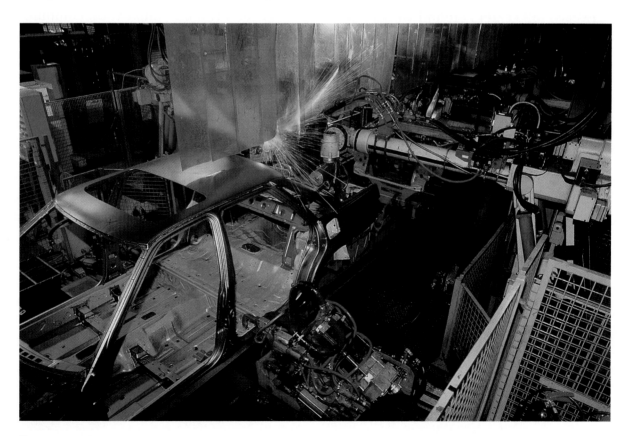

▶ **Fig. 16-7** The workstation cell is the central unit of the flexible manufacturing system.

CAD/CAM saves time in manufacturing. It allows the manufacturer to go directly from the design and testing of a product to the making of the product. The need for drawings and program writing is greatly reduced.

Flexible Manufacturing System (FMS)

Often manufacturers produce a variety of products that differ slightly in design. Athletic shoes are a good example of this. The manufacturer of your favorite sneakers might produce a dozen similar styles. A **flexible manufacturing system (FMS)** is a system that allows the manufacturer to make a variety of designs using the same machines.

Every FMS centers around workstations called "cells." A *cell* consists of many machines that are controlled by computers to perform a variety of operations. Fig. 16-7. When a change in a product's design is needed, the manufacturer simply reprograms the computers to adjust to the new design.

Automobile makers use flexible manufacturing to create different models of the same car. Code numbers on the body, chassis, and other parts tell the computers which assembly program should be used.

FMS speeds up manufacturing and produces parts and products at lower costs. However, a flexible manufacturing system can cost millions of dollars.

IMPACTS

Computer control systems have made manufacturing more efficient. One example of this efficiency is *just-in-time manufacturing*. In such manufacturing, the necessary parts are delivered to the factory just before they are needed. This has several advantages. The manufacturer does not have money invested in parts that are stored unused in the warehouse.

Because parts are delivered to the factory just before they will be needed, the factory is able to adjust its order so that it receives only the needed number of parts. This procedure also ensures that the parts ordered will actually be used. Computer control systems can aid in the use of just-in-time manufacturing. Fig. 16-8.

One unfortunate impact of computer control systems has been their impact on the job market. Many have lost their jobs because of the introduction of computer control systems in manufacturing. However, some have stated that the installation of computer control systems has created jobs in new areas.

THE FUTURE

One key feature of the Information Age has been the rapid transfer of technical information. The increase in communication has allowed others to quickly adopt good ideas. This has been especially true in manufacturing. The manufacturing of items in one country for sale in another is common. It has led to *standardization*, or the manufacturing of items to a set standard. In the case of machine parts, for example, the parts must be interchangeable. Computer control systems have aided the growth of standardization. As international trade increases, the standardization of products will become more important. Computer control systems will play an important part in this.

▶ **Fig. 16-8** The engine blocks in the foreground were manufactured through a computer-controlled process.

Apply What You've Learned

Design and Build Computer Numerical Control Coordinates

State the Problem

Check the coordinates assigned to the drilling locations for a computer numerical control program. Plot each drill location using the Cartesian coordinate listed.

Develop Alternative Solutions

Evaluate methods for locating and plotting each of the following coordinates along the X axis and Y axis.

(-3,4)
(3,4)
Line Ends
(3,4)
(3,-4)
Line Ends
(3,-4)
(-3,-4)
Line Ends
(-3,-4)
(-3,4)
Line Ends
(-2,-3)
Drill Point #1
(2,3)
Drill Point #2

Select the Best Solution

Select the procedure for plotting the coordinates on graph paper.

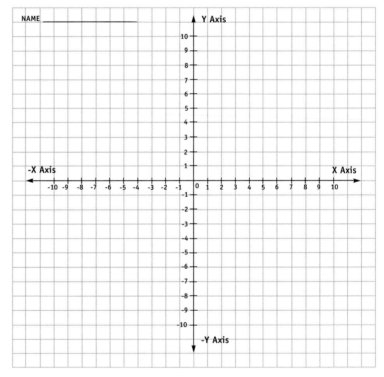

Fig. A

Fig. B

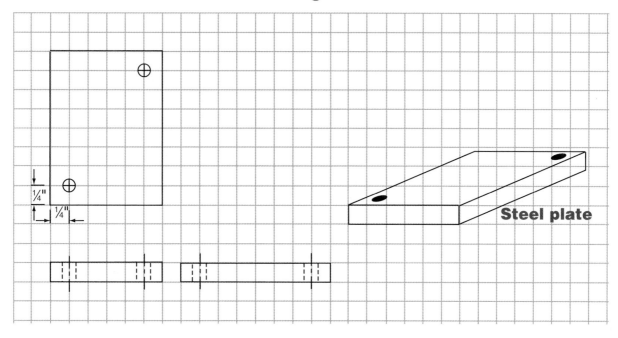

Steel plate

Implement the Solution
1. Lay out the numbered X and Y coordinates on graph paper. Fig. A.
2. Plot the ordered pairs of numbers given as coordinates.
3. Connect the points with line segments. Start a new line after the
words "Line Ends."

Evaluate the Solution
Check that the locations have been accurately plotted. Compare the
locations you plotted with the locations on the drafter's drawing. Fig. B.

CAREERS IN
Computer Systems Control

COMPUTER NUMERICAL-CONTROL OPERATOR

Experienced CNC operator needed to run horizontal lathes and turning centers. Willing to train applicant with good math and mechanical aptitude blueprint reading skills. Health and dental insurance, profit sharing, and attendance bonuses. Further education funding available. Apply weekdays from 9:00 a.m. to 3:00 p.m. to: General Die Casters, 6212 Harrisburg Road, Peninsula, Pennsylvania.

TOOL AND DIE MAKERS

Small automatic screw machine product manufacturer looking for tool and die maker with tool grinding experience and ability to design fixtures. Must be able to operate numerically-controlled lathes and mills. Full- or part-time. Send resume to: Eber Manufacturing Company, P.O. Box 6650, Independence, MO 68402.

NUMERICAL-CONTROL ROUTER OPERATOR

Electronics manufacturing company needs operator to run numerical-control router machine. Will load control media, select cutting tools, and monitor machine in cutting individual printed circuit boards. Apply to Ampex Industries, 5300 Smith Road, Brookpark, NC 30406.

INDUSTRIAL ENGINEER

Automated manufacturer is seeking an Industrial Engineer to conduct operation research and design manufacturing and information systems. Our goal is to produce high quality products as efficiently as possible. Bachelor's degree required with excellent communication skills. Send resume to: Diverse Industries, 3480 West Market Street, Fairview, AL 36400.

NUMERICAL-CONTROL DRILL OPERATOR

Growing linear motion products manufacturer has immediate opening for drill operator. Must be computer literate. Will load control media, attach depth collars to drill bits and measure bit depth, monitor machine, compare display data to specifications. We offer competitive wages, overtime and health insurance. Non-smoking facility. Apply to: Nook Industries, 4950 East 23rd Street, New Bedford, Indiana.

CAM TECHNICIAN

Creative problem-solver needed to work as technician for manufacturer of compressors. Should have experience with CAD/CAM for drafting of electromechanical components. Strong analytical and communication skills needed to work in team environment. Competitive salary plus benefits. Forward resume to: Coppland Corporation, P.O. Box 699, Hilliard, OH 43565.

Linking to the WORKPLACE

Were you aware of careers in computer numerical control before now? How many jobs do you think there are in this field in all of the United States? There are over 20,000 different job titles that exist in the United States! Did you guess correctly?

Write down the names of 20 different jobs. It is important for you to explore as many job titles as possible before deciding on your career. With 20,000 possibilities, don't limit yourself to only a few careers.

Chapter 16 Review

SUMMARY

▶ Detailed and accurate drawings called working drawings are needed to produce a product.

▶ CAD software enables a designer to produce three-dimensional as well as two-dimensional drawings.

▶ Computer numerical control (CNC) is a system that uses computers to control the operation of machines during production.

▶ CNC machines require only three axes to locate any point on a workpiece.

▶ The letter-number combinations that locate points are referred to as Cartesian coordinates.

▶ CAD/CAM is a combination of two computer control systems: computer-aided design and computer-aided manufacturing.

▶ A flexible manufacturing system allows the manufacturer to make a variety of designs using the same machines.

CHECK YOUR FACTS

1. How has the use of computers in manufacturing changed the way products are produced? Give three specific examples.

2. What type of drawing software would an engineer use to design a part for a product?

3. Give an example of how computers are used in product design.

4. Explain the process of computer numerical control.

5. Describe how a CNC machine locates a point on the surface of a material.

CRITICAL THINKING

1. Identify three types of products that could be manufactured using a flexible manufacturing system. Give reasons for your selections.

2. What impacts do you think computer use in manufacturing has had on employees? List one positive and one negative impact.

3. Describe the changes in education and training that you think must take place to help people, at all levels, enter the manufacturing industry. Be specific.

Robotics

OBJECTIVES

▶ identify technological developments that led to modern robotics.

▶ explain how the stepper motor is used in robotics.

▶ define the work envelope.

▶ explain how feedback control is used.

KEY TERMS

automation

degree of freedom

feedback control

robot

robotics

work envelope

A duck runs quacking across the floor, flapping its wings, and spreading its feathers.

Finally it dives into a pond. Stretching out his hand, a man offers food to the duck. The duck gently eats the pieces of grain from the man's hand.

Not so remarkable is it? You need a little more information. The duck is a mechanical device made from metal.

Still not so remarkable? The duck was made by a French inventor named Jacques de Vaucanson in the 1700s. That's pretty remarkable!

We think of robots as marvels of modern technology. However, the idea of machines designed to imitate human actions existed over 3,000 years. The ancient Egyptians made puppets on strings.

HOW ROBOTICS DEVELOPED

The duck you just read about was a crude robot. A **robot** (ROW-bot) is a machine made to act like a living thing. **Robotics** is the study of robots. Robotics is a control system technology. Through input data and a variety of processes, people have learned to control the output of machines. Today, robots are designed to do many of the tasks humans used to do.

Ancient attempts at robotics were not control system technologies. Crude robots could not react to changing conditions as modern robots can. Many technological events had to take place before true robotics could develop. Designers and engineers had to learn how to transfer forces. They did this through gears and levers. The development of punch cards, computers, and automation was also important.

Modern Robots

Robots are used today in a variety of ways. For example, during some joint replacement operations in humans, a hole must be drilled into the bone to accept the artificial joint. The accuracy of this hole is critical. Some surgeons now use robotic systems to position the drill and make the actual hole. The accuracy and steadiness of the robotic arm are hard to beat.

Mobile robots can crawl into the most unusual places. The Mermaid is an underwater robot that searches the sea bottom for unexploded mines. It can hold onto the sea floor in the roughest current. When the robot locates the mine, it self-destructs by blowing itself up. This detonates the mine. Fig. 17-1.

Walking robots are capable of moving over rough ground and even up stairs. Police departments in large cities use walking robots to retrieve explosives from buildings.

▶ **Fig. 17-1** The Mermaid, an underwater research robot. Notice the robotic arm in the foreground. It has a gripper that allows it to pick up items with extreme precision.

Deep beneath the dark Atlantic Ocean robotic divers searched for the sunken ocean liner *Titanic*. The cameras and sonar on board the robotic diver sent signals to the computer on the mother ship. Suddenly, observers could see the hull of the great ship resting on the floor of the ocean.

Early Mechanisms

During the eighteenth and nineteenth centuries, mechanical robotics were built for entertainment purposes. These robots used springs, gears, levers, and pulleys to perform many tasks. These robots were called *automatons*. Some of them could play musical instruments, write letters with pen and ink, and even perform magic tricks.

In the early 1800s, Joseph-Marie Jacquard, a French weaver, invented a punch card system. He used it in his factory to produce fabric patterns. Holes were punched in stiff paper cards. These holes corresponded to patterns woven in fabric. Some rods were pushed through the openings in the cards. Other rods were held back. This arrangement of rods represented the color and pattern of threads that the machine had to weave. The punch card process laid the foundation for modern computers.

Punch Cards and Computers

Herman Hollerith invented the punch card tabulating machine. The machine was used to tally data (information) during the 1890 United States census. Data was punched into the card. The machine inserted rods through the holes in the card. The rods then made contact in small cups of mercury, completing an electrical circuit. The electrical connection made the hands on a dial move one space. In this way data was recorded and added.

Hollerith had created an electrical scanner and sensor. This technology was essential to later developments in computer control technology. Hollerith's company later joined other companies to form IBM.

The first computers were built during the 1940s. They were huge calculating machines that took up an entire room. In 1948, a team of scientists invented the transistor. This reduced the size of the computer. It also multiplied the speed at which the computer could make calculations. Fig. 17-2.

Automation

Automation (auto-MAY-shun) is a technique that is used to make a process automatic. The word was first used in the 1940s. It described work that had been

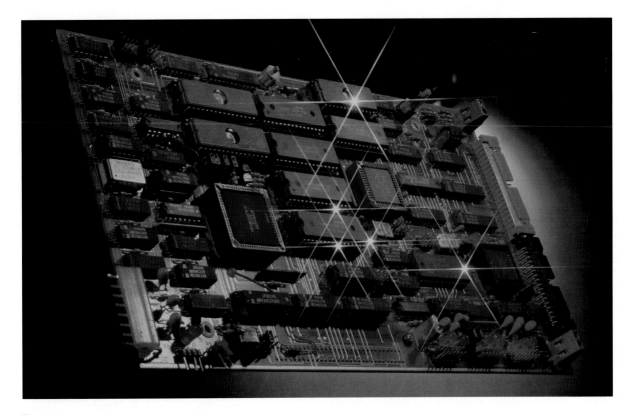

▶ **Fig. 17-2** The motherboard of a personal computer. It is a giant leap from the punch card tabulator to transistors, microchips, and personal computers. New technologies build upon old technologies. The personal computer created the foundation for modern robotic control.

done by people that was now being done by machines.

Automated factories used machine tools that were computer controlled. The most accurate machines at the time were numerical control machines. The system used numbers to describe the shape of a part and the tool's movement through the material. A punched paper roll (punched tape) had these movements recorded on it. This roll was fed through the computer. The machine tool would then automatically cut, grind, or drill the part to shape.

MODERN ROBOTIC SYSTEMS

Machine tools are not robots. Robots are more accurate. They are also more flexible and can make decisions. The word *robot* was first used in 1922 by Karel Capek, a Czechoslovakian writer. He wrote a play about mechanical humans, or robots. They worked in factories, where they replaced human workers. The Czech word *robota* means "slave labor." In Capek's play, the robots finally rebel against their masters—and take over the world!

Robots represent great power in today's workforce. By the 1970s, Japan had over 7,000 robots in automobile factories. These robots moved materials and parts. They welded, assembled, and painted automobiles on the assembly line.

Robotic systems are designed to have humanlike movements. In many ways, robotic systems model human systems. Just as your brain sends commands to your arms and legs, a computer sends instructions to a robot. Computers are the brains of modern robotic systems. They control the movement of mechanical robotic arms. Fig. 17-3.

Robotic Arms

There are sixteen joints in the human arm, wrist, and fingers. These joints provide us with forty degrees of freedom. A **degree of freedom** is the ability of the robot to move in a direction. Movement in a combination of joints allows human, as

well as robotic, arms to move in any direction. The flexibility of these joints allows robots to handle a variety of materials in a variety of shapes. This flexibility gives robotic arms, or manipulators, the ability to move in any direction and grasp a variety of items.

Robotic Hands

Human hands are very flexible. They can grip almost any object in a variety of positions. Robotic hands have a much more difficult time. Often a variety of robotic hands must be used as an assigned task changes. Robotic hands, known as *end effectors*, can quickly be attached to a robotic wrist as a task may change.

The Work Envelope

The place where two moving parts of a robot are connected is called a joint, or axis. An arm robot (manipulator) moves at its waist (or base), shoulder, elbow, and wrist joints. Each degree of freedom in a robotic arm is provided by combining the movements of these joints.

Linking to SCIENCE

Joints and Motion. The joints of the fingers, wrist, and elbow allow freedom of movement in many, but not all, directions. Explore the ability of your joints to move in many directions. Attempt to move your fingertip, thumb tip, entire finger, entire thumb, hand, and forearm in (1) a pivoting (circular) motion, (2) a back-and-forth motion, (3) an up-and-down motion. Hold the body part above the joint to immobilize it. This will make it easier to identify the directions in which the joint allows movement.

Fig. 17-3 Robotic arm and effectors. Robotic arms use cables, motors, gears, and pneumatic cylinders to move within a space. The arms receive directions from a computer much like our arms and hands receive instructions from our brain. Robotic arms move through space by rotating a combination of joints. A robot's degree of freedom is based on the number of joints it has and the degree of movement the joints allow.

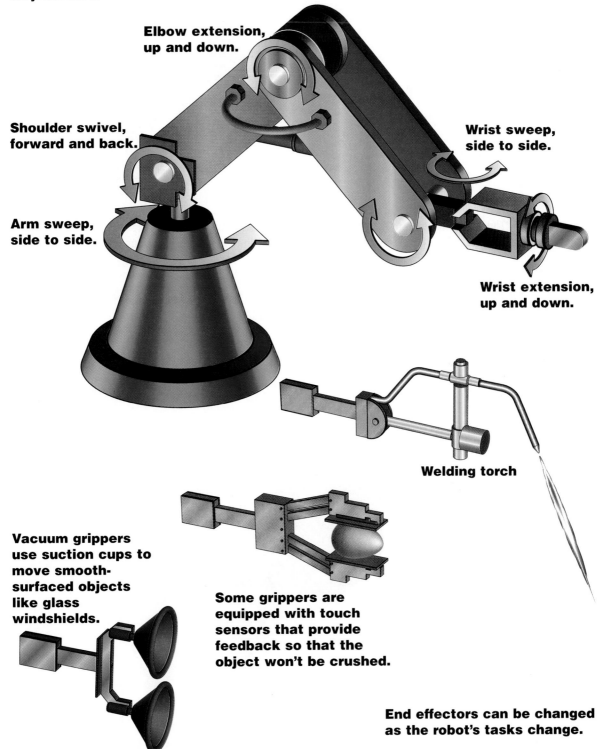

Elbow extension, up and down.

Shoulder swivel, forward and back.

Wrist sweep, side to side.

Arm sweep, side to side.

Wrist extension, up and down.

Welding torch

Vacuum grippers use suction cups to move smooth-surfaced objects like glass windshields.

Some grippers are equipped with touch sensors that provide feedback so that the object won't be crushed.

End effectors can be changed as the robot's tasks change.

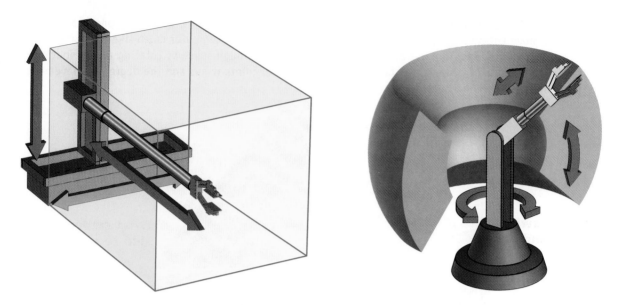

▶ **Fig. 17-4** The area a robot moves within is called its work envelope. The size of the robot and the degrees of freedom determine the size of its work envelope. What is the size of your work envelope?

The space a robotic arm moves within is called its **work envelope.** The design, or architecture, of the robotic arm will determine the size and shape of its working envelope. Fig. 17-4.

POWER FOR ROBOTIC MOVEMENTS

Each moving robotic part can be powered in a variety of ways. The selection of a power source depends upon

Electrical signal from computer interface.

The shaft of the motor can be made to rotate fractions of a degree by the computer. It is this controllability that allows the robotic arm such a high degree of accuracy.

▶ **Fig. 17-5** Stepper motors provide the force to move the robotic arm. Gear trains and chain drives transfer mechanical energy from the motor to the moving part of the robotic arm. Gears are used to adjust the speed of the motors.

what the robotic arm has to do. An electric motor known as a *stepper motor* is commonly used as an actuator, or power source, for robotic movement. Fig. 17-5.

One complete rotation of a stepper motor can be divided into hundreds of individual steps. Each step represents a fraction of a degree of movement. A stepper motor can rotate a small amount or a step each time an electrical signal is sent to it. Waist, shoulder, elbow, and wrist joints may each be powered by separate motors. The motor shaft transmits the mechanical energy through gears, shafts, and pulleys to the robotic joint. Robot programmers control the precise movements of each joint by controlling the steps of the motor.

Electric stepper motors power robotic arms with speed and accuracy. At times, robotic arms lift heavy objects. Pneumatic and hydraulic power supply the extra force. Pneumatic and hydraulic actuators use compressed air (pneumatic force) or hydraulic fluids (hydraulic force) to transfer power to the joints and grippers. Fig. 17-6.

Linking to MATHEMATICS

The Work Envelope. What is the volume of your work envelope? Since you reach forward, backward, to the left and to the right, you reach a full three-dimensional circle, which is a sphere.

The formula for the volume of a sphere is $V = 4/3 \, \pi \, r^3$. Radius (r) is the reach of your arm forward or up or any direction. You are the center of your sphere of reach. For example, if your little sister's reach is 10 inches, her work envelope is:

$V = 4/3 \, \pi \, r^3$
$V = 4/3 \times 3.14 \times 10^3$
$V = 4186.66$ cubic inches

Have a classmate measure your reach. You should then figure your work envelope. ▲

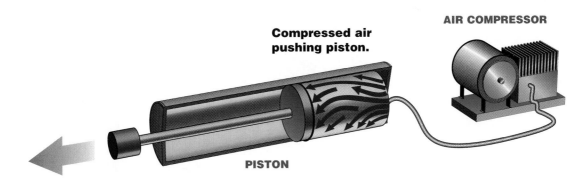

COMPRESSED air pushing piston.

AIR COMPRESSOR

PISTON

▶ **Fig. 17-6** In this pneumatic system, a computer-controlled valve allows air to enter the cylinder. The air applies force to the piston, which pushes on the shaft controlling the gripper.

Explore

Design and Build a Feedback Control Game

State the Problem

Your eyes send feedback to your brain about the movements of your hand. This feedback allows your brain to make constant corrections to your hand movements. This helps you to form letters, lines, and shapes. What would happen if your brain received the wrong feedback signals? In this activity you will design and build a device that will fool your brain.

Develop Alternative Solutions

You will need to construct a device that will fool your brain and give it incorrect feedback. The device will contain a vision blocker that will keep you from directly seeing the image your hand will trace.

The only visual feedback your brain will receive regarding your hand movements will come from what you see in the mirror. As you know, mirror images are backwards or the reverse of the real image. How do you think this will affect your ability to trace the printed image?

Design the game so it is large enough to accommodate your hand comfortably on the base. You may need to create several possible designs. One possible design is shown in Fig. A.

Collect Materials and Equipment
base material (wood, plastic, cardboard)
mirror, glass or acrylic
cardboard (for vision blocker)
paper with printed images
material processing tools and machines

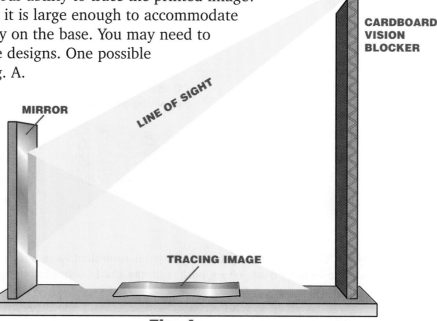

MIRROR

LINE OF SIGHT

CARDBOARD VISION BLOCKER

TRACING IMAGE

Fig. A

Select the Best Solution

Select the design that you think is most effective.

Implement the Solution

1. Build the game, using your design sketch as a guide. Be sure the vision blocker is large enough to block your line of sight. Be sure the mirror is placed to reflect the reversed image to your eyes.
2. Place the game on your desk. Sit behind it so the vision blocker prevents you from directly seeing the image printed on the paper.
3. Position yourself so that you can see the printed image only through the mirror.
4. With a pencil, trace the printed image using only the mirror as your feedback mechanism.

Evaluate the Solution

1. After playing the game for a while you will probably be able to trace the image more accurately. How can this be explained?
2. Are some images more difficult to trace than others? Why?

Pneumatic and hydraulic systems are made up of cylinders and pistons, much like a doctor's syringe. The piston pushes on the fluid in the cylinder. A second piston on the other end of the cylinder moves as the fluid presses on it. The moving piston can make the robotic arms move forward and back. The pistons are controlled by electrical switches connected to computers. The switches open and close valves controlling air and hydraulic fluids.

CONTROLLING ROBOTIC SYSTEMS

Binary Code

Cables of wire travel from the computer to the robotic interface. The *interface* links the robotic motors to the computer. Fig. 17-7. Inside the interface are electronic switches that turn the motors on and off. Electrical signals travel through the cable as coded information. The code consists of

▶ **Fig. 17-7** The job of a programmer is to change flowchart information into a language that the computer can understand. The computer sends electrical signals, or codes, as pulses of electrical energy.

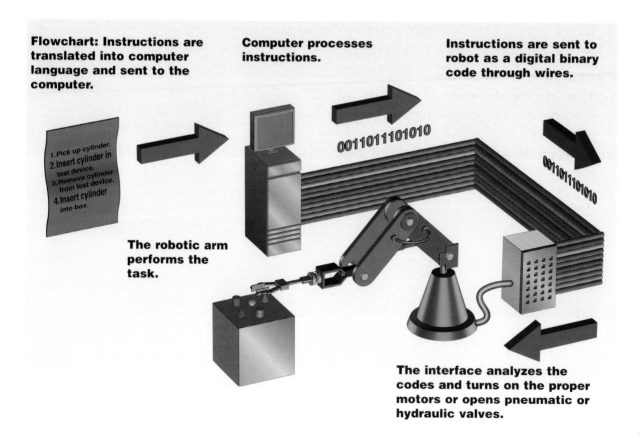

Flowchart: Instructions are translated into computer language and sent to the computer.

Computer processes instructions.

Instructions are sent to robot as a digital binary code through wires.

1. Pick up cylinder.
2. Insert cylinder in test device.
3. Remove cylinder from test device.
4. Insert cylinder into box.

0011011101010

0011011101010

The robotic arm performs the task.

The interface analyzes the codes and turns on the proper motors or opens pneumatic or hydraulic valves.

bursts of electrical current. When written, the code is represented by a series of 0's and 1's. This code is commonly referred to as the *binary code*.

Computer Control

How does a robotic arm know what movements to make?

How much pressure should the gripper apply?

How can a robotic arm "remember" the patterns needed to paint an automobile?

Just as your brain controls your every movement, so do computers control the movement of robotic systems. The computer uses a series of instructions known as a *program*. Robotic programs are very complex. They must list in logical order all the steps needed for the robot to perform a task.

Imagine listing each command the brain sends to muscles when you pour milk from a container into a glass. Robotic software (program) designers prepare flowcharts that list the basic movements of the robot. These movements are then broken down into finer detail. They are then written in a language that computers can understand.

Computers send instructions to the actuators that power the movement of the joints. The instructions tell the robotic arm how far to travel, how much pressure to apply, and how to move a tool to perform a task. The instructions are sent as electrical signals that rotate stepper motors or open pneumatic valves, causing the pistons to travel.

Robotic programmers write the computer instructions or software to control the robotic hardware. Instructions

Linking to COMMUNICATION

Word Origins. The English language has taken words from many languages. In this chapter you learned that "robot" is a Czech word. With the help of your language arts teacher and/or librarian, compile a list of foreign words used commonly in English. Hint: Think of words associated with food, music, and clothing.

can be created by guiding the robotic arm through a sequence of movements and programming the computer to remember the pattern of motion. Teaching a robot in this manner is called lead-through programming. Robots can also be programmed using keyboards or teach pendants. The pendant and keyboard give the robot direct instructions to move up, down, left, and right. Each movement is remembered by the computer and repeated as often as required.

Feedback Control

How does a robot know where an item is? Humans have sensory organs such as eyes, skin, and ears. These allow us to track changes in our environment. Robots also have sensors so they can keep track of what's going on around them.

Imagine that you are going to touch the handle of a saucepan on a stove. Your brain sends signals to the muscles and tendons in your hand to grasp the pot handle. Information is quickly sent to your brain through nerve bundles. The message is that your hand has grasped the handle and is ready for the next command.

Robots

Some jobs are better done by robots. They are used, for example, for welding jobs in automobile assembly. They are also used for drilling precise holes in hip-replacement surgery.

The way machines interact with people has become an important dimension of technology. In the future, control by voice commands will become more widespread. Cog is the name of the humanoid robot shown across these two pages. It is programmed to mimic human movements and senses.

ROBOT ANTS

These one-inch robots contain tiny motors for locomotion and grasping. When it detects food, the robot ant sends out an electronic signal, which summons other ants. Robots that cooperate in imitation of social insects could be ideal for collecting toxic waste for disposal.

SECURITY ROBOT

This security robot is programmed with a detailed map of the museum it guards. Its detectors give warnings of smoke, fire, and increases in humidity. It can also detect movement.

ROVER SOJOURNER

This solar-powered robotic vehicle was used by NASA to study the composition of soil and rocks on Mars. Controlled by computer, it could steer itself to avoid obstacles. It provided regular location updates.

▶ **Fig. 17-8** The feedback control process. On robotic systems, sensors monitor what the system is actually doing. If the input to the system does not match the output from the system, the system is adjusted.

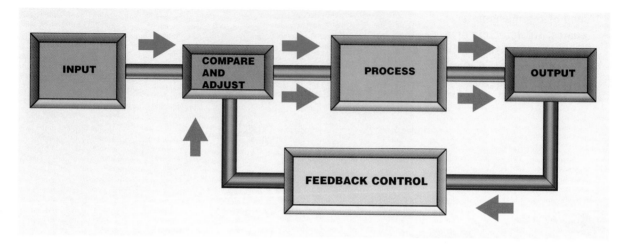

What if the handle is too hot? If the handle is too hot, signals quickly return to your brain. There they are translated as pain signals. Your brain sends new signals to your hand and your grip is released. The process of sending signals, interpreting received signals, and adjusting through signals is called **feedback control**. Robots use feedback control constantly. It allows them to know where they are and to adjust their actions. Fig. 17-8.

Your fingers have over 17,000 sensors (nerve endings) that send data to your brain as you touch things. Robots use touch sensors known as contact sensors. These feed information back to the computer about a task it is working on. Contact sensors send electrical signals to the computer. The data might include information on the shape of an object and how much pressure the grippers are placing on it. The computer can then adjust the actions of the robot if changes are needed. Fig. 17-9.

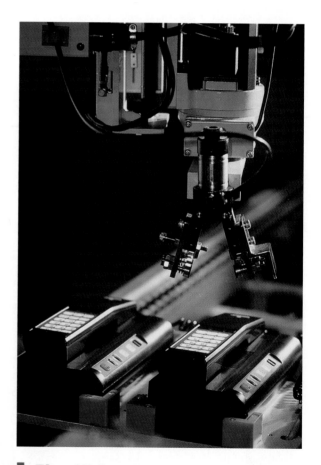

▶ **Fig. 17-9** Advances in computer control, electronics, and mechanics enable robots to perform more intricate work.

A robot may be equipped with cameras so it can view objects in the work envelope. The image picked up by the camera is input. It is sent to the computer for analysis. Using that data, the computer outputs directions to the robot.

Imagine freshly baked cookies moving down a conveyor line. Using its attached camera, the robotic arm looks for burnt and broken cookies. When one is sighted, the computer instructs the grippers to remove the bad cookie from the line.

Robotic arms that perform detailed work like welding and painting use angular or optical sensors. These track the arm's movement. These sensors, which are shaped like disks, have markings. The disks are placed at each joint in the robotic arm. As the joint moves, optical scanners (like cameras) read the code. They send this information to the computer. The computer interprets the information, calculates the angle of the joint, and outputs needed commands to the arm.

Probably the most famous robot arm, seen hard at work by millions of people, is the Remote Manipulator System (RMS) used aboard the Space Shuttle. Fig. 17-10. The RMS is a jointed robotic arm. It uses cameras and angle sensors to tell the

▶ **Fig. 17-10** In the darkness of space, the Hubble Space Telescope (HST) is lifted by the remote manipulator system (RMS) from its berth in the cargo bay of the Earth-orbiting Space Shuttle Discovery. The orbiter uses electric motors to position its robotic arm. Video cameras help astronauts view the arm as they manually move it towards its target.

computer its position. The arm can be used to release satellites into space as well as retrieve items already in orbit. The RMS can also serve as a remote work platform for doing repair work on space vehicles. Robots may also use microphones to sense sounds. They may use sonar to measure distances. They use sensors to detect poisonous materials in the work envelope. These sensors are called non-contact sensors.

Robot Generations

The first generation of robots was designed by industry to perform a variety of tasks. Known as steel collar workers, these robots did simple tasks that were dangerous or unpleasant for human workers. Early robots were used to handle hot metal, weld metal parts, spray-paint, move parts, and load pallets. These early robots were large and not very flexible.

The second-generation robots used today can perform tasks more complex than the tasks performed by early robots. Today's robots are flexible. Fig. 17-11. They can quickly be taught to do several

▶ **Fig. 17-11** IT (Interactive Technology), is demonstrated here as an interactive robot. It mimics the human emotions of happiness, sadness, surprise, boredom, sleepiness, and anger.

different operations. With movements accurate to a fraction of a millimeter, robotic arms can assemble intricate electronic circuits. They can solder wires as thin as a human hair.

IMPACTS

Are there negative impacts to the use of robot technology?

What if your friend worked in a factory that assembled automobiles? She performs her job with the greatest accuracy and never misses work. Her supervisor often tells her that she is the most productive and reliable employee in the company. One day when your friend goes to work, she walks onto the assembly line floor to find a shiny robotic arm in her spot. The arm works twice as fast as your friend. It takes no breaks, and works twenty-four hours a day.

Your friend has just been displaced. "Displaced" is a term used to describe a person whose job has been taken over by automation or new technology. Many experts tell us that robotic technology may cause increased unemployment as companies switch to automation.

Others say that being displaced is not the same as being dismissed. Displaced employees usually find new work within the same company or with other companies that are not yet automated.

THE FUTURE

The use of robots in business and industry is part of the automation revolution. Automation is the process by which computers control a series of tasks in manufacturing. Automated factories can operate with very few people. Automated machines can usually work faster, at lower cost, and more accurately than human workers. Remember that robots are not paid a salary, are never late, never call in sick, never need health insurance, and never take vacations.

Robotic automation now allows manufacturers to produce products more cheaply. This allows products to be sold for less. This allows the manufacturer to become more competitive in the world market. In the future, the use of robots will increase.

▶ **Fig. 17-12** Welding is commonly done by robots. Weld placement can be precisely controlled.

Apply What You've Learned

Design and Build a Pneumatic Control Device

State the Problem

Design and build a device that will convert the pneumatic force of the balloon's motion into linear (straight) or rotary (circular) motion.

Develop Alternative Solutions

Using the materials listed, design a device that can convert the pneumatic force of the balloon's expansion into linear or rotary motion. Sketch several.

Select the Best Solution

Choose the design that you think will be most effective.

Implement the Solution

1. Attach the balloon to the plastic tubing, using a rubber band. Attach the other end of the tubing to the squeeze bottle by forcing it through the hole in the cap. A tight fit is needed.

Collect Materials and Equipment
1/4" plastic tubing
squeeze bottle
cardboard
dowels of assorted diameters
wood scraps
tape
glue
string
rubber bands
material processing equipment
balloon

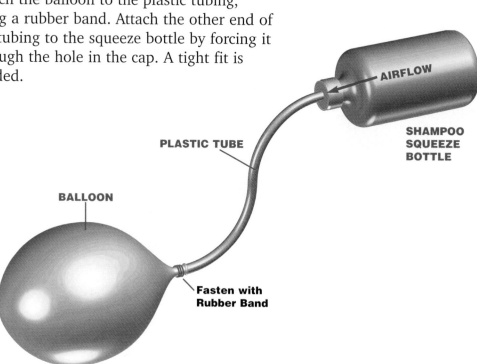

AIRFLOW

PLASTIC TUBE

SHAMPOO SQUEEZE BOTTLE

BALLOON

Fasten with Rubber Band

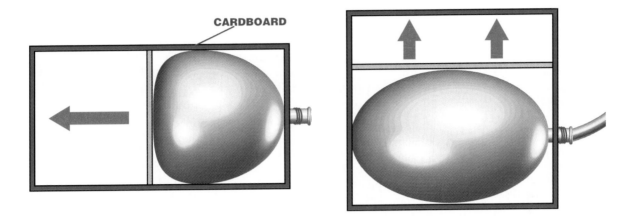

CARDBOARD

2. Experiment with the pneumatic system. Get a feel for its power and how the balloon expands. Try lifting a stack of books.

Evaluate the Solution

1. In a real device, what mechanism would replace the balloon as a power source?
2. What effect would a larger squeeze bottle have on your system?
3. Make a sketch showing how the mechanism could be used as a gripper arm.
4. Explain what makes the balloon expand.
5. If the air in the bottle was replaced with water, what kind of system would it be?
6. Can you think of a use for your design? Can you find everyday devices that use a mechanism similar to yours? List them.

CAREERS IN
Robotics

ROBOTICS TECHNICIAN

Manufacturing firm needs robotics technician to program and monitor robotics equipment. Must be able to perform precise work and have knowledge of hydraulics, electronics, and programming. Post-high school training required. Responsible for setup, operation and preventative maintenance. Please send resume to: Southland Manufacturing Company, 7223 Heyward Drive, Lansing, LA 44208.

MAINTENANCE MECHANIC

Manufacturing firm has immediate opening on the evening shift for a maintenance mechanic with knowledge of circuit board functions. Responsible for locating and replacing faulty printed circuit boards in robot controller. Work in newly renovated facility. Salary plus overtime opportunities. Submit resume to: Personnel, Brooks Manufacturing, Inc., 70 Ackerman Place, Indianapolis, IN 52201.

MECHANICAL ENGINEER

Engineering firm needs mechanical engineer with background and knowledge in design. Work with engineering team to design and develop robots for use in production operations. Bachelor's degree is required. Will be actively involved in all phases of development. Excellent working conditions. Submit resume to: Robotics Engineering Design Group, 1313 Henderson Drive, Whittier, CA 90020.

INDUSTRIAL ROBOT OPERATOR

Experienced operator needed for first or second shift. Good working knowledge of MS-DOS and computer literacy a plus. Must be able to diagnose and solve problems without direct supervision. Fast-paced direct marketing company offers competitive wages and excellent benefits. Please send resume to Human Resources Manager, HKN Direct Marketing, 5590 Cass Avenue, Minneapolis, MN 41024.

SALES ENGINEER

Robotics company needs sales engineer to act as liaison between company engineers and sales and marketing staff. Responsible for training sales staff on products and applications and communicating new design ideas to engineering staff. A degree in engineering or computer science is required. Must have excellent communication skills to explain highly technical terminology. Forward resume to: Human Resources, Robotics International, 8820 Burlington Avenue, Greensboro, NC 34421.

Linking to the WORKPLACE

Have you considered a possible career? If you have, what jobs do you think a robot could perform in your workplace? Robots are designed to perform repetitive tasks that require a high degree of accuracy. If you could have a robot perform three jobs for you every day, what would they be?

Chapter 17 Review

SUMMARY

▶ A robot is a machine made to act like a living thing.

▶ Robotics is the study of robots.

▶ The transistor and computer are important in robot technology.

▶ The space a robot moves within is known as its work envelope.

▶ The stepper motor is a common power source for robotic movement.

▶ Robots depend on feedback control, which allows them to adjust their actions.

▶ Uses for robots have expanded as technology has developed.

CHECK YOUR FACTS

1. In what ways are industrial robots similar to humans?

2. What important technological developments had to take place before computer control robotics could become a reality?

3. How does a robotic arm achieve degrees of freedom?

4. Explain how the stepper motor is used in robotics.

5. Define the work envelope of a robotic arm.

6. List a variety of end effectors commonly used on robots and describe how they work.

7. Describe two ways a robot can be taught a task.

8. Explain how feedback control is used to adjust robotic movements.

CRITICAL THINKING

1. You are the owner of a company that manufactures small appliances. You have decided to add robots to the assembly process. Describe the plans you have for the workers that will be displaced by your actions.

2. Prepare a list of tasks usually performed by people in their home that could be done by robots.

3. Prepare sketches of a robotic arm that could be used to turn pages in a book for a disabled person. Label the parts.

SECTION 7

Integrated Technologies

CHAPTER 18 *Lasers and Fiber Optics*

CHAPTER 19 *Engineering*

CHAPTER 20 *Applied Physics*

Some projects are simple. Others are complicated. Consider, for example, the building of a spacecraft. This is a complicated engineering project. To complete the project successfully, engineers must draw on the resources of several different technologies. They must integrate, or bring together, the principles of different technologies. For this reason, engineering is known as an integrated technology. Several different technologies are integrated into the engineering process.

The chapters in this section discuss integrated technologies and their uses.

Technology and Society

Rocket Going Up, Glider Coming Down

An airplane is an airplane, relying on air to make its wings work. A spacecraft is a spacecraft, at home where there is no air. A vehicle such as the Space Shuttle, which travels between the two environments, has to combine elements of an airplane and a spacecraft. This leads to some loss of efficiency. A successor to the Shuttle is now being designed. This new craft will be sleeker. Its design is intended to provide cheaper access to space.

Less Clunky, Less Costly

The Reusable Launch Vehicle (RLV), has been dubbed VentureStar. It will launch vertically and land on a runway. Its fuel tanks will be built-in. Keeping these tanks internal creates design challenges. Internal fuel tanks affect vehicle weight, cargo capacity, and flying characteristics.

VentureStar will save weight by taking advantage of breakthroughs in materials technology. The Space Shuttle used ceramic tiles for heat-shielding.

VentureStar will use honeycombed panels made from a nickel-based superalloy. This superalloy has been designed to be lightweight. It has high strength at temperatures up to 1,650° F. The metallic skin of VentureStar will require far less maintenance. This will help reduce the cost per launch.

Automation of systems is a key to reducing launch costs. Costs for the Space Shuttle exceed half a billion dollars per flight. Much of this cost relates to the large number of engineers needed to tend the subsystems.

Aerospike: An Engine for All Altitudes

One key to the integration of RLV technologies is an engine that performs well at both high and low altitudes. The Aerospike engine is designed to operate efficiently at low altitudes and high altitudes.

Linking to the COMMUNITY

Get acquainted with Facts on File as a reference resource at your local public library. Look for the latest on the X-33 prototype and the VentureStar.

CHAPTER 18

Lasers and Fiber Optics

OBJECTIVES

▶ tell what the term laser stands for.

▶ explain the difference between laser light and ordinary light.

▶ describe how a laser works.

▶ identify uses for lasers in communication, manufacturing, construction, medicine, and business.

KEY TERMS

coherent light

directional light

electromagnetic wave

holography

laser

monochromatic light

photon

Have you ever thought about how important light is? Light is all around us. It enables us to see. It creates rainbows and makes it possible for us to enjoy the colors created by nature and technology. Without the sun, there would be no light and no life on Earth.

Fire was the first human-made light source. Later came candles, oil lamps, gas lamps, and then, electric lights. Engineers continue to work to find new and more efficient light sources. In our own century, they have introduced the laser.

THE NATURE OF LIGHT

Take a look at an electric lightbulb. If the glass is clear, you'll be able to see a thin wire inside. This wire is called a *filament*. When you turn on a lightbulb, its filament heats up as electricity flows through it. Each tiny unit of energy given off by the filament is called a **photon** (PHOH-tahn). Millions of photons produce the light we see. All light sources, from the sun to a pocket-size flashlight, produce light energy. This happens as a result of atoms being excited and giving off photons.

Light travels in waves much like the waves we see on a lake or ocean. Light is a type of electromagnetic wave. An **electromagnetic wave** is a wave produced by the motion of electrically charged particles. Scientists often call these waves *electromagnetic radiation* because they radiate from the particles. Visible light waves are one form of electromagnetic radiation. Other forms include radio waves, X-rays, and gamma rays. These waves are classified by wavelength. Fig. 18-1.

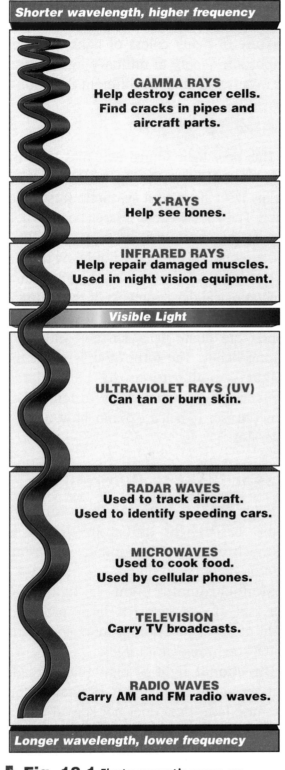

Fig. 18-1 Electromagnetic waves are classified by wavelength on the electromagnetic spectrum.

FASCINATING FACTS

Insects do not perceive light the same way we do. Most insects see ultraviolet light. Some beetles see infrared light. Although many insects perceive more light than we do, they cannot focus their eyes well. It is hard for them to see objects that are not nearby.

Though it looks white, ordinary light such as light from a bulb or the sun is a mixture of many colors of light. Another important quality of ordinary light is that its rays travel in many different directions.

LASERS

This new light source was first built by Theodore Maiman, an American scientist. In his laser he used a synthetic (artificial) ruby. The ruby created a laser beam when an intense beam of ordinary light was flashed on it. Maiman did not know he had created one of the most important technological developments of the century.

The **laser** (LAY-ser) is a light source that sends out light in a narrow and very strong beam. The term *laser* is short for "Light Amplification by Stimulated Emission of Radiation." Try to remember this phrase. It helps explain how a laser operates.

Laser Light Is Different

Laser light is different from ordinary light. Laser light differs in that it is monochromatic, directional, coherent, and bright.

Monochromatic light is light that consists of one color. A laser produces light of only one color, or monochromatic (MON-o-chrome-attic) light.

Directional light is light that spreads out very little compared to ordinary light. Because laser light does not spread out, it can be focused on a small spot by a lens or mirror.

Laser light is also coherent (coh-HEAR-ent). **Coherent light** is light in which all of the light waves have the same wavelength. They are also "in phase."

Waves that are in phase have their peaks and valleys aligned. In other words, all the waves of laser light are in step with each other.

Waves of laser light are like the members of a marching band, where everyone moves in step. In contrast, waves of ordinary light are incoherent. The waves of incoherent light are like many people leaving a store, with each person then walking in a different direction. Fig. 18-2.

Laser light is bright. The light waves from a laser work together to produce a high-energy beam that is brighter than any ordinary light source. A 60-watt lightbulb produces enough light for reading. A 60-watt laser can melt metal.

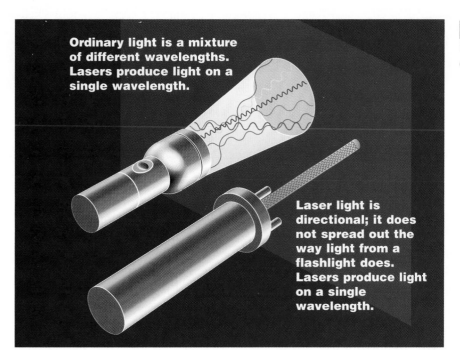

▶ **Fig. 18-2** Laser light compared with ordinary light.

Ordinary light is a mixture of different wavelengths. Lasers produce light on a single wavelength.

Laser light is directional; it does not spread out the way light from a flashlight does. Lasers produce light on a single wavelength.

Laser Safety

Lasers are safe when used properly. Follow the safety guidelines included with the laser you use. Check with your teacher if you have questions.

Linking to COMMUNICATION

Figures of Speech. Writing and speech can be made more interesting and informative through the use of figures of speech.

"Light waves are like ocean waves."

"The waves of laser light are like the members of a marching band, where everyone moves in step."

"Waves of incoherent light are like many people leaving a store, with each person then walking in a different direction."

These comparisons are called similes. A *simile* (SIM-uh-lee) is a comparison using "like" or "as." Similes help us to understand ideas more clearly. Write a simile for each of the following terms: fiber optic, CD, and hologram.

Lasers are grouped into four classes according to the hazard they present. *Class I lasers* produce no known hazard to people. The lasers used in supermarket checkout scanners, laser printers, and CD players are Class I lasers.

Class II lasers can cause eye damage if not used properly. The helium-neon lasers usually used in technology education are Class II lasers.

Safety Note
Do not stare at a laser light or aim it at another person!

Class III lasers produce a powerful beam that can damage the eyes. Special glasses must be worn when using Class III lasers.

Class IV lasers are more powerful than Class III lasers. They can burn the skin. Special glasses are also required with Class IV lasers.

Explore

Design and Build a Prism System

State the Problem

Determine the colors in light by duplicating an experiment performed by Isaac Newton in 1666.

Develop Alternative Solutions

You will need to shine light through a prism onto a piece of white cardboard. This will show the colors in light. As light sources, you will be using a flashlight and a laser. Draw some simple sketches showing how you might place the items needed to perform this experiment.

Select the Best Solution

Evaluate the various plans. Select the plan that you think will be the most effective.

1/8" SLIT

PRISM

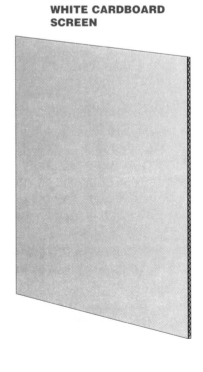

WHITE CARDBOARD SCREEN

Implement the Solution

1. Cut a circle of black construction paper slightly larger than the lens of the flashlight.

2. Fold the circle in half. Cut along the crease.

3. Using masking tape, attach both pieces of paper to the front of the flashlight. Leave a slit of about 1/8" in the center, between the pieces.

4. Place the sheet of white cardboard to create a screen in a darkened area of the room.

5. Turn on the flashlight. Shine a narrow beam of light through one of the prisms so that a rainbow of colors shows on the white cardboard screen. You may need to adjust the distance between the prism and the cardboard. How many colors do you see? Make a diagram that shows the colors in the order they appear. Label the colors.

6. Set up the second prism. Direct the light from the first prism through the second prism. What do you see now on the cardboard screen?

7. Repeat the first experiment. This time use the laser as the light source. How many colors do you see? Why?

Evaluate the Solution

1. What difference, if any, did you notice when you directed the light through the second prism?

2. What were the color differences between the prism light produced by the flashlight and the prism light produced by the laser?

3. Write a brief report (150-200 words) on this experiment. In your report, describe the outcomes of the various stages of your experiment. Be sure to discuss any differences in the bands of light produced by the different light sources and prisms. Write your report in complete sentences. It should contain no spelling errors.

Compact Discs

Compact discs (CDs) have changed the way we listen to music and obtain information. CDs hold information in a spiral track. When you play a CD, a laser beam reads the track. It converts the information into electrical signals. These signals then reproduce the sound, pictures, and text on the disc.

INPUT

Electronic pulses are converted to short bursts of strong laser light. These cut pits in the surface of the spinning disc. The smooth spaces between pits are called flats.

MASTER DISC

CD STATISTICS

Disc diameter: 12 cm

Disc speed: 500 turns per minute

Recording track width: Narrower than a human hair

Recording track length: More than 5 kilometers

Number of pits: Billions

Width of a pit: About one-hundredth the breadth of a human hair

As the CD turns, a low-power laser is focused from below on the spiral track of the CD.

The reflected light falls on light sensors. The presence (or absence) of light causes the sensors to give out an on/off digital signal. This digital signal is then converted into sound, images, and words. The result is a multimedia CD.

OUTPUT

The flats reflect the laser light. When light hits a pit, it scatters.

How Lasers Work

The laser is a system designed to produce a special kind of light. Its subsystems include an excitation mechanism, an active medium, and a feedback mechanism. Fig. 18-3.

All lasers need an energy source. The energy can be produced by electricity, a chemical reaction, or it can be light from another source. This energy comes from the *excitation mechanism* of the laser. This is one of the basic subsystems of a laser.

The *active medium* changes the energy from the excitation mechanism to light and amplifies (strengthens) it. The active medium is made of a material that can absorb and release energy. It can be a solid, a liquid, a gas, or a semiconductor.

The *feedback mechanism* usually consists of two mirrors placed at each end of the active medium. The mirrors are used to build the strength of the laser beam. One mirror is made to allow some of the light to escape the active medium. This mirror is called an *output coupler*.

Since light travels at 186,000 miles per second, a laser process occurs in a fraction of a second. The laser process is shown in Fig. 18-4.

Types of Lasers

The light that comes from a laser can be continuous or released in short pulses. Pulsed lasers are powerful because their energy is very concentrated.

The strength of a laser is measured in watts. The laser you use in technology education produces less than 1/1000 of a watt. Some lasers being used for research can produce millions of watts of energy. These lasers fill an entire building.

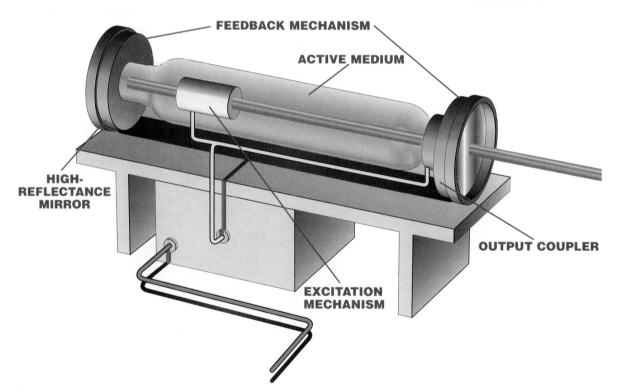

Fig. 18-3 A basic laser system.

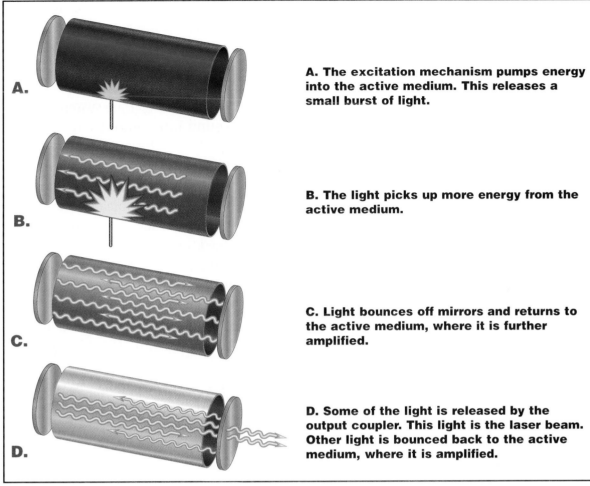

A. The excitation mechanism pumps energy into the active medium. This releases a small burst of light.

B. The light picks up more energy from the active medium.

C. Light bounces off mirrors and returns to the active medium, where it is further amplified.

D. Some of the light is released by the output coupler. This light is the laser beam. Other light is bounced back to the active medium, where it is amplified.

Fig. 18-4 The laser process releases energy as a burst of light.

Lasers are usually named after their active medium. For example, a solid piece of ruby is the active medium in a ruby laser. Carbon dioxide gas is the active medium in a carbon dioxide laser. Semiconductor lasers use electronic components as their active medium. The helium-neon (HeNe) laser uses a mixture of the two gases. This is the type of laser that is usually used in technology education classes.

FASCINATING FACTS

In 1969, astronauts from the Apollo 11 Moon Mission left a mirror on the Moon. Later, scientists on Earth aimed pulses of laser light through a telescope at the mirror. They measured the time it took for the pulse to be reflected off the mirror and return to Earth. Because light travels at a constant speed, scientists could measure the distance from the Earth to the Moon very accurately.

Laser Applications

Lasers are bringing about dramatic changes in technology. Engineers are finding new uses for existing lasers. They are also working to develop new kinds of laser systems. Lasers are now used in many different applications. They are widely used in communication, manufacturing, construction, medicine, and business.

Communication

The most important use of lasers in communication is in fiber optics. In 1880, four years after inventing the telephone, Alexander Graham Bell proved that light could carry sound from one place to another. Today, fiber-optic systems are used to transmit audio, video, and data information. This information is sent as coded light pulses.

Laser fiber-optic systems are replacing conventional metallic-cable communication systems. They are less expensive and more efficient. They are also less subject to interference (static).

In a fiber-optic telephone system, a laser changes sound input into a series of light pulses. Fig. 18-5. These pulses are called *bits*. They travel along optical fiber at 90 million bits per second. Along their route, the pulses are electronically amplified. When the pulses reach a receiver, they are changed back into sound.

Holography (hole-OG-ruh-fee) is a photographic process. It uses a laser as well as lenses and mirrors to produce three-dimensional images. These images are called *holograms* (HOLE-o-grams). Fig. 18-6.

Holograms are used in art, advertising, business, and industry. Most credit cards now have embossed holograms on them.

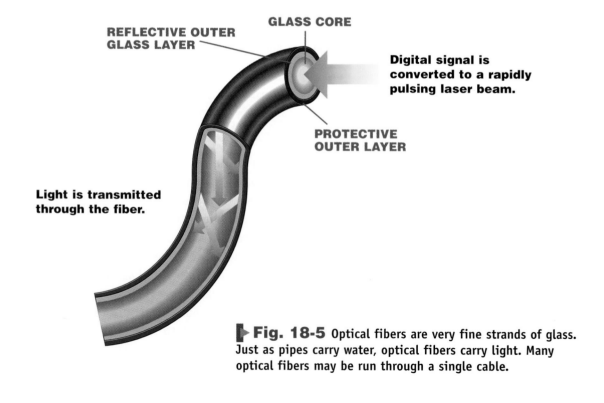

GLASS CORE

REFLECTIVE OUTER
GLASS LAYER

Digital signal is
converted to a rapidly
pulsing laser beam.

PROTECTIVE
OUTER LAYER

Light is transmitted
through the fiber.

Fig. 18-5 Optical fibers are very fine strands of glass. Just as pipes carry water, optical fibers carry light. Many optical fibers may be run through a single cable.

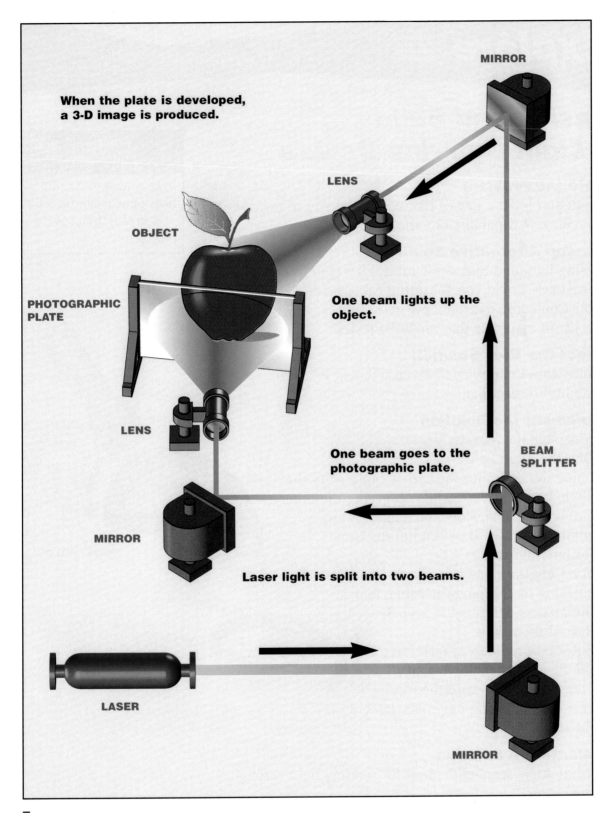

When the plate is developed, a 3-D image is produced.

MIRROR

LENS

OBJECT

One beam lights up the object.

PHOTOGRAPHIC PLATE

LENS

One beam goes to the photographic plate.

BEAM SPLITTER

MIRROR

Laser light is split into two beams.

LASER

MIRROR

▶ **Fig. 18-6** The holographic system. This system is used to make holograms.

Explore

Design and Build a Light-Carrying Device

State the Problem
Design and build a device that will allow you to test the light-carrying capability of various materials.

Develop Alternative Solutions
In using light and fiber-optic cables, try to consider alternatives. Could you substitute cable made of string or wool? Could you use a hollow item, such as a straw? Draw up a list of materials that might be tested.

Select the Best Solution
Identify those materials that you will test for their capability to carry light.

Implement the Solution
1. Assemble the parts as shown.
2. Dim the room lights.
3. Shine the light into the end of cable. Vary the angle from 0° to 90° while observing the other end of the cable. How much light, if any, is coming from the other end of the cable? Record your results.
4. Try tying the cable into simple knots. Avoid cracking or crimping the cable. Repeat the process described in Step 3. Record the results.
5. Substitute a different material for the fiber-optic cable. Repeat Steps 3 and 4. Record the results.
6. Test the remaining materials. Record the results.

Evaluate the Solution
1. What angle seemed the best for sending light through the cable?
2. Did bending the cable have any effect?
3. Compare the results for the other materials tested. What conclusions can you draw?

light source
fiber-optic cable (at least 3' in length)
protractor

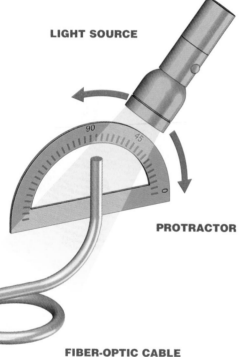

LIGHT SOURCE

PROTRACTOR

FIBER-OPTIC CABLE

The embossed holograms are designed to discourage forgers from trying to make copies of the cards. Holograms can also be used to check for hidden cracks in aircraft parts.

Manufacturing

Lasers have now become valuable manufacturing tools. They are being used for material processing and measurement.

Lasers can drill holes in most materials, including rubber, wood, metal, diamonds, and paper. What's more, they can drill holes more smoothly than standard drill bits can. Laser-drilled items include aerosol spray nozzles, and contact lenses.

Industrial carbon dioxide lasers can cut heavy steel plate more accurately than any saw. Lasers are used in meters. Fig. 18-7.

Automated laser welding systems are used in making automobiles and ships.

Construction

Surveying is the measurement of the boundaries of a piece of land. A laser

Linking to MATHEMATICS

Estimating. You are the purchasing agent for a clothing manufacturer. One bolt of men's suiting material contains 15 yards of material. Each man's suit requires 3 1/2 yards of material. It takes 1 1/2 hours to align the material and prepare it for cutting. The laser cuts the material for 500 suits at one time. It takes the laser 30 minutes to make this cut.

How many suits can be made from one bolt?

How many bolts are needed for one laser cutting?

How many suits can be cut in an 8-hour day?

How many bolts of material will you need to have in inventory for the next 8-hour day? ▲

beam can be used as a straight line in surveying. Distances are measured by timing a light pulse from the laser to a mirror and back to a detector near the laser. Laser beams are being used to align water and sewer pipes and tunnels.

▶ **Fig. 18-7** This velocity meter uses an industrial laser to measure the speed of this escaping gas.

Fig. 18-8 The laser transit allows extremely accurate measurements in surveying.

Lasers can guide the equipment used to level a construction site. For example, a rotating laser sends out a signal that is picked up by a receiver on a bulldozer. In this way the bulldozer operator receives information that the land level is low, high, or on grade (level). The blade of the bulldozer is automatically adjusted by the laser. Similar systems are used to prepare farmland for planting. Laser transits are used in surveying. Fig. 18-8.

Medicine

One of the best uses for lasers is in surgery. Lasers have revolutionized many traditional operations. They have led to the development of new surgical techniques. Advantages of laser surgery include less blood loss, reduced infection risk, and less patient discomfort.

Lasers are particularly useful for eye surgery. They cut more accurately than a scalpel. There is also less damage to nearby tissue. Often the surgery can be performed in the doctor's office. The patient can resume nearly normal activities the same day. Fig. 18-9.

Physicians use lasers to treat skin diseases, for cosmetic surgery, and to remove tumors and birthmarks. Recently, lasers have been used to open clogged arteries. Lasers are being used by dentists for drilling and cleaning teeth and for treating gum disease.

Fig. 18-9 A thin laser is used in delicate eye surgery, where absolute precision is essential.

Business

Your local supermarket is where you are most likely to see a laser in action. Most of the items sold there have a Uniform Product Code (UPC) printed on the package. A UPC symbol consists of a bar code, parallel lines of varying widths.

A laser scanner at the checkout counter reads each symbol. As each item is moved across the scanner, information from the item's bar code is sent to a computer. The computer identifies the item. It then signals the cash register to print the name and price of the item. At the same time, the information is used to update the supermarket's inventory. This helps managers pinpoint when it is time to reorder.

Many offices now have a laser printer. Laser printers use a low-power laser to form images on a rotating drum. Powdered ink adheres (sticks) to the image formed on the drum. This image is transferred to paper. Laser printers print quickly and quietly. They produce high-quality copies.

IMPACTS

Lasers have changed modern life in many ways. The military uses laser guidance systems to aim handheld weapons and deliver bombs and missiles. In museums, lasers are used to clean valuable paintings and sculptures. Some concerts include laser light shows choreographed to the beat of the music. Fig. 18-10.

THE FUTURE

Research and development continue to produce new applications for lasers. The advantages of laser surgery are encouraging doctors to refine laser-surgery techniques. Most stores will switch to laser checkout systems. Such systems allow them to improve customer service and inventory recordkeeping. Since lasers are becoming less expensive, we can expect to see them more in everyday use. For lasers, the future is bright!

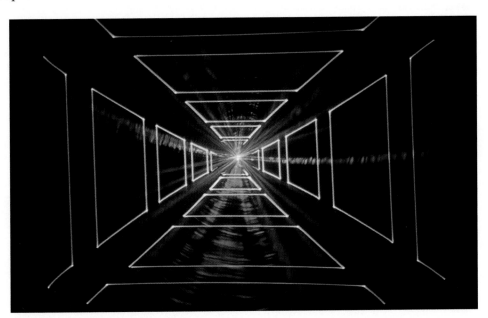

▶ **Fig. 18-10** The versatility of lasers suggests uses in entertainment.

Design and Build a Lighted Monogram

State the Problem

Design and build a lighted monogram, or letter, using the principles of fiber-optic light transmission.

Develop Alternative Solutions

Develop a series of sketches for the letter. Avoid tight bends and multiple pieces. Develop a method to secure the letter to a base or wall plaque.

Select the Best Solution

Select the most pleasing monogram design.

Implement the Solution

1. Using graph paper, prepare a full-scale drawing of the monogram. This drawing will be the pattern on which you will shape the plastic rod.

Collect Materials and Equipment
Monogram
plastic rod, round 1/4" - 1/2" diameter
plastic solvent
Light source
two "D" cells, lamp and socket
cardboard tube
Mounting plaque
3/4" wood
1/4" dowels
Bending form
3/4" plywood
1/4" dowel pins
Equipment
plastic material processing equipment
plastics oven

A. Bending Form

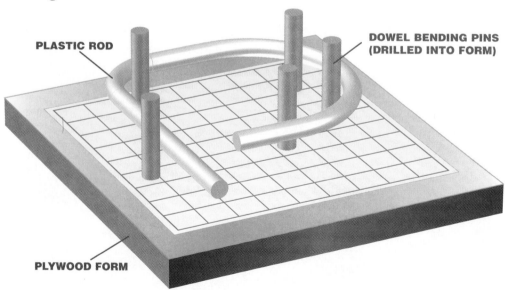

PLASTIC ROD

DOWEL BENDING PINS (DRILLED INTO FORM)

PLYWOOD FORM

B. Monogram Display

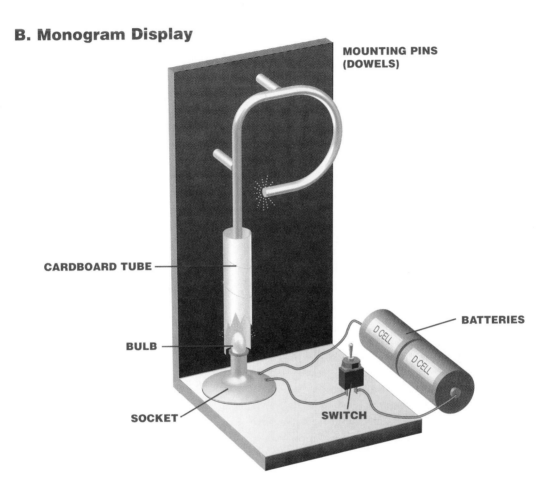

MOUNTING PINS (DOWELS)

CARDBOARD TUBE

BATTERIES

D CELL

D CELL

BULB

SOCKET

SWITCH

2. Tape the graph paper to a piece of 3/4" plywood. Place dowel pins (1/4") at strategic locations where the rod has to be bent.

3. Using string or a flexible ruler, calculate the length of rod you will need for the monogram.

4. Heat the rod to the proper temperature.

5. Using gloves, remove the now flexible rod from the oven. Bend it around the form you created. Hold in place until cool.

6. Mount the monogram to the stand or plaque.

7. Install the light source.

Evaluate the Solution

Does the light flow through the entire plastic tube? Does the light flow through some areas better than others?

CAREERS IN
Lasers and Fiber Optics

LASER-BEAM WELDERS

Growing controls manufacturer needs welder to operate laser-beam machine to weld metal components. Monitor machine activities and perform general maintenance. Must have necessary skills to plan and lay out work from drawings, blueprints, or other written specifications. Salary and benefits. Submit resume to: Merstech Industries, 6770 Snoville Road, Rockford, IL

COMMUNICATIONS EQUIPMENT MECHANIC

Norstand is a full-range provider of integrated voice, video, and data communication business products. Immediate opening for a mechanic for remote troubleshooting of calls for equipment service and repair. Computer or telephone background and strong analytical skills preferred. Send resume to: Support Central Manager, Norstand, 8832 West 49th Street, Sylvania, MD 22011.

SERVICE SALES REPRESENTATIVE

Computer company has opening for service sales representative with excellent technical and communication skills. Successful rep will be involved in initial sales and continued service. Will be responsible for equipment setup and maintenance. Will assist in cable installation for network systems. To apply, forward resume to: Computer Source, 5454 Spring Street, Boulder, CO 70348.

FIELD TECHNICIAN

Scantron Service Group, a growing laser service company, is seeking field service technicians for the Charlotte area. Technicians are needed to respond to service requests and perform preventive maintenance on laser equipment. Excellent communications/customer service skills and a clean driving record are essential. Competitive salary and company vehicle. Please mail resume to: Phil Adams, P.O. Box 778, Charlotte, NC 32002.

FIBER OPTIC TECHNICIAN

Opening for technician to assist in the research and development of fiber optics technology. Work in laboratory setting performing detailed, precise work. Two-year degree required with knowledge of fiber optics. Good technical skills preferred. Please send resume to: Palmer Fiber Optics, Inc., 348 Westside Drive, San Mateo, CA 90550.

Linking to the WORKPLACE

Lasers are used in a variety of ways. List workplace locations where lasers are used. Consider the following locations: hospitals, supermarkets, banks, factories, and military bases. Can you think of other places where lasers are being used? Identify possible uses for lasers in the locations you have listed.

Chapter 18 Review

SUMMARY

▶ Light is a type of electromagnetic wave. Electromagnetic waves are classified by wavelength.

▶ The term *laser* is short for "*L*ight *A*mplification by *S*timulated *E*mission of *R*adiation."

▶ Basic subsystems of the laser are an excitation mechanism, an active medium, and a feedback mechanism.

▶ Laser light differs from ordinary light. Laser light is monochromatic, directional, coherent, and bright. Ordinary light is a mixture of many colors. It also spreads out, is incoherent, and is less bright.

▶ Lasers are grouped into four classes according to the hazard they present.

▶ Lasers are usually named after their active medium.

▶ Lasers are used in fiber-optic communication systems. They are used to read information contained on compact discs. They are used in holography to create three-dimensional images.

▶ Lasers are used in manufacturing, construction, medicine, and business.

CHECK YOUR FACTS

1. What does the term *laser* stand for?

2. List four characteristics of laser light.

3. How does laser light differ from ordinary light?

4. Name the type of laser that is usually used in technology education.

5. Describe several ways lasers are being used in communication, manufacturing, construction, medicine, and business.

6. How can lasers be used for material processing?

7. List two advantages of laser surgery.

CRITICAL THINKING

1. Name the three basic subsystems of a laser. Explain how these subsystems work together to produce laser light.

2. Identify some ways in which lasers might be used in the future.

Engineering

OBJECTIVES

▶ describe the basic job of an engineer.

▶ discuss the specific jobs performed by different types of engineers.

▶ identify and explain the basic steps in the engineering process.

KEY TERMS

engineer

engineering

engineering process

standard

Have you ever made a paper airplane? Have you ever straightened out a paper clip—and then reshaped it to what you thought was a better design? Have you wrapped a present when there was just enough paper to cover the box? Have you ever figured out the best way to squeeze the last bit of toothpaste from a toothpaste tube? If you have done any of these things, you have solved an engineering problem.

WHAT IS ENGINEERING?

What everyday item do you think represents excellence in design and engineering? If you ask an engineer this question, you might be surprised by the answer. For example, he or she might suggest something as simple as a paper clip.

Everybody understands how the paper clip works. Instructions are not necessary. Paper clips can be used without looking at them—simply by feel. They are lightweight, durable, and reusable. They are also easy to make and inexpensive.

You might wonder how something so simple could represent the complex process of engineering. Often, we take such everyday items for granted. We sometimes fail to recognize the beauty of a simple design.

The paper clip is an engineering marvel, but it didn't happen by chance. The common paper clip is a product of the engineering process. Fig. 19-1.

Engineering is a process used to develop solutions to problems. Solutions come in the form of products, methods, or designs used to satisfy a need. An **engineer** is a person who uses his or her knowledge of science, technology, and mathematics to solve technical problems.

Look around. Most of the things that make up our people-made world were influenced by engineers. Even the making of this textbook was influenced by engineers and the engineering process. *Computer engineers* designed the software and hardware used to lay out the text and graphics on each page. *Mechanical engineers* helped to design the

machines that printed the pages. *Chemical engineers* developed the inks that transferred the printed images to the paper. They also helped plan the

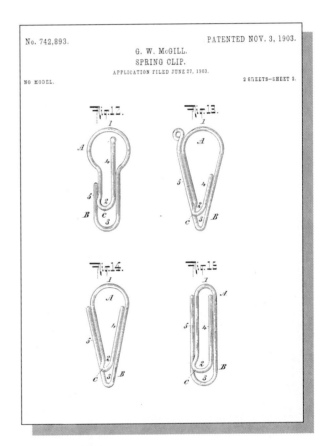

▶ **Fig. 19-1** These drawings were filed with the application for a paper clip patent. Even simple things like a paper clip go through many design changes until the best possible design is found.

composition of the paper used to print the book. Fig. 19-2. *Systems engineers* oversaw the entire publication process. This required printing, binding, and shipping the textbooks.

TYPES OF ENGINEERING

The engineering profession changes as new needs develop. The driving force behind this change is technology. As new technologies emerge, new types of engineering evolve.

"Classic" Engineering

Civil, mechanical, mining, chemical, and electrical engineering are referred to as "classic" engineering. Formal programs designed to train these professionals began in the 1700s. The Industrial Revolution fueled the need for people who could design machines and create the *infrastructure* (roads, bridges, tunnels) that a growing technological society needs.

Civil engineering is the oldest engineering profession in the United States. The word "civil" refers to public works. *Civil engineers* work with architects to design and build government buildings, airports, water-treatment plants, highways, bridges, and other structures used by the public.

Machines and machine parts are designed by *mechanical engineers*. Among other things, mechanical engineers play a vital role in the development of robots used in factories.

Within a branch of engineering, there might be different specialty areas. For example, *mining engineers* design and manage the processes used to remove minerals from the ground. Some mining engineers specialize in processing metals or refining petroleum.

▶ **Fig. 19-2** The work of engineers influences the design of every people-made product. Engineers also play a role in the design of the tools, machines, and processes used to create products.

Engineers often work together on projects. *Chemical engineers*, who design the chemical reactions that produce products, might work with mechanical engineers to design the vessels in which the chemical reactions take place.

Electrical engineers (and *electronics engineers*) design circuits and power supplies for electrical and electronic devices. They are concerned with producing electrical energy and developing more efficient motors, lighting, and machines that use electricity. Fig. 19-3.

A New Generation of Engineering

The twentieth century brought new technologies. With these technologies came a new generation of engineers. Many of these engineers focus their design efforts on manufacturing, transportation, energy sources, computers, and the environment.

Manufacturing engineers try to find the best way to produce a product. Their goal is to increase productivity. *Productivity* is the relationship between the number of hours worked and the quantity of products made. Manufacturing engineers design and organize the placements of machines on the assembly line. They also develop the sequence of processes used to transform materials into products.

The people responsible for designing and building new types of aircraft are called *aerospace engineers*. These people

▶ **Fig. 19-3** Products ranging from blow dryers to air conditioners require circuits to control the flow of electricity. Electrical engineers design these electrical circuits.

focus their design work on engine design, the frame of the aircraft, the guidance and control systems, and other flight-related systems.

New technologies require new materials. *Materials engineers* are responsible for developing materials with specific properties that meet a manufacturer's needs. Properties such as strength, conductivity, and hardness can be engineered into a new material.

Environmental engineering came about as a result of the rapid growth in manufacturing. The congestion of city life and the reckless use of our natural resources increased land, air, and water pollution. *Environmental engineers* work for the government or private industry. Their job requires them to design cleaner production methods that will meet strict government regulations. Unfortunately, a great deal of pollution was caused before society really understood the dangers. Environmental engineers also design systems to clean up polluted sites. Fig. 19-4.

Computer engineering is a spin-off of electronics engineering. Some *computer engineers* specialize in hardware or computer design. Others specialize in software design. Companies hire computer engineers to design, install, and maintain large computer systems.

There are other engineering specialties. As new technologies emerge, new kinds of engineers are always needed.

▶ **Fig. 19-4** Environmental engineers design the processes used to clean up polluted areas.

Architects and Industrial Designers

Not all designs are made by engineers. Many companies employ architects and industrial designers. These professionals also use their imagination and knowledge to create designs. *Architects* design buildings. They are experts on how buildings are constructed. Architects may specialize in designing houses, skyscrapers, schools, churches, or hospitals. Fig. 19-5.

Industrial designers, sometimes called product designers, specialize in improving designs that already exist. Their attention is focused on how the product looks and works and how it might be improved.

Fig. 19-5 Architects use engineering skills to design buildings.

Explore

Design and Build a Jack-in-the-Box

State the Problem

Design a Jack-in-the-box with a mechanism that allows the spring-loaded "Jack" to pop up after each turn of the crank.

Develop Alternative Solutions

How many ways are there to release a hatch? Are all hatches spring-loaded? Do you think that all Jack-in-the-box toys operate in the same way? Prepare several possible designs for the Jack-in-the-box hatch release. Discuss the problem with your instructor and within the group. Make certain the assignment is clearly understood. Most of the parts for one possible design are shown in Fig. A. Note that the release mechanism is not shown.

Select the Best Solution

Compare your various designs. Select the one that you think will be most effective.

Implement the Solution

1. Build the Jack-in-the-box.
2. Install the release mechanism. Follow the design you have chosen.
3. Test the finished model.
4. Present your finished model to the class.

Evaluate the Solution

1. Does the crank work properly?
2. Does the hatch-release mechanism work?
3. How might the mechanism be improved?

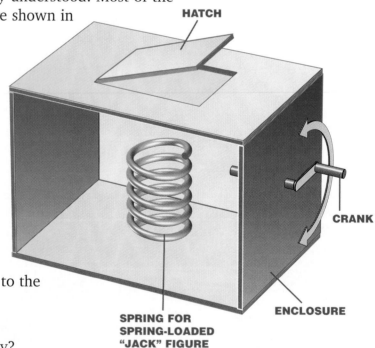

HATCH

CRANK

ENCLOSURE

SPRING FOR SPRING-LOADED "JACK" FIGURE

Fig. A

THE ENGINEERING PROCESS

As you have seen, engineers do many things. They build bridges, design airplanes, run factories, improve machines, and clean up toxic-waste sites. Regardless of the type of engineer, all engineers basically do the same thing— they design. Even engineers working in operations, maintenance, or construction are designers.

All engineers solve problems. After thinking of a possible solution, the engineer has to turn the idea into a product, process, or system. Engineers follow an established path in solving a problem. They perform many different tasks on the way to a design solution.

The **engineering process** is a problem-solving process and a design process. This process requires the engineer to state the problem, gather information, develop alternative solutions, choose the best solution, implement the solution, and evaluate the solution.

State the Problem

The first step in the engineering process is to *state the problem*, or identify the need. The engineer must identify the problem. He or she must also understand the true nature of the problem. For example, a civil engineer might be asked to redesign an entrance ramp to a highway. What are the safety needs? How can access to the ramp be made easier? The engineer must be able to answer these and other questions before he or she can think about a solution.

Gather Information

As the engineer develops ideas, new questions and concerns will arise. The second step in the process is to *gather information*. This involves collecting materials and equipment. For example, an electrical engineer is asked to develop a cordless headset for a portable stereo. The first thing he or she might do is to investigate the solutions others used to solve a similar problem.

The engineer might search books and trade magazines for new technologies that might offer solutions to the design needs. He or she might perform some experiments or tests. Fig. 19-6. Gathering information requires that questions be asked and answers be found.

▶ **Fig. 19-6** The testing of aircraft parts in wind tunnels helps engineers evaluate the strength of those parts. Here, a wing section is being prepared for testing.

Develop Alternative Solutions

The third step in the process is to *develop alternative solutions*. There are several ways to develop solutions. *Brainstorming* is thinking of ideas that may solve the problem, but deferring judgment concerning their worth. When people brainstorm they express ideas out loud. A team member acts as a recorder, writing down every suggested idea.

The goal is to record as many ideas as possible within a certain time period. The most important rule in brainstorming is to keep the ideas flowing. Every idea should be allowed. No idea should be criticized during the brainstorming process.

Freehand sketching is another technique that can be used to generate design ideas. Drawing shapes and building upon them with new shapes can spark a great design idea.

Categorizing is another technique used to develop design ideas. The technique requires making a chart on a sheet of paper. The top of the chart carries headings that match the design criteria. *Criteria* are specifications. The headings are the idea categories. When the process begins, items that fit into each category are listed. The process is finished by connecting strings of ideas, one from each category. You can quickly see that many combinations of ideas are possible.

Having evaluated all of the possible solutions, the engineer selects the best idea for a solution.

Before deciding on the solution, the engineer may find that he or she may need to make trade-offs. A trade-off is giving up one thing to get another. Engineers must stay within budgets and use certain materials. In addition, they must deal with safety issues. Perfect solutions are hard to find.

For example, a computer hardware engineer may be asked to develop a handheld game for children. The engineer is told that the game's selling price cannot exceed $20. The engineer's first choice is to use a color display. However, this will raise the selling price of the game to $35. Therefore, a black-and-white display must be used. The color display would make the game more interesting, but more expensive. The engineer has to trade a color display for a lower selling price.

Select the Best Solution

After all ideas have been presented, you need to select the best solution. To select the best solution, you must compare the ideas generated with the criteria assigned to the product. The solutions that match the criteria are the best ideas. Note, though, that the best solution is not always the most appropriate solution. For example, a solution that meets all the design criteria might be unsafe or too difficult or expensive to manufacture. A design checklist is one way to arrive at design decisions.

Implement the Solution

After a solution to the problem has been selected, the next step is to *implement the solution*. Implementing the solution will require you to prepare careful drawings.

An *instrument drawing* is an accurate, precise drawing of a design idea. Engineers or drafters prepare the drawings showing the size, shape, and details of the object. The drawings are often prepared using computer-aided design and drawing software. Some computer systems can animate the drawing so designers can actually see the product in motion.

You may also need to build a model. Engineering models have many purposes. Models help the designer work out details and test the solution idea. Models also help to communicate the design solution to other people who may also be working on the project. Modeling design ideas can be accomplished in a variety of different ways.

A *working model* may not look like the final product, but it is constructed to work like the final design solution. A working model will give the designer a good feel for the product's shape and size. The final product, however, might end up being made from a different material. Fig. 19-7.

A *prototype* is a working model of a design idea. Usually handmade, it looks and works like the finished product. A prototype is evaluated as a final test before manufacturing takes place.

Evaluate the Solution

The next step in the engineering process is to *evaluate the solution*. This evaluation judges how well the solution worked. In other words, it checks its effectiveness and efficiency. It determines if the solution solves the problem. The results of the evaluation are compared to standards. A **standard** is a rule or model against which a product, action, or process can be compared. Standards serve as a gauge for the measure of criteria such as quantity, weight, extent, or value.

▶ **Fig. 19-7** Testing is an important step in choosing materials for a product.

Explore

Design and Build a Shelf for a School Locker

State the Problem

Design a shelf suitable for a school locker. The shelf must be easy to install and strong enough to support several textbooks. It must be inexpensive and fairly easy to make. (You will again be playing the role of an engineer. Each engineering team in your classroom, as determined by your teacher, will be responsible for its own design.)

Develop Alternative Solutions

Investigate the interior of a locker. Ask yourself questions such as: Will the locker shelf be one piece? Can a shelf be designed for easy installation and removal? Is it possible to make the shelf so that it fits lockers of different sizes? Can a two-tiered shelf be made? Sketch your ideas on paper.

Select the Best Solution

Evaluate the various sketches. Select the design that you think will be most effective.

Implement the Solution

1. Build your proposed shelf.
2. Build a mock-up (a model just to test size and appearance). Make changes as desired.
3. Build prototype (actual working model).
4. Test the prototype.
5. Present your finished model to the class.

Evaluate the Solution

1. Does the shelf install easily?
2. Does the shelf support several textbooks?
3. Can the shelf be removed without damage?

Collect Materials and Equipment
Materials
sheet metal or other available material for shelf
fasteners and other hardware for mounting shelf
Equipment
miscellaneous hand tools

The results of the evaluation will determine whether the design needs to be changed and tested again. Fig. 19-8.

USING THE ENGINEERING PROCESS: A CASE STUDY

What would it be like to work through the engineering process and actually design a product? Imagine that you are an engineer for the National Aeronautics and Space Administration (NASA). You are working at a spaceflight center. A group of scientists wants to conduct an experiment on board the Space Shuttle Orbiter. The scientists want to know if, unlike on Earth, oil and water will mix (and stay mixed) in a microgravity. You have been assigned the job of designing the experiment.

NAME: _____ AGE: _____ HEIGHT: _____		
GRADE: _____ TEACHER: _____		
	Yes	No
Was the chair the correct height for you?	✓	
Was the chair seat comfortable?	✓	
Was the chair back comfortable?	✓	
Was the chair too heavy to lift?		✓
Did the chair glide on the floor easily?	✓	
Did the chair scratch the floor?		✓
Did your book bag fit under the chair?		✓
Was the chair comfortable?	✓	

▶ **Fig. 19-8** This checklist helps engineers evaluate the design of a chair.

Linking to SCIENCE

Density. Oil floats on water because of a difference in density. Oil has less mass per unit volume than water. You can demonstrate this difference using water and vegetable oil. Half-fill a clear glass container with water. Pour a small amount of vegetable oil into the glass. The oil will float on top of the water.

OIL

WATER

Will *oil* and water layer in space as they do on Earth and how *do* we test it?

State the Problem

After studying the scientists' requirements, you determine that the following questions need to be examined:

- How will the items for the experiment be stored until the astronauts are ready to conduct the experiment?
- How will the two liquids be mixed?
- How will the results of the experiment be recorded?

Gather Information

After speaking with other engineers and checking the specifications for shuttle payloads (cargo), you identify new problems. You didn't know that the temperature in the cargo bay ranges from below 0°C [32°F] to above 100°C [212°F]. How will you keep the liquids from freezing and boiling?

Linking to MATHEMATICS

Cubic Inches. The cargo compartment is 300 cubic inches. The depth of the compartment is 10 inches. The width of the compartment is 5 inches. What is the height of the compartment? ▲

Space and power are also in short supply on the Shuttle Orbiter. The equipment for the experiment must fit into a 300-cubic-inch compartment in the cargo bay. Electrical engineers have told you that your experiment can use no more than 5 ampere-hours of electricity. Fig. 19-9.

Develop Alternative Solutions

You have put together a team of engineers to help develop design solutions for the problems you have uncovered. The team is made up of a thermal engineer, a mechanical engineer, and an electrical engineer. As lead engineer, it is up to you to evaluate their ideas.

Choose the Best Solution

After you have evaluated all possible solutions, you will need to choose the best solution. In choosing the solution, you will need to make certain it meets all the criteria.

Implement the Solution

NASA requires that you build a physical model of your experiment. You must use the actual materials and equipment that will be used in the final design. The following are the recommendations made by each of the design team engineers. There were many trade-offs, but they feel the solutions will work. Drawings will

▶ **Fig. 19-9** Here, an astronaut (lower left) spacewalks in the open cargo bay area of the Space Shuttle Orbiter as it passes over Australia. Engineering for spaceflight presents interesting problems. There is no gravity in space. Also, the temperature swings from freezing to boiling many times each day. NASA is very strict about placing hazardous materials aboard the orbiter. Any materials taken on board must be approved by safety engineers.

have to be submitted to the technicians so the Experimental Module can be built.

After the model of the Experimental Module is completed, it will be plugged into a ground-based simulation unit. The unit will provide power and control identical to that found on board the shuttle. The Module will be tested many times and its performance evaluated.

> ▶ **Fig. 19-10** Mechanical engineering provided this sketch. It shows the possible arrangement of the parts needed for the oil/water experiment.

Mechanical Engineering Preliminary Report

1. The two liquids will be kept inside a plastic container (module container) of the type already used on many shuttle missions.
2. The two liquids will be injected at the same time into an empty container by a syringe-like plunger. A solenoid will be used to push the plunger. This will mix the fluids sufficiently.
3. A digital still camera will take one picture every 5 minutes for a period of 1 hour. This will record the settling process. Fig. 19-10.

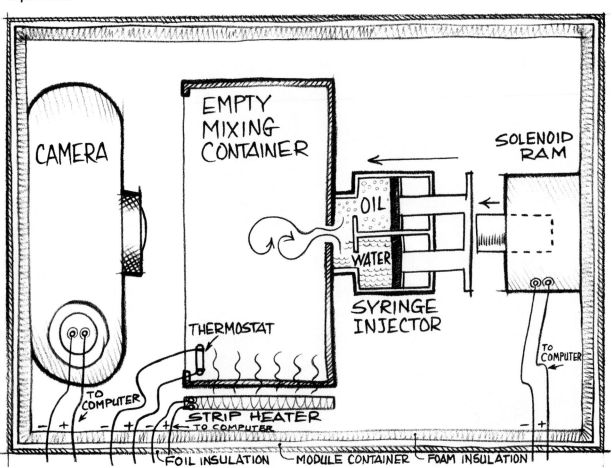

Thermal Engineering Preliminary Report

1. Insulation will be used as a passive means to keep the liquids from freezing.
2. A heater, controlled by a thermostat, will serve as a backup in case the temperature gets too low.
3. Cooling the liquids will take more energy than is available. Insulation will be used to prevent the materials from boiling.

Electrical Engineering Preliminary Report

1. NASA will provide a Module Electronics Unit (MEU). The MEU contains a microprocessor that can be programmed to control the camera and take the sequence of photographs needed.
2. Backup mini-heaters and the camera will use 3 amp-hours of electricity, which is below the specified maximum.

Evaluate the Solution

Now you must make your final recommendations. You must evaluate the solution. You must then report the results of your modeling and testing. Make sure that your report is both well prepared and well presented.

IMPACTS

Engineering is a process that applies to most of the topics discussed in this book. Every product and service had its birth in an idea. Engineering was needed to make the product or service attractive and useful.

Because engineering affects activities in many different areas, its impacts are enormous. One goal of good engineering is to add value to a service or product. It is the use of the product or service that determines its impact on society. Can you think of examples of well-engineered products that draw criticism?

THE FUTURE

Advances in technology have promoted the introduction of many new products. Expanding technologies have driven the need for more engineers. Anyone interested in engineering should take math, science, and technology classes while in high school. This will help prepare the student for advanced courses in college.

Engineers are trained at four-year colleges. Upon completion of the engineering program, the graduate is awarded a bachelor's degree.

Presently, women and minorities are underrepresented in the field of engineering. Only 17% of the students enrolled in engineering schools today are women. The percentage of minorities is even lower. Some people predict that the United States will experience a shortage of engineers.

Apply What You've Learned

Design and Build a Toothpick Dispenser

State the Problem

Design and manufacture a new type of toothpick dispenser. The dispenser should dispense toothpicks one at a time (much like the type seen in restaurants). It should also hold business cards for customers to take. Characteristics of the dispenser should include:

- Sanitary storage of toothpicks.
- Capacity of at least 100 toothpicks.
- Easy to load and dispense toothpicks.
- Able to hold at least 25 business cards.
- Attractive.
- Small enough to fit near cash register on counter.
- Simple to use.

Develop Alternative Solutions

Prepare several designs for a toothpick dispenser. How many moving parts do they have?

Select the Best Solution

Select the design that you think will be most effective.

Implement the Solution

1. If your dispenser has movable parts, build the mechanism. Does it work?
2. Build the case for the toothpick dispenser.
3. Assemble the components.
4. Test the dispenser. Improve performance if possible.
5. Present the dispenser to the class and instructor.

> **Collect Materials and Equipment**
>
> matteboard
> foamcore board
> stiff cardboard
> plastic
> wood
> Easy-to-cut materials will probably prove most satisfactory. Ask your instructor if you need a material that is not immediately available.

Evaluate the Solution

1. Does the dispenser work properly several times in succession?

2. Is the dispenser attractive?

3. Can a consumer easily use the dispenser the first time?

4. Is the dispenser suitable for a counter location?

5. Will the dispenser store at least 100 toothpicks in a sanitary manner?

6. Is the dispenser easy to load?

7. Can 25 business cards be held by the dispenser?

CAREERS IN
Engineering

AEROSPACE ENGINEER

Leading aircraft manufacturer needs engineer to design, construct, analyze, and test aircraft. Bachelor's degree required with specialization in structural design. Good communication skills, CAD experience and excellent math ability needed to work as part of team. Competitive salary. Submit resume to: Dobbins Aircraft Corporation, 5500 Airport Way, Houston, TX 66002.

MINING ENGINEER

Operational coal mining company seeks engineer to ensure environmentally sound, safe, and economical operation of mines. Examine plans and oversee construction of mine shafts and tunnels. Excellent opportunity for engineer interested in field work. Competitive salary and benefits. Send resume to: Blackstone Coal Company, Inc., Route 4, Box 218, Taylorville, PA 30422.

CIVIL ENGINEER

Civil engineering firm seeks engineer experienced in highway and bridge design. Experienced in design and detailing using CAD systems. Salary commensurate with qualifications. Attractive benefits and performance incentives. Submit resume to: Bard Engineering, Inc., 5 East Long Street, Seattle, WA 88642.

CHEMICAL ENGINEER

A leader in the professional beauty industry is seeking a chemical engineer with well-developed skills to plan, design, and evaluate pilot and plant experiments. Should be able to direct major/complex projects of both a chemical and engineering nature. Will also provide training and technical support for plant operations. Bachelor's degree in chemical engineering required. If interested, please send resume to: Matrix, Inc., Dept KL, 10603 Carter Street, Springfield, IL 62704.

MATERIALS ENGINEER

Aircraft parts manufacturing company has opening for materials engineer to lead research team. Responsible for conducting the testing and evaluation of composite materials used for production. Work in laboratory setting. Engineering degree required. Salary plus benefits package. Submit resume to: Airways Manufacturing, Inc., 3355 Airway Drive, St. Louis, MO 62204.

Linking to the WORKPLACE

Let's take a trip into your future. Imagine that you are thirty years old and working in one of the jobs listed above. Close your eyes and imagine what your day will be like. Ask your older self these questions: How much education did you complete? Where are you living? How do you get to work? What hours do you work? What tasks do you perform at work each day? Do you enjoy your job? What changes are you planning for the future?

Now think of a job that you are interested in, and answer these questions again. Can you see how the work you perform has an impact on your daily life?

Chapter 19 Review

SUMMARY

▶ Engineering is a process used to develop solutions to problems.

▶ Solutions come in the form of products, methods, or designs used to satisfy a need.

▶ The engineering profession changes as new needs develop. The driving force behind such change is technology.

▶ All engineers basically do the same thing—they design.

▶ The engineering process is a problem-solving process and a design process.

CHECK YOUR FACTS

1. Describe the basic job performed by all engineers.

2. Discuss the specific jobs performed by different types of engineers.

3. What type of engineer would be employed to design a new delivery system for a vending machine?

4. What type of engineer would be employed to determine the power requirements for a portable radio?

5. Identify and explain the basic steps in the engineering process.

6. Why are many solutions developed at the start of a design problem?

7. What does the term "trade-off" mean, and why do engineers make trade-offs?

8. List two sources an engineer may use to gather information about a topic.

CRITICAL THINKING

1. Develop a series of sketches for the Experimental Module from the Space Case Study. Base your sketches on the engineering team's input and your own ideas. Be sure your sketches are labeled and detailed.

2. You have been asked to design a new book bag. Using the engineering process, explain how you would proceed to solve your design problem. Be specific.

Applied Physics

OBJECTIVES

▶ list the steps in the scientific method.

▶ define the term *applied physics*.

▶ explain the forces that are involved in motion.

▶ state and explain each of Newton's three laws of motion.

▶ explain the difference between work and power.

▶ explain how sound and light waves are different.

KEY TERMS

applied physics

force

friction

gravity

inertia

machine

mechanical advantage

power

work

Do you want to know why things happen and what makes things work? If so, you have the curiosity of a good scientist or engineer. Curiosity is important. Leonardo da Vinci was very curious. In the late 1400s and early 1500s, he questioned the world around him. For example, Leonardo studied the flight of birds. He observed and sketched their movements as they flew. He studied their anatomy to make detailed drawings of their muscles, wings, bone structure, and feathers. Leonardo used the knowledge he gained about flight to create designs for flying machines. He prepared designs for a helicopter, a plane, and a parachute.

MOTION

Science is the study of the natural world. Scientists use the scientific method to gain facts about nature. Fig. 20-1.

Science is divided into many branches, depending upon the topics being studied. *Biology* is a branch of life science that focuses on living things. Chemistry and physics are branches of physical science. *Chemistry* is the study of what substances are made of and how these substances combine. *Physics* is the study of energy and the laws of motion.

This chapter discusses how the laws of nature are used in technological design. Engineers and designers base their design ideas on these principles. This chapter will focus on the study of applied physics. The word "applied" means "to put into action."

Applied physics is the study of the ways in which objects are moved and work is done. Applied physics makes use of the principles that apply to motion and energy. Objects move only when a force is applied to them. The study of motion is a good place to start our exploration of applied physics. Motion involves force, friction, and gravity.

FASCINATING FACTS

Some floating cranes can lift over 5,500 tons. The most powerful floating cranes are used for work on offshore oil rigs. These cranes are used in assembling the rigs. They are mounted on semisubmersible vessels.

Scientific Method
1. State the Problem.
2. Gather Information.
3. Form a Hypothesis.
4. Perform Experiments to Test Hypothesis.
5. Record and Analyze Data.
6. State Conclusion.

▶ **Fig. 20-1** Scientists investigate the laws of nature in a systematic manner. The system they use is known as the scientific method. The basic steps in the scientific method are listed here.

Design and Build a Crane

State the Problem
Design and model a tabletop crane that can be used to lift panels into place. The crane must increase your mechanical advantage and ability to do work.

Develop Alternative Solutions
Gather photos showing cranes used to lift heavy objects. Also gather information on how to increase mechanical advantages. Determine the size of the crane you will build. Sketch possible design solutions. Include the frame and body of the crane. Sketch the pulley combinations you will use. Show how they will connect to the frame of the crane.

Select the Best Solution
Select the design that you think will be most effective.

Implement the Solution
1. Build the crane.
2. Hang a 1000-gram weight from the crane. Lift the weight 10" off the tabletop.
3. Calculate the work accomplished by your crane.

Evaluate the Solution
1. Was the crane able to lift the weight?
2. Was the crane able to support the weight?

Collect Materials and Equipment
1/8" x 1/8" balsa strips
string
pulleys
cardboard
dowels
glue
modeling materials
material processing tools and machines
1000-gram weight

Force

A **force** is a push or pull applied to an object. Forces affect how an object moves by giving energy to it or taking energy from it. Once an object is moving, a force can cause it to slow down, speed up, or change direction.

Friction

Friction is a force that opposes motion. It acts in the direction opposite to the direction of the motion. Friction is the force that brings an object to rest.

Friction occurs when two surfaces come in contact with each other. There are different types of friction. *Surface friction* occurs when one surface slides over another surface. *Rolling friction* takes place when one surface rolls over another surface. *Fluid friction* takes place when an object moves through a fluid. Remember that both air and water are considered to be fluids.

Friction and Technology

How can you make the force of friction work for you?

When you operate the hand brakes on a bicycle, you use the force of surface friction to slow and stop the motion of the bike. Operating the hand brake presses a rubber brake pad against the rim of the turning wheel. The brake applies a force that slows and eventually stops the motion of the wheel. Fig. 20-2.

▶ **Fig. 20-2** Friction can be a helpful force. Here the friction of one of the bicycle's brake pads against the wheel rim slows the bicycle. This brake pad would be more effective if it were not worn down. The metal clamp that holds it is also contacting the wheel rim, producing sparks.

Inline skates are affected by rolling friction. You've probably noticed that the wheels on inline skates spin very easily. This is because there are small ball bearings between the axle and the wheels. These bearings reduce the friction between the wheels and the axle.

When automobiles and airplanes move at high speed, they encounter *drag*. Drag is a form of fluid friction. Molecules of air bump into the moving vehicle. This friction slows the vehicle. Creating a shape that allows molecules of air to slip easily over the surface of a moving vehicle reduces surface friction. Such a shape is described as being *aerodynamic*.

Gravity

Gravity is the force of attraction that exists between two objects. The amount of gravitational force depends on the size of the objects and the distance between them. For example, the gravitational force between you and the classmate sitting next to you is almost nonexistent. This is because your masses are small and equal to each other. On the other hand, the gravitational force between you and planet Earth is quite great. In fact, the force of gravity actually holds you to the surface of the planet. Fig. 20-3.

Your weight is a measure of the amount of force placed on you by gravity. You

On Earth, an astronaut has a weight of about 30 newtons (N).

On Jupiter, she would weigh 80 N. Jupiter's gravitational force is 2.65 times stronger than on Earth.

On the moon, she weighs 5 newtons (N). The moon's gravitational force is 1/6 as strong as the Earth's.

▶ **Fig. 20-3** Scientists use the *newton* (N) as a unit to measure force. Weight is a force created by gravitational pull. The newton is used to express the force of weight.

would weigh less on the surface of the moon because the mass of the moon is one-sixth that of the Earth. Therefore, the moon's gravitational force is also one-sixth of the force on Earth.

Gravitational Force and Technology

Gravity pulls all objects on our planet towards the center of Earth. You may have heard the old saying, "What goes up, must come down." How can a rocket travel beyond the reach of Earth's gravitational pull? To do this, the rocket must have engines powerful enough to produce enough forward thrust, or force, to reach escape velocity.

Escape velocity is the minimum speed at which an object must travel to escape the gravitational pull of a planet. The escape velocity needed to overcome Earth's gravitational pull is 40,000 km/h. The rocket must reach this speed while moving in the direction opposite that of the gravitational force.

NEWTON'S LAWS OF MOTION

Dynamics is the study of the movement of an object and the forces acting on it. Isaac Newton, an English scientist, studied this relationship in the 1600s. He developed three laws of motion.

Newton's First Law of Motion

Newton's *first law of motion* states that an object at rest will stay at rest and an object in motion will stay in motion. For example, if an object is on a table, it will take an unbalanced force to make it move. An *unbalanced force* is a force strong enough to cause an object to move or change direction. Fig. 20-4. This law holds true for an object that is moving. A

▶ **Fig. 20-4** Newton's first law of motion. A body stays at rest unless an unbalanced force has enough strength to cause the object to move or change direction.

moving object will continue to move unless an unbalanced force is applied to it. The tendency of an object to continue to move or to stay at rest is called **inertia** (in-ERSH-uh).

Newton's first law can be used to explain the effectiveness of an automobile's seat belt system. A car moving down a highway has inertia. Its tendency is to continue to move in a straight line. If the driver slams on the brakes, the car begins to slow down.

Do the passengers in the car also slow down? No, they don't. They also have inertia. They continue to move forward in a straight line until something stops them. Unfortunately, this can be the steering wheel, dashboard, or windshield. If the passengers are wearing seat belts, the belts hold them in place, stopping their forward motion and preventing injury. Fig. 20-5.

Linking to COMMUNICATION

Sayings. The purpose of a saying is to quickly and simply explain a more complicated concept. The English language contains many sayings. A saying is also called an adage, maxim, saw, or proverb. The saying, "What goes up, must come down," is a simple explanation of the force of gravity. Other examples of sayings are, "The early bird gets the worm," "Haste makes waste," and "What goes around, comes around."

Interview your parents, grandparents, teachers, and friends. Record a list of common sayings. Compile a more complete list as you share your findings with your study group. Try writing a complete explanation of one of your examples to discover how useful and true sayings are.

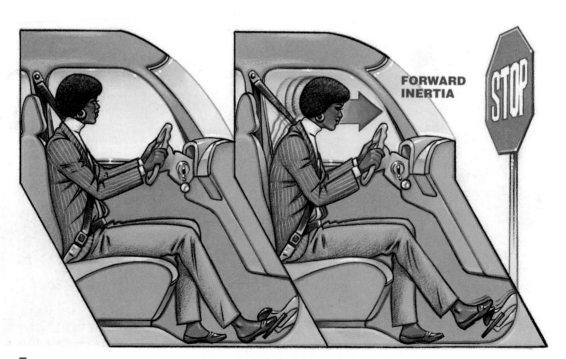

▶ **Fig. 20-5** The brake slows the car, but the driver continues forward because of inertia. The safety belt stops the passenger's forward motion.

Newton's Second Law of Motion

Newton's *second law of motion* explains the relationship between force, mass, and acceleration. The rate at which an object moves depends on two things: the size of the unbalanced force acting on it and the mass of the object. This relationship is stated in the following formula:

Force = Mass × Acceleration
or
$F = M \times A$

Suppose you applied the same force to an empty wagon and to a wagon of the same size filled with bricks. Which wagon would move with greater acceleration? Because the mass of the empty wagon is less, it would move with greater acceleration. The wagon filled with bricks would require a greater force to accelerate at the same rate. That force can be calculated. Multiply the wagon's mass by the desired change in velocity. What do you think would happen to the force required to move a heavy object as its mass is reduced?

Newton's Third Law of Motion

All forces occur in pairs. Newton's *third law of motion* states that for every action there is an equal and opposite reaction. For every force exerted, an equal force is exerted in the opposite direction. Suppose you were standing absolutely still on a skateboard. If you were to push a basketball away from you, in what direction would the skateboard move? Remember that the forward motion of the ball would be the action. What would be the direction of the reaction force? Fig. 20-6.

▶ **Fig. 20-6** Newton's third law describes forces as always being in pairs. While such forces work in opposite directions, they have equal strength.

A tennis racket also demonstrates the principles of Newton's laws. A tennis racket is designed in an oval shape. This shape allows the racket frame to withstand the force placed on it by the stretched strings. Why are the strings stretched so tightly? The answer is explained by Newton's third law of motion.

Imagine that you are on a tennis court. Your opponent serves the tennis ball to your side of the court. The ball is moving at a high velocity. If you simply hold out your racket to meet the ball, the ball will return at the same force applied to it by your opponent. This is action and reaction.

If you swing the racket towards the oncoming ball, you will meet it with *additional* force. This will cause the ball to stretch the flexed strings even further. The stretched strings try to return to their original position. In doing so, they multiply the force applied to the ball. This causes the ball to return across the net at an even higher velocity.

WORK, POWER, AND MACHINES

Work

What does the word "work" mean to you? You might think of it as cutting the grass or cleaning your room. However, work has a different meaning to a scientist or engineer. To them, **work** is the application of force to cause an object to move in the direction of the force. If you push a heavy box two meters across the floor, you are doing work. The amount of work accomplished can be measured using the formula:

Work = Force × Distance
or
$W = F \times D$

In the metric system, the unit of measure for force is the newton (N). Distance is measured in meters (m) in the metric system. The metric unit of measure for work is the *newton-meter* (*N-m*). The force of one newton pushing or pulling an object a distance of one meter equals one newton-meter (N-m) of work. Scientists sometimes use the term *joule* (jewel) to describe work. One newton-meter equals one joule (J).

Assume that the box you pushed along the floor weighed 700 N. As mentioned, you moved the box two meters. How much work have you accomplished? (To convert newtons into pounds, divide newtons by 4.5.) Fig. 20-7.

Suppose a crane lifts a steel beam weighing 1400 N to a height of 20 m. How much work did the crane perform?

$$W = F \times D$$
$$W = 1400 \text{ N} \times 20 \text{ m}$$
$$W = 28,000 \text{ J}$$

Power

In science, **power** is the rate at which work is done or the measure of how much work is accomplished in a certain period of time. To calculate power, use the following formula:

Power = Force × Distance ÷ Time
or

$$P = \frac{F \times D}{T}$$

In the example previously given, you performed 1400 J of work by pushing the box along the floor. If it took you 10 seconds to perform the work, how much power was used?

$$P = \frac{1400 \text{ J}}{10 \text{ seconds}}$$

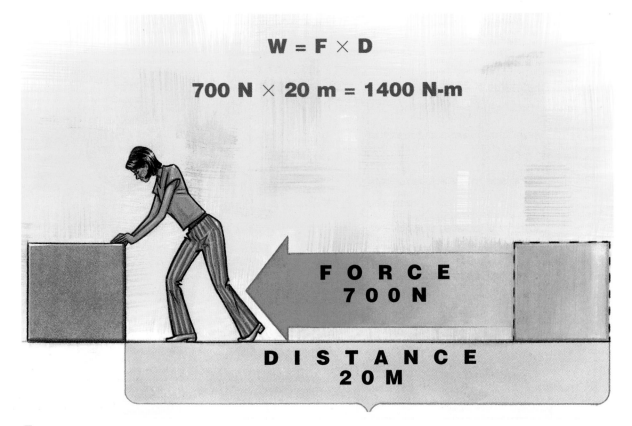

W = F × D

700 N × 20 m = 1400 N-m

FORCE 700N

DISTANCE 20 M

▶ **Fig. 20-7** Work is accomplished when an object is moved a certain distance by an applied force. To calculate the work accomplished, multiply the force by the distance.

You used 140 J/sec (joules per second) of power. The watt (W) is also a unit of power measurement. One watt (W) equals 1 J/sec.

It took 50 seconds for the crane to lift the beam 20 meters. What is the crane's power? Fig. 20-8.

$$P = \frac{W}{T}$$
$$P = \frac{28,000\ J}{50\ seconds}$$
$$P = 560\ J/sec,\ or\ 560\ W$$

At some point a crane will not be able to lift an object because the weight of the object will exceed the crane's lifting capacity. How can the crane's manufacturer increase a crane's lifting capacity?

Machines

A **machine** is a device designed to obtain the greatest amount of force from the energy used. A machine is used to make work easier and faster. For example, by using a can opener, you can open a metal container in seconds. A car jack can help you easily lift a car weighing thousands of pounds. We use machines all the time to accomplish tasks more quickly.

Machines help us accomplish work by increasing a force or changing its direction. A machine may look very complicated. However, a complex machine really consists of a combination of simple machines. Fig. 20-9.

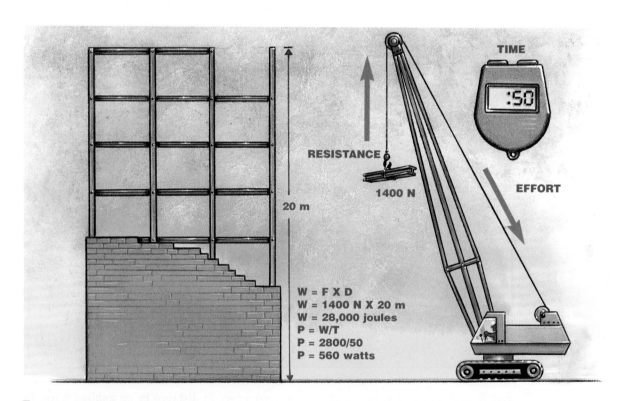

RESISTANCE

1400 N

20 m

W = F X D
W = 1400 N X 20 m
W = 28,000 joules
P = W/T
P = 2800/50
P = 560 watts

TIME

:50

EFFORT

▶ **Fig. 20-8** The steel beam weighs approximately 311 pounds. To convert newtons into pounds, divide newtons by 4.5. The crane shown here uses 560 watts of power to lift the beam 20 meters in 50 seconds.

Some machines multiply a force more effectively than others. The number of times a machine multiples a force is called the **mechanical advantage**.

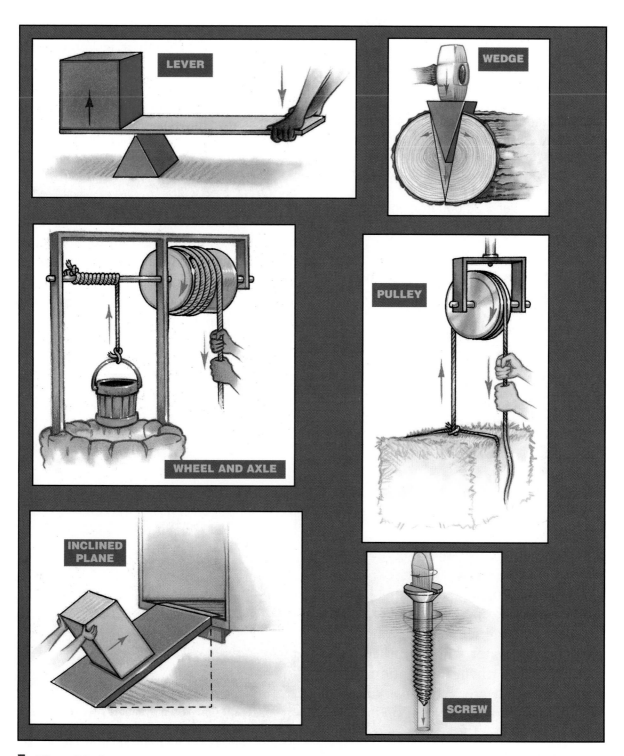

▶ **Fig. 20-9** The six simple machines. Complex machines are made up of simple machines. When a machine is used, two forces are always involved. The force applied to the machine is called the "effort force." The force working against the machine is called the "resistance force."

We know that waves can travel through liquids like water. They can also travel through solids and gases such as metal and air. A liquid, solid, or gas that allows waves to pass through it is known as a *medium*. Water is the medium for ocean waves. Air is a good medium for sound waves. Energy is transferred by a medium because the particles that make up the medium vibrate.

Sound Waves

Sound is an energy form that causes the molecules within a medium to vibrate. As the molecules vibrate, they compress neighboring particles and a sound wave is formed. The sound energy is transferred through the medium as a wave.

SOUND WAVES AND LIGHT WAVES

What happens when you drop a pebble into a bowl of water? Can you describe what you see?

The pebble creates waves. These waves move out from the point where the pebble hits the water. Did you know that sound and light also move through the air in waves? Fig. 20-10.

When you drop a pebble into a bowl of still water, the energy from the falling pebble is transferred to the water. The particles of water are disturbed. The energy places them in motion. Each particle of water bumps into a neighboring particle of water and a wave is formed. A *wave* is a disturbance that transfers energy from one place to another through matter or space.

Fig. 20-10 When a pebble hits the water, it transfers its energy to the water. A wave pattern spreads out from the point of contact, as particles of water are pushed away. Light waves and sound waves move in a similar way.

Some mediums or materials transfer sound better than others. The molecules within a solid material transmit sound waves more easily than liquids and gases. The particles are packed more closely together. The transfer of the wave energy is more efficient.

Have you ever shouted in a large empty room and heard an echo? Echoes are heard because sound energy bounces off a hard, flat surface back to our ears. The repeated sound is an echo.

Sonar is a device used on ships to measure the depth of water and to locate underwater objects. Sonar equipment on the ship sends pulses of sound energy through the water in the form of sound waves. These sound waves strike the ocean floor or an underwater object and echo back to the ship. The time it takes for the sound wave to travel down and back is measured.

Sound travels through water at a speed of 1500 meters per second. By knowing the speed of sound and the time it takes for the sound wave to bounce back to the ship, we can calculate the distance the wave traveled. Fig. 20-11.

Ultrasound is sound that is so high pitched that we can't hear it. Ultrasound machines are used in the medical field. For example, they can produce images of babies before birth. The ultrasound machine works in the same way as sonar. It sends sound waves that bounce off the baby inside the mother. The echo is measured by the machine, which produces an image of the baby on its screen.

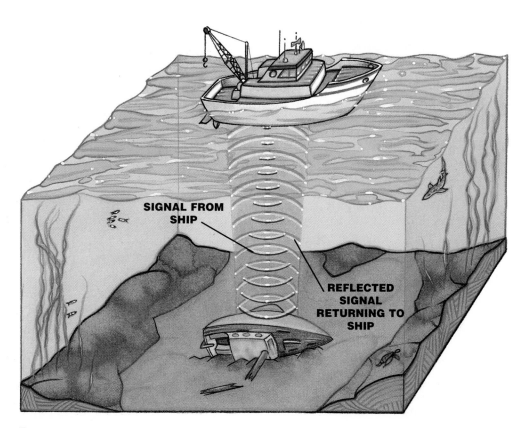

SIGNAL FROM SHIP

REFLECTED SIGNAL RETURNING TO SHIP

▶ **Fig. 20-11** Sonar is used to locate objects in the water and to determine depth.

Light Waves

Like sound, light also travels in waves. However, light waves and sound waves are very different. Unlike sound energy, light energy can be transmitted with or without a medium. For example, sound waves cannot travel in a vacuum (the absence of air). Light waves can.

The source of all light energy is the atom. An atom consists of a central nucleus. The nucleus is surrounded by a

Explore

Design and Build a Sound Wave Tester

State the Problem

When architects design concert halls, theaters, and auditoriums, they must carefully consider how sound waves will move through the space. Conduct an experiment to determine how the placement of materials affects the movement of sound waves.

Develop Alternative Solutions

1. On a separate sheet of paper, make a data sheet. The data sheet should include space to list the materials. It should also include space to describe their sound transmission qualities.
2. Prop or brace the plastic sheet upright on the table.
3. Place the two tubes on the table at angles to each other and the plastic. Be sure the plastic is close to the tubes but not touching.
4. Place the ticking clock inside the outer end of one of the tubes.
5. Place your ear near the outer end of the second tube.
6. Record descriptions of what you hear on your data sheet. Take notes on the volume, pitch, and crispness of the sound.
7. Place your hand or a book over the inside end of the tube containing the clock. Repeat Step 6.
8. Repeat Steps 2 - 7, first covering the plastic with cardboard, then fabric, and then foam. Note whether each material absorbs or reflects sound.

Collect Materials and Equipment
2 cardboard tubes about 20" long, 2"-3" in diameter
smooth cardboard 15" × 15"
plastic sheet, 15" × 15"
piece of fabric, 1" × 15"
piece of foam, 15" × 15"
ticking watch or clock

cloud of negatively charged particles called *electrons*. Electrons position themselves around the nucleus of the atom in energy levels. At times, electrons absorb additional energy and step up to a higher energy level. This change in level is only temporary. The electron soon loses its extra energy and returns to its original level. At that moment, the electron releases its extra energy as photons.

9. Change the angles of the tubes. Repeat the experiments. Record how the sound is affected on the data sheet.

Select the Best Solution

Analyze the data you have collected. What types of materials would you use to cover the walls of a new movie theater? How should the materials be placed?

Implement the Solution

Think of a place that has poor acoustical conditions.

Evaluate the Solution

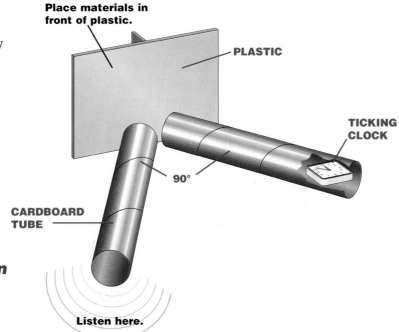

Place materials in front of plastic.

PLASTIC

TICKING CLOCK

90°

CARDBOARD TUBE

Listen here.

Be sure to include the types of acoustical materials used on the walls and the layout of the walls. Prepare a sketch showing how you would place materials to improve the sound quality. Refer to the information learned from your experiment. Will your suggested placement of materials improve sound quality?

Light is a stream of packets or bundles of light energy called *photons*. Photons move in waves. As the waves of photons move, they produce electrical and magnetic fields around them. For this reason light waves are called *electromagnetic waves*. Electromagnetic waves do not need a medium in which to travel.

We can see some forms of light, but we can't see the patterns in which light waves travel. To describe a light wave, we need to talk about four important properties of light. They are speed, wavelength, frequency, and amplitude.

All light waves travel at the same speed, 300,000 kilometers per second. The wave moves up and down. The *wavelength* is the distance between the crests of the wave. The *frequency* is the number of waves that pass a particular point each second. Millions of light waves go by each second. The height of the light wave is called its *amplitude*. A bright light has a high amplitude.

There are many forms of light. Some light waves are visible to our eyes and some are not. The visibility, color, and energy of light waves depend on their wavelength, frequency, and amplitude. Fig. 20-12.

Light Waves and Technology

There are many sources of light. Most of our natural light comes from the sun. Artificial light is created in several ways. The most common source of artificial light is the incandescent lamp. An incandescent lamp produces light from heat. Inside the glass lightbulb is a filament, or very thin wire. The filament is made of tungsten. Electricity runs through the filament when the light is turned on. Because the filament is very thin, it resists the flow of electrons. This causes heat to build up. The filament glows and gives off photons of visible light. Fig. 20-13.

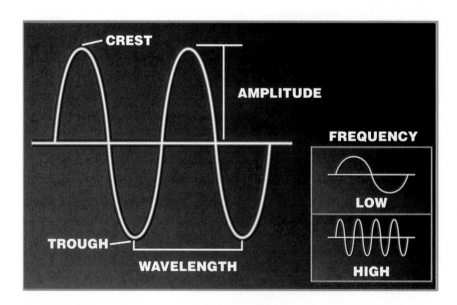

▶ **Fig. 20-12** Whether we can see light, as well as the color of that light, depends on the characteristics of the light wave. The height of the wave, as well as its frequency, determines the kind of light.

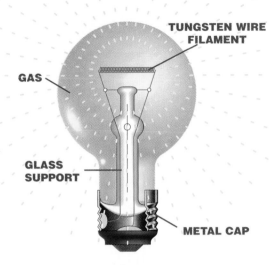

PHOTONS OF
VISIBLE LIGHT

TUNGSTEN WIRE
FILAMENT

GAS

GLASS
SUPPORT

METAL CAP

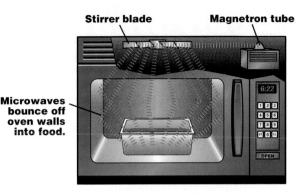

Stirrer blade   Magnetron tube

Microwaves
bounce off
oven walls
into food.

MICROWAVE OVEN

▶ **Fig. 20-14** The operation of a microwave oven.

▶ **Fig. 20-13** The filament of a lightbulb is made of tungsten wire. This metal wire creates heat by resisting the flow of electricity. When the filament is white hot, it gives off photons of visible light.

Lasers are another source of light energy. Unlike an incandescent bulb, lasers give off coherent light. Coherent light is light made up of only one frequency. This means that all of the light waves produced by a laser move in the same direction at the same time. A laser beam is a narrow, high-energy beam of parallel light rays. Laser light is a very powerful source of single-color light.

Did you know that microwaves are a form of electromagnetic waves, or light waves? The food in a microwave oven absorbs the energy of the microwaves. The water or sugar molecules in the food vibrate. This vibration causes the molecules in the food to heat up. This causes the food to cook. Fig. 20-14.

IMPACTS

Applied physics has been used in the planning and completion of many of the great technological advances of our century. Its impact is seen in the design of items in every area of technology. Its laws are used to help predict the stability of such structures as skyscrapers and bridges.

THE FUTURE

In our world, there is constant change. The laws of physics provide a standard against which new ideas can be evaluated. They also provide a means of reassessing old ideas. The applications of physics must be carefully considered in considering any complicated project, especially a building project. As such, the laws of physics will continue to show the close relationship between science and technology.

Apply What You've Learned

Design and Build a "Reaction" Rocket Racer

State the Problem

Rocket and jet engines are "reaction" engines. Newton's third law of motion explains how these engines work. Hot gases escaping from the rear nozzle of the engine are the "action" force. The "reaction" force, called thrust, propels the rocket or jet in the opposite direction.

Design and build a "reaction" rocket racer that will travel a measured distance.

Develop Alternative Solutions

Determine what factors influence the speed at which an object will move across the floor when powered by the escaping air of a balloon. List these factors in a table. Use these factors to help you prepare several designs for a rocket racer. Incorporate the factors that produced a faster vehicle into your final design. One possible design is shown on the opposite page.

Select the Best Solution

Select the design that you think will be the most effective.

Implement the Solution

1. Construct the test rocket racer. Follow your design.
2. Using the tape measure, lay out a 20' test track on the floor of your classroom.

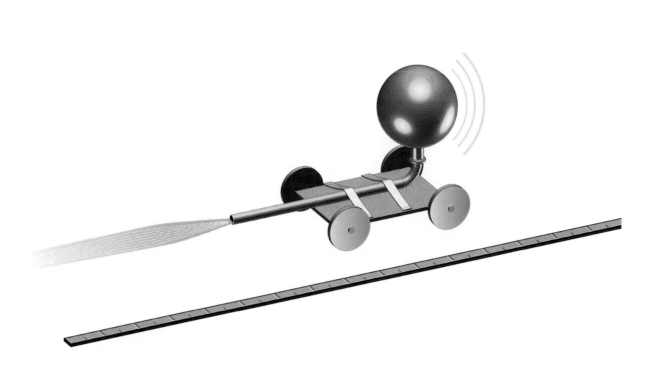

3. Test the rocket racer. Time how long it takes for your vehicle to travel down the track.

4. Make at least five test runs. Average the run times.

Evaluate the Solution

1. Compare the success of your vehicle with the success of other vehicles in your class. What additional modifications might you make to increase speed?

2. Weight is important. Can you think of any modifications that will reduce the weight?

CAREERS IN
Applied Physics

BIOPHYSICIST

Research laboratory has opening for biophysicists to investigate dynamics in the damage to cells and tissues caused by X-rays and nuclear particles. Prefer previous experience in area of nuclear particles for cancer treatment. Will be working in new facility. Ph.D. required. Send resume to: Durham Research Laboratories, 2201 Duke Avenue, Raleigh-Durham, NC 32088.

PHYSICS INSTRUCTOR

Small, private, liberal arts university has opening for physics instructor. Will prepare materials and lecture to students. Will also serve as advisor to students majoring in physics. Flexible schedule according to teaching load. Send vitae and list of references to: Physics Department Chair, Private University, 7540 University Place, Enid, TX 50229.

ASTRONOMER

University research position in astronomy department. Will be involved in project planning, conducting experiments, analyzing results and formulating theories. Also responsible for coordinating activities at university observatory. Please send vitae and list of references to: Astronomy Search Chair, State University, 800 South Edge Street, Omaha, NE 62035.

PHYSICIST

Department of Energy has opening for physicist in research and development. Will work as part of team with engineers in a laboratory setting. Primary area of research is in nuclear energy. Must have good communication skills. Master's degree in physics required. Proposal and report writing experience helpful. Please send resume to: Department of Energy, Laboratory Building, 1380 State Street, Albuquerque, NM 72033.

ELECTRO-OPTICAL ENGINEER

Engineering firm has position open in research and development of gas and solid state lasers and other light-emitting and light-sensitive devices. Requires a degree in physics or engineering with project management experience a plus. Will conduct application analysis and write technical reports. Competitive salary and benefits offered. Forward resume to: Caswell Engineering, 21165 Livingston Avenue, Gary, IN 63301.

Linking to the WORKPLACE

Advances in aeronautics and space technology can result in new discoveries in physics that can impact the workplace. Bring in an article from a newspaper or magazine that discusses a recent discovery. List ways that you think jobs may be affected by this discovery in the future. Write a short report and share it with the class.

Chapter 20 Review

SUMMARY

▶ Applied physics is the study of the ways in which objects are moved and work is done.

▶ Gravity is the force of attraction that exists between two objects.

▶ Newton's first law of motion states that an object at rest will stay at rest and an object in motion will stay in motion.

▶ Newton's second law of motion explains the relationship between force, mass, and acceleration.

▶ Newton's third law of motion states that for every action, there is an equal and opposite reaction.

▶ Work is the application of force to cause an object to move in the direction of the force.

▶ Power is the rate at which work is done.

▶ A machine is a device designed to obtain the greatest amount of force from the energy used.

CHECK YOUR FACTS

1. List the steps in the scientific method.

2. Define the term *applied physics*.

3. Explain how force, friction, and gravity relate to motion.

4. Name and describe three types of friction.

5. On what two factors does the amount of gravitational force depend?

6. State and explain each of Newton's three laws of motion.

7. Explain the difference between work and power.

8. Define the term *mechanical advantage*.

9. How is a sound wave different from a light wave?

10. What is a photon?

CRITICAL THINKING

1. What forces are acting on a person parachuting from an airplane?

2. Describe the principle of physics that explains how we hear the sounds from a guitar.

3. Describe the principle of physics that explains how an applied force, like snow on the roof of a building, can cause the building to collapse.

4. You have been pushing and pushing, trying to move a refrigerator in your kitchen, but you have not been able to move it. How much work have you done?

Glossary

A

aerodynamics. The study of the forces of air on an object moving through it.

aerospace. The study of how things fly.

ailerons. Wing flaps on an airplane that change the shape of the wing, increasing and decreasing the amount of lift the wing creates.

airfoil. A shape, like the wedge-shaped wing of an airplane, that is designed to speed up the air passing over the top surface.

amplitude modulation (AM). A way of transmitting radio broadcasts. In this type of radio transmission, the amplitude (strength) of the carrier wave changes.

animation. The creation of the illusion of movement in a series of still images.

applied physics. The study of the ways in which objects are moved and work is done.

audio. What we hear on a telecast or broadcast.

automation. A technique that is used to make a process automatic.

B

battery. A device that converts chemical energy into electrical energy.

Bernoulli Effect. States that a fast-moving fluid exerts less pressure than a slow-moving fluid.

bio-related technology. A technology that improves the techniques used to nurture life.

biomechanical engineering. The use of engineering principles and design procedures to solve medical problems.

bioprocessing. A process that uses living microorganisms or parts of organisms to change materials from one form to another.

brainstorming. A process in which group members suggest ideas as they think of them. It is a group problem-solving method.

building codes. Codes enacted by state and local governments that specify the methods and materials that can be used for each aspect of construction.

buoyancy. The upward force a fluid places on an object placed in it.

C

CAD/CAM. A process that combines computer-aided drawing and computer-aided manufacturing. Output from the computer used to design a product is used to operate the machines that manufacture it.

Cartesian coordinate system. A system that allows you to plot points on a drawing. The system is based on an imaginary grid.

Cartesian coordinates. The letter-number combinations that locate points. The letters identify imaginary lines running through an object.

CD-ROM. Short for "compact disc read-only memory." It can store many kinds of information, including music, video, and animation.

chromosome. A tiny particle that carries the genes that pass on inherited characteristics. It is located in the nucleus of a cell.

circuit. In electricity, the pathway through which electrons travel.

coherent light. Light in which all of the light waves have the same wavelength. They are also "in phase."

command. An instruction that directs the software program to perform a specific task (such as drawing).

communication. The process of giving or exchanging information.

computer network. A communication system created by connecting many computers.

computer numerical control (CNC). A control system that uses computers to control the operation of machines during production.

computer-aided design (CAD). Software that provides tools for drawing and dimensioning a product.

computer-aided design/computer-aided manufacturing (CAD/CAM). A combination of two separate systems. A computer can be used to design a product, and that same computer might then be used to control the machines used to manufacture the product.

computer-aided drafting (CAD). The process of using a computer to create drafted documents.

computer-aided manufacturing (CAM). Using computers to manufacture a product.

continuous production. A type of production in which products are mass produced, usually on an assembly line.

controlled environment agriculture (CEA). An agricultural system that recreates the conditions a plant needs to grow. It allows plants to thrive by providing the perfect growing environment. Humidity, temperature, lighting, watering, and feeding are five of the conditions controlled by this technology.

coordinate pair. A set of two numbers that will locate a point on a grid. The first number is the X coordinate. The second number is the Y coordinate.

current. The flow of electrons in a wire or other conductor.

custom production. A type of production in which products are made to order.

D

degree of freedom. The ability of a robot to move in a certain direction.

design. A plan for making something.

design brief. A statement of the problem that is to be solved. It should include all the information that the designer needs to understand the problem.

design process. A process that uses problem solving to arrive at the best solution, or design. It is also used to improve products and services that are already in use.

desktop publishing. The use of a computer and special software to produce documents that combine text and graphics (photographs and drawings).

diagnosis. The process of examining a patient and studying the symptoms to find out what illness the patient has.

diesel engine. An internal-combustion engine that burns fuel oil by using heat produced by compressing air.

digital camera. A camera that can produce electronic images that can go directly into a computer.

directional light. Light that spreads out very little compared to ordinary light.

domain. The name of the Internet provider. In e-mail, the part after the @ but before the period, or dot (.), is the domain.

drafting. The process of representing three-dimensional objects in two dimensions.

drag. The force of fluid friction on moving objects.

E

e-mail. Messages sent over the Internet.

electricity. The flow of electrons through a pathway that conducts electricity.

electromagnetic wave. A wave produced by the motion of electrically charged particles.

electromagnetism. The relationship between electricity and magnetism.

engineer. A person who uses his or her knowledge of science, technology, and mathematics to solve technical problems.

engineering. A process used to develop solutions to problems.

engineering process. A problem-solving process and a design process. This process requires the engineer to identify the need, gather information, develop alternative solutions, choose the best solution, implement the solution, and evaluate the solution.

ergonomics. The study of designing equipment and devices that fit the human body, its movements, and its thinking patterns.

F

feedback. Information about the output of a system. It is used to monitor how a system is working.

feedback control. The process of sending signals, interpreting received signals, and adjusting through signals.

flexible manufacturing system (FMS). A system that allows the manufacturer to make a variety of designs using the same machines.

fluid. Any substance that flows.

fluid power. The use of pressurized liquids or gases to move heavy objects and perform many other tasks.

force. A push or a pull that transfers energy to an object. Forces on a structure can be external or internal.

foundation. The part of the structure in contact with the ground.

frequency modulation (FM). A way of transmitting radio broadcasts. In this type of radio transmission the frequency of the carrier wave changes.

friction. A force that opposes motion. It acts in the direction opposite the direction of the motion. It is the force that brings an object to rest.

G

gasoline piston engine. An engine that creates power by burning fuel inside the engine.

generator. A device that changes mechanical energy into electrical energy. A generator uses electromagnetic induction to force electrons from their atoms.

genetic engineering. The process of changing the genetic materials (genes) that make up living organisms.

genetics. The science that studies the laws of heredity.

gravity. A force that pulls objects towards the center of the Earth.

H

heredity. The study of the passing on of certain traits from parents to offspring.

holography. A photographic process that uses a laser as well as lenses and mirrors to produce three-dimensional images.

human factors engineering. A design process that is used in product design. It gives special attention to the strengths and limitations of the human body.

hydraulic systems. Fluid power systems that use oil or another liquid.

hydroponics. The process of growing plants in a soilless environment.

hyperlinks. "Links" that can be attached to text, buttons, or graphics on web pages. When you click on them, they take you to other pages.

hypermedia. Used by CD-ROM programs to make them interactive. It gives you the chance to choose the information you want and provides more than one way to get it. Two kinds of hypermedia are hypertext and hotspots.

hypertext. The system of connecting documents through hyperlinks.

I

immunization. An action that protects the body against a disease.

industrial materials. Materials that are used to make products.

inertia. The tendency of an object to remain still or to continue to move in the same straight line unless an outside force acts on it.

innovation. A change created by improving an existing technology. It occurs when something new is introduced.

input. Something that is put into a system. In some systems, the input is a combination of the seven resources of technology.

invention. The process of designing new products.

J

job-lot production. A type of production in which a specific quantity of a product is made.

K

key frame. A frame that shows a beginning or ending point in an action sequence.

L

laser. A light source that sends out light in a narrow and very strong beam. The term is short for Light Amplification by Stimulated Emission of Radiation.

leading. In desktop publishing, the spacing between lines.

load. An external force on an object.

M

machine. A device designed to obtain the greatest amount of force from the energy used.

maglev. A term that is short for MAGnetically LEVitated. Trains are levitated or floated above a guideway (track) and propelled (moved forward) by magnetic fields. The magnetic fields are created by large electromagnets.

manufacturing. The changing of materials into usable products.

mechanical advantage. The increase in force gained by using a machine; also, the number of times a machine multiplies a force.

model animation. Animation that involves the use of three-dimensional figures called puppets.

modem. Changes the signals from your computer into signals that can be transmitted over telephone lines.

monochromatic light. Light that consists of one color.

multimedia. The combination of several kinds of communication, such as text, pictures, sound, video, and animation.

O

Ohm's law. The law that states that when voltage is applied to a conductor, the current that moves through the conductor is directly proportional to the applied voltage.

output. That part of a system that is the result of the system's process. If the system is working properly, the output will be the desired product or result.

P

page layout programs. Software programs that combine text and graphics in a document.

Pascal's Principle. The principle that states that when force is applied to a confined liquid, the resulting pressure is transmitted unchanged to all parts of the liquid.

persistence of vision. The blending of individual images into one image that seems to move.

photon. A tiny unit of energy given off by the filament of a light bulb.

photosynthesis. The process plants use to convert sunlight to energy.

physical enhancements. Replacement body parts.

plans. Drawings that show the builder how to construct the structure.

pneumatic systems. Fluid power systems that are based on the use of air or another gas.

potential difference. The force that causes electrons to flow.

power. The rate at which work is done or the measure of how much work is accomplished in a certain period of time.

pressure. The force on a unit surface area (such as a square inch).

primary processes. Processes that change raw materials into industrial materials.

primitives. Computer-generated basic geometric shapes used in modeling.

process. That part of a system during which something is done. It is the "action" part of the system.

prosthesis. A people-made device used to replace human body parts.

prototype. A full-size model of a product. It looks and works like the actual product.

publishing. The process of presenting material in printed form.

R

random-access memory (RAM). Temporary memory in a computer.

raw materials. Materials as they occur in nature.

read-only memory (ROM). Permanent memory in a computer. It usually includes the computer's operating system.

resistance. The opposition to the flow of electrons.

resolution. The number of dots per inch (dpi) of ink on printed images.

robot. A machine made to act like a living thing.

robotics. The study of robots. Robotics is a control system technology.

S

scanners. Devices that change images such as photographs into an electronic form that computers can use.

secondary processes. Processes that turn industrial materials into finished products.

selective breeding. A technique used to combine the traits of one animal with those of another animal.

simulation software. Software that can test a product design by simulating the environment in which the product must be able to work.

smart materials. Materials that have built-in sensors to warn of unsafe conditions.

sound. A form of energy produced by vibrations that act on the ear so that we can hear.

specifications. Written details about materials and other project-related concerns.

standard. A rule or model against which a product, action, or process can be compared.

storyboard. A series of sketches used as the "blueprint" for making the film.

structural member. A building material used to make up the frame of a structure.

structure. Something that is constructed, or built.

superstructure. The part of a structure above the foundation.

system. An orderly way of achieving a goal.

T

technology. Using knowledge to develop products and systems that satisfy needs, solve problems, and increase our capabilities.

transistor. A semiconductor device that amplifies and acts as an electronic switch.

transportation. The process by which people, animals, products, and materials are moved from one place to another.

typeface. A set of letters, numbers, and symbols that look the same.

U

UHF (ultra-high frequency). Channels that have a frequency between 470 and 806 megahertz.

Uniform Resource Locator (URL). The special address system used by the World Wide Web. Every web site has its own URL.

V

vessel. A water vehicle that transports people and products.

VHF (very high frequency). Channels that have a frequency between 54 and 216 megahertz.

video. The part of a telecast that you see.

W

work. The application of force to cause an object to move in the direction of the force.

work envelope. The space within which a robotic arm moves.

Index

A

action/reaction, 196, 397, 398
active medium, of laser, 358, 359
activities, design and build (Apply)
  casein glue manufacturing system, 282-283
  catamaran, 224-225
  continuity tester, 304-305
  dispenser, 386-387
  gameboard, 242-243
  home page, 110
  keyboard, 260-261
  lighted monogram, 366-367
  newsletter, 80-81
  numerical control coordinates, 322-323
  orthographic projection, 62-63
  pneumatic control device, 344-345
  product packaging, 154-155
  profile map, 26-27
  propeller, 200
  rocket racer, 408-409
  small structure, 176-177
  spreadsheet, 42-43
  tin can telephone, 132-133
  zoetrope, 94-95
activities, design and build (Explore), 22, 32-33, 38, 56, 58-59, 70, 74, 88, 92, 102, 108, 120-121, 128-129, 146-147, 150-151, 172-173, 191, 192-193, 208-209, 214-215, 231, 238-239, 251, 256-257, 268, 275, 292, 312, 316-317, 334-335, 354-355, 362, 376, 380, 392, 404-405
activities, of technology, 18-19, 20-21, 29
actuator(s), 232, 333, 337
aerodynamics, 187-189
  defined, 187-188, 412
aerodynamic shape, 394
aeroponic systems, 276, 277 illus.
aerospace, *defined*, 184, 412
aerospace engineers, 310, 373-374
Aerospike engine, 349
aesthetics, 39
aggregate, 276, 277
aggregate systems, 276, 277 illus.

Agricultural Age, 15
agriculture, 240
  controlled environment, 273-274, 278, 285
  and environmental technologies, 265-278
ailerons, 193 illus., 203
  defined, 192, 412
air bags, 13
aircraft, 238, 373-374
air cushion vehicle(s), 222, 238
airfoil(s), 192, 194, 203, 221 illus., 311
  defined, 190, 412
airplanes, 189-197, 203
  controlling, 192-193
air pressure, and lift, 190 illus., 203, 311
algae, 279
alloy, 140
alternating current (ac), 295
aluminum, as element, 289 illus.
American National Standards Institute (ANSI), 236
Americans with Disabilities Act, 257-258
ampere (amp), 296
amplifier, 118, 119, 303
amplitude, 115, 406
amplitude modulation (AM), *defined*, 117, 412
AMTRAK, 213
angiogram, 250 illus.
animation, 93, 97
  clay, 89
  computer, 84-97
  defined, 85
  types of, 86-87, 89-91, 93
animation programs, 90
animation software, 89, 311
animators, 87, 90
annealing, 145
aorta, 250
appearance models, 36
applied physics, 390-411
  defined, 391, 412
aquaculture, 278
arch, as structural shape, 166
arch bridge(s), 164 illus., 166
Archimedes' principle, 220
architect(s), 36, 48, 166-167, 174
  and designers, 375

architectural design program, 90
area, figuring of, 241 *mathematics*
armature, in generator, 295
artificial body parts, 249 illus., 250, 258, 259
artificial intelligence, 328
assembly line, 149, 405
atomic charges, 291
atomic structure, 290-291
atom(s), 289, 290 illus., 291, 404-
  audio, 135
  defined, 115, 412
  and video and multimedia, 114-135
audio console, 117
automated factories, 329, 343
automation, 153, 313, 328-329
  defined, 328, 412
  effect on employment, 152, 343
automation revolution, 343
automatons, 328
automobiles, 213. *See also* cars.
autorotate, 194
aviation, careers in, 202
axes, X, Y, Z, 52-53 illus., 55 illus., 313, 315-317, 315 illus.
axis, 315, 330

B

bacteria, 253, 259 illus., 272, 279
bacterium, 270
  dividing of, 253 *mathematics*
barges, 219, 220
batteries, 223, 294-295
battery, 293, 297, 307
  defined, 294, 412
beam (ray), laser, 356, 407
beams (construction), 54, 163 illus.
Bell, Alexander Graham, 360
Bernoulli effect, *defined*, 190, 412
bicycle, 206, 207, 311, 393
binary code, in robotics, 336-337
binary number system, 303
bioengineers, 253, 259
biological control, of pests, 271
biology, 391
biomechanical engineering, 263
  defined, 249, 412
biomechanical engineers, 249, 251, 259

bionics, 249 illus.
bioprocessing, 282, 283, 285
    defined, 279, 412
bio-related technologies, 19, 265,
    281. *See also* environmental
    technologies
bio-related technology, 249, 279,
    285
    defined, 265, 412
biosphere, 281
bits, in fiber optics, 360
body replacement parts, 249, 250,
    251
boosters, rocket engine, 198 illus.
Boyle's Law, 217 illus., 230
brainstorming, 35-36, 45, 378
    defined, 35, 412
brazing, 144
bridges, 163, 164-165 illus.
broadcasting, radio, 116-118
browsers, web, 104, 105, 107
building, 171
    as a system, 159
    of a residential structure, 166-
    173
building codes and building
    permits, *defined,* 174, 412
building inspections, 174
building(s), 160 illus., 161, 163,
    170 illus., 175, 375
buoyancy, *defined,* 220, 412

C

CAD, 48-65, 309-311
    advantages of, 49-50
    See also computer-aided design;
    computer-aided drafting
CAD/CAM, 61, 319-320, 325
    defined, 61, 319, 412
    See also computer-aided
    design/computer-aided
    manufacturing
CAD drawings, 49-50, 53-54
CAD system(s), 36, 65
    components of, 50-52
    using, 57, 60
calculations
    for power, 399, 400
    for work, 398, 399
CAM, *defined,* 319
cam, 214
cameras, 71, 75, 122-124, 125,
    341
cantilever, 165
career, selecting, 64

Careers in
    applied physics, 410
    audio, video, and multimedia,
    134
    aviation, 202
    computer-aided drafting
    (CAD), 64
    computer animation, 96
    computer systems control, 324
    design and problem solving, 44
    desktop publishing, 82
    electronics, 306
    engineering, 388
    environmental technologies,
    284
    fluid power, 244
    health technologies, 262
    Internet technology, 112
    lasers and fiber optics, 368
    manufacturing, 156
    robotics, 346
    structures, 178
    technology, 28
    transportation, 226
carpenter, 239
cars, 223, 309
Cartesian coordinates, 315-317,
    325
    defined, 315, 412
    plotting, 52 *mathematics,* 322-
    323
Cartesian coordinate system, 52-
    53, 54, 55 illus., 65
    defined, 52, 412
Cartesian geometry, 52
cartoon(s), 84, 85, 89
casting, 142 illus., 143
categorizing, 378
CAT scan, 254 illus.
CD-ROM, 50, 126-127, 130
    defined, 412
CD-ROM drives, 131
CD-ROMs, in desktop publishing,
    69, 75
CDs, 356-357
ceiling, 171, 172
cel animation, 86-87
cell(s) (electricity)
    and batteries, 294-295
    photovoltaic, 297
cells (natural), 272, 273 illus.
celluloid, 87
cell (workstation), 320
cels, 87 illus.
cement, 171
central processing unit (CPU), 50

    defined, 52
ceramics, 140
channel, 115, 124
charges, 291, 294
Charles' law, 217 illus.
chat, on Internet, 106-107
checklist, design, 379, 381 illus.
chemical conditioning, 146
chemical engineers, 373
chemical processes, 141
chemical reaction, in converting
    energy, 294 illus., 295
chemistry, 391
chip removal, 143, 144 illus.
chlorophyll, 266, 267
chromosome, 273 illus.
    defined, 272, 412
CIM, 19, 152
circuit(s), 293, 307, 373
    defined, 297, 413
    electrical, 297-300
    in fluid power system, 236, 237
civil engineers, 48, 372
classic engineering, 372-373
clay animation, 89
climate-control systems, 171
clip art, 69, 75
cloning, 259
closed-loop system, 24
CNC. *See* computer numerical
    control
CNC data input, 314-315
CNC machine(s), 313-314, 315,
    318, 325
CNC programmers, 317-318
coating, 145
code(s) (information), 314, 318,
    320, 336-337
codes (laws), building, 174
coherent light, 407
    defined, 352, 413
color(s), 78 illus.
    in animation, 90
    in desktop publishing, 75, 78
    in light, 352, 354-355
    and safety messages, 153
color vision, 79
column(s), 163
combining, 144-145
combustion, in engine, 216
command(s), CAD, 58-59, 60, 65
    defined, 57, 413
commercials, 89, 90
communication, 320
    defined, 99
    electronic, 115-116, 135

uses of lasers in, 360-361
Communication, Linking to, 19, 35, 61, 72, 93, 105, 127, 149, 162, 194, 213, 234, 252, 274, 301, 319, 337, 353, 375, 396
communication software, 101
communication system(s), 99, 126
communication technologies, 19
communication technology, 109
compact discs (CDs), 356-357
complex circuits, 298-300
composite material, 140
compounds, 290
compression
    as force, 161 illus., 179
    video, 131
compression stroke, 218, 216
compressor, 232
computer-aided design (CAD), 36, 309-311
    defined, 309, 413
    See also CAD
computer-aided design/computer-aided manufacturing (CAD/CAM), 61, 151-152
    defined, 151, 413
    See also CAD/CAM
computer-aided drafting (CAD), 48-65, 75, 175
    defined, 49, 413
    See also CAD
computer-aided drafting/computer-aided manufacturing (CAD/CAM), 61. See also CAD/CAM
computer-aided manufacturing (CAM), defined, 319, 413
computer-aided manufacturing (CAM) systems, 319-320
computer animation, 84-97
computer control, 151-152, 308-325, 337
computer control systems, 308-325
computer engineers, 371, 374
computer-integrated manufacturing (CIM), 19, 152
computer modeling, 310-311
computer network, 113
    defined, 100, 413
computer numerical control (CNC), 19, 313-318, 325
    defined, 311, 413
computer numerical control machines, 61, 152, 313-314

computer(s)
    in animation, 89-90, 91
    for desktop publishing, 69
    development of, 301, 328
    Internet, 109
    in manufacturing, 151-152, 308-325
    on moon mission, 318
    operation of, 303
    and product design, 309-310
    in publishing, 67
    systems controlled by, 175
    testing on, 36, 54, 310-311
    types of memory of, 50
    use with robots, 328, 330, 331 illus., 336, 340, 347
computer simulations, 36, 310
computer technology, 100
concrete, 140, 171
conditioning, 145-146
conduction, of heat, 24
conductors, 293 illus., 297, 301
construction, 174, 239, 363-364. See also structures
construction equipment, 229, 239
construction technologies, 18-19
construction workers, 174
consumer products, 146
consumers, 40
containers, 205-206, 219
continuous production, 149-150
    defined, 149, 413
control
    computer numerical, 311, 313-318
    feedback, 337, 340-342, 347
    of pests, 270, 271
controlled environment agriculture (CEA), 265, 273-274, 276, 278, 285
    defined, 273, 413
control systems
    computer, 308-325
    for construction quality, 173-174, 179
control system technology, 327
control valves, 232
conveyor, 205, 274
coordinate pair(s), 56, 58, 59
    defined, 52, 413
coordinate system. See Cartesian coordinate system
cost, as design consideration, 39
costs, comparing, 90 mathematics
crane, 137, 206, 391, 399, 400

criteria, 378, 379
crop defenses, 270-271
crops, 278
    genetically improved, 270, 271
    steps in growing , 265-266, 265 illus.
cross-pollination, 269, 271
current, 295, 298, 299, 300, 303
    defined, 293, 413
    and resistance and voltage, 293-294, 296
custom production, defined, 149, 413
cycles, of gasoline engine, 216 illus.
cylinder(s), 216, 217, 218, 232

D

dams, 240-241, 278 illus.
data, 328
    display of, 39 mathematics
    in CNC, 314-315, 318
da Vinci, Leonardo, 390
dead load, 162
degree of freedom, defined, 330, 413
density, 382
design
    computer-aided, 36, 309-311, 319
    defined, 31, 413
    effect of, 35
    and engineering, 147-148
    and engineering process, 381-385
    of environments, 256-258
    ergonomic, 255-258
    for physically challenged, 257-258
    of parts, in CNC, 318
    and problem solving, 30-45
design brief, 35 illus., 45
    defined, 34, 413
designer(s), 34, 36, 309, 310, 318, 327, 375 , 391
design guidelines, for newsletters, 79
designing, for people, 263
designing documents, 71-78, 83
design process, 31-41, 34-35 illus., 249, 377
    defined, 31, 413
    using, 41
designs, testing of, 310-311
desktop publishing, 66-83

defined, 67, 413
what it is, 67-68
detail drawings, 168 illus.
diagnosis, 252, 253-254
defined, 253, 413
diagram(s)
block, in electronics, 302 illus.
of fluid power system, 236, 237
illus.
dialogue, 86, 93
dies, in forging, 143
diesel engine(s), 217, 227
compared with gasoline, 218
illus.
defined, 218, 413
digestion, process in treating
waste, 279
digital analysis, 309
digital camera(s), 75
defined, 71, 413
digital video editing, 130-131
digitizer, 50
digitizing tablet, 50, 57
dimensions, in CAD, 57, 309
diode(s), 292, 301, 302 illus., 307
direct current (dc), 295
directional light, *defined*, 352, 414
disease, prevention of, 252-253
disk, floppy, 309, 314
disk drives, hard, 50
Disney, Walt, 85, 87
distance, figuring, 210
mathematics
distance learning, 119
distribution and sales, in
manufacturing, 150-151
DNA, 270, 271, 273
documents, 67-68, 71-78
domain(s)
defined, 104, 414
top-level (TLDs), 103
doping process, 301
drafter(s), 61, 318
drafting
computer-aided (CAD), 48-65
defined, 49, 414
drag, 192, 220, 222, 394
defined, 188, 414
drawing, to scale, 169
mathematics
drawing commands, in CAD, 57,
58-59, 60
drawing programs, 69, 75
drawing(s)
in animation, 86-87, 89, 90, 91

CAD, 49-50, 52, 53-55, 65,
309, 311
in desktop publishing, 72-73
detail, 168 illus.
instrument, 379
residential, 167-169 illus.
section, 168, 169 illus.
system, 168
working, 167, 309, 325
drugs, in treatment of illness, 255
dry cells, 294
drywall, 172
dynamic loads, 162
dynamics, 395

E

echo, 403
economy, 152, 153
Edison, Thomas, 31
editing
digital video, 130-131
of images, 69, 71
editing commands, in CAD, 60
effectors, robotic arm and, 331
illus.
efficiency, in manufacturing, 320
Eiffel Tower, 378
elastic limit, 166
electrical circuit, 297-300
electrical energy, 223, 259, 288,
291, 294, 295
electrical engineers, 303, 373
electrical systems, 171
electric cars, 223
electric energy, 297
electricity, 289-294
defined, 289, 414
and electronics, 288-307
sources of, 294-296
used by maglev trains, 213
electric power, 229, 240
electric railroad, 206 illus.
electrocardiograph (EKG), 254
electrodes, 294, 295
electroencephalogram (EEG), 254
electrolyte, 294, 295
in maglev trains, 212, 213
electromagnetic induction, 295
electromagnetic radiation, 351
electromagnetic wave(s), 116, 351
illus., 369, 406, 407
defined, 351, 414
speed of, 124 *mathematics*
electromagnetism, *defined*, 296,
414

electromagnet(s), 295 illus., 296
electromotive force (emf), 293.
See also voltage
electron cloud, 290 illus., 291
electronic communication, 115-
116, 135
electronic cottage, 14
electronic mail, 104. *See also*
e-mail
electronics, 288-307, 300-304
electronics engineers, 48, 373
electronic signal(s), 122, 135, 301
electronic switches, 301-302
electronic systems, 303
electron(s), 289, 290
control of, 301
energy levels of, 291, 405
flow of, 291, 293, 294, 296,
298, 300
valence, 291, 293, 301
elements (matter), 289-290
elements (parts), of systems, 207
elevation drawing, 168 illus.
elevators, 218
e-mail, 107
defined, 104, 414
employees, displaced, 343
employment, in construction, 174
end effectors, 330
end mill, 313
energy
biomass, 183
converting of, 117, 294
electrical, 288, 291, 294, 295,
297
geothermal, 182
heat, 216
kinetic, 189
light, 404-406, 407
magnetic, 212, 254
mechanical, 216, 218, 295
nuclear, 183
for photosynthesis, 274
in plant growth, 266
potential, 189, 293
and power, 182-183
solar, 182
sound, 259, 402, 403
sources of, 182-183
in transportation, 207, 223,
227
energy levels, and electrons, 291
energy sources, primary, 294
engineering, 263, 370-389
biomechanical, 263
computer, 249, 272, 374

defined, 371, 414
and design, 147-148
genetic, 259, 265, 273, 281
human factors, 249
types of, 372-375
engineering models, 379
engineering process, 377-379,
    381, 389
    case study of, 381-385
    defined, 377, 414
engineer(s), 109, 309, 310, 318,
    319, 350, 377, 391
    biomechanical, 249, 251, 259
    civil, 48, 372
    in communication, 117, 122
    defined, 371, 414
    design skills of, 206
    electrical, 303, 373
    electronics, 48, 373
    in fluid power, 229, 236, 241
    human factors, 255, 256
    preparation for becoming, 385
    in robotics, 327
    team of, 383
    types of, 371, 372-374
engine operation, scientific
    principles in, 217
engine(s), 214, 402
    diesel, 217, 218, 222, 227
    gasoline piston, 216-217, 227
    heat, 216
    jet, 197 illus.
    reaction, 194, 196-197, 408
    rocket, 198-199 illus.
English Channel Tunnel
    (Chunnel), 210, 223
enhanced senses, 259
entity, in CAD, 57
environment, 270, 276, 284
    bioprocessing and, 279
    controlled, in agriculture, 265,
        273-274, 276-278
    effects on, 223, 281
    growing, 273, 274 illus.
    impacts on, 152, 240-241
    simulating of, 310
    water, 278
environmental engineers, 374
environmental technologies, 264-
    285
environment design, 256-258
ergonomic design, 255 illus., 257
ergonomic features, for persons
    with disabilities, 258
ergonomics, 45, 255-258, 263

defined, 39, 255, 414
escalators, 218
escape velocity, 395
estimating, 363 *mathematics*
exhaust, vehicle, 223
exhaust stroke, 217
exposure sheets, 86
exterior finishing, 173
exterior walls, 171
external-combustion engines, 216
external force(s), 161, 162, 179
extrusion, 143 illus.

F

factories, 139, 320, 329, 330
factory-made houses, 175 illus.
FAQ area, in Internet newsgroup,
    107
farmers, 265, 270, 272, 281
farming, 266, 272, 279, 281
fasteners, mechanical, 144
feature film production, 91, 93
feedback, 79, 167, 207
    defined, 24, 414
    as part of a system, 159
feedback control, 337, 340-342,
    347
    defined, 340, 414
feedback control process, 340
    illus.
feedback-control system, 24
fiber optics, lasers and, 350-369
fiber-optic systems, 360
filament, 351, 406, 407 illus.
film, production of, 91
films, animated, 85, 87, 89, 90,
    97
finishing, of a structure, 172-173
firefighters, 196-197
firing, in making ceramics, 140
fission, nuclear, 183
flatcar, railroad, 206, 211
flexibility
    in manufacturing, 150
    of joints in robots, 330
flexible manufacturing system
    (FMS), 325
    defined, 320, 414
flight, 184-203
flip-book, 85, 87
floor, framing of, 171
floor plan, 32, 167 illus.
    drawing to scale, 169
    mathematics
floppy disk, 50, 309, 314

flowchart, 314, 317 illus., 336,
    337
flow regulators, 232, 234
fluid friction, 189, 393, 394
    defined, 188
fluid power, 228-245
    defined, 229, 414
fluid power systems, 230, 233
    illus.
    in aircraft, 238
    components of, 232-234, 245
    types of, 229, 232-236, 245
fluid power system safety, 230,
    232
fluid(s), 230, 393
    defined, 229, 414
    in fluid power systems, 232
fluid science, 229-230
food production, 272, 279, 281
footings, 171
force, 411
    buoyant, 220, 221 illus.
    calculating of, 235
        mathematics, 397
    defined, 161, 185, 393, 414
    electromotive (emf). *See*
        voltage
    of gravity, 185, 186, 394, 395
    of magnets, 213
    and mass and acceleration, 188
        mathematics, 397
    to move electrons, 296, 291
        illus.
    multiplying of, 234-235 illus.
    of potential difference, 293
    transfer of, 234-235 illus.
    unbalanced, 395
forces, 185-187, 196
    on aircraft, 187 illus.
    on beams, 163 illus.
    and flight, 187, 193 illus., 203
    and motion, 188-189 illus.
    that oppose motion, 186-187,
        203
    pairs of, 397
    on structures, 161-162, 179
    transfer of, in robotics, 327
Ford, Henry, 13
forging, 142 illus., 143, 146
forming, 142-143
fossil fuels, 182, 213, 223
foundation, 170 illus.
    defined, 171, 414
four-stroke cycle, of engine, 216-
    217, 216 illus.

frame, members used in, 163
framing, 171
free electron, 291
freighters, 219, 220
freight trains, 211
frequency
    in broadcasting, 124
    of light, 406, 407
    of sound, 115, 118
frequency modulation (FM),
    defined, 117, 414
friction, 187-189, 220, 393-394
    defined, 186, 393, 414
    and technology, 393-394
fuel(s), 198, 203, 213, 216, 218,
    223
fulcrum, 22
Fulton, Robert, 221
fuse, in a circuit, 299, 300 illus.
future
    of applied physics, 407
    of CAD, 61
    of computer control systems,
        320-321
    of desktop publishing, 79
    of electricity and electronics,
        303
    of engineering, 385
    of farming, 281
    of fluid power systems, 241
    of health technologies, 259
    of Internet, 109
    of lasers, 365
    of manufacturing, 153, 320-
        321
    of multimedia, 131
    of robotics, 343
    of structures, 175
    of technology, 25
    of television, 125
    of transportation, 223

G

gas(es), 217, 223
    molecules of, 230
    as state of matter, 229, 289
gasoline engines, compared with
    diesel, 218 illus.
gasoline piston engine, 216-217,
    227
    defined, 216, 414
gene, 272
gene farming, 281
generator(s), 209, 210, 240, 295,
    297, 307

defined, 295, 414
gene-splicing, 270 illus.
genetically altered plant, 270, 276
genetically improved crops, 271
genetic engineering, 265, 273,
    281
    defined, 259, 272, 414
genetic information, 269
genetics, 272, 285
    defined, 267, 414
geometry, Cartesian, 52
global economy, 153
Goodyear, Charles, 147
governmental system, 206
grafting, 265, 272
graphics, 68, 72-73, 75, 79, 83
gravitational force, 394, 395
gravity, 186-187, 203, 394-395,
    411
    defined, 186, 394, 414
greenhouses, 274, 276
gross domestic product (GDP),
    152
ground-effect vehicles, 222
guideway, 212, 213

H

hands, robotic, 330
hard disk drives, 50
hardening, 145
hardware
    CAD, 50-51, 65
    for desktop publishing, 69, 71
    and software, 68-69, 71, 100-
        101
hardwood, 139
harvesting, 266
health care, 240
health care technology, 252-255
health technologies, 248-263
hearing, 258, 259
heat, changing of
    into light, 406
    into mechanical energy, 216
heat conduction, 24
heat engines, 216
helicopters, 194, 195 illus., 203
heredity, *defined*, 267, 415
hertz, 118
holes, in semiconductors, 301
Hollerith, Herman, 36, 328
holograms, 360-361, 363
holographic system, 361 illus.
holography, 360-361, 363
    defined, 360, 415

home-heating system, 23-24
home page, 103, 110, 111 illus.
horsepower (hp), 402
    mathematics
hotspots, 126, 127
houses, 163, 175 illus.
Hovercraft®, 222, 223 illus.
hull design, and buoyancy, 220
human factors engineering, 263
    defined, 249, 415
human factors engineers, 255,
    256
human systems, model for robotic
    systems, 330
human technology resources, 249-
    250
humidity, in plant growth, 274
hydraulic lines, 232
hydraulic power, 333
hydraulic system(s), 229 illus.,
    234-235, 245
    components of, 232-234
    defined, 232, 415
    in robotics, 336
    safety in using, 230, 232
hydrofoils, 222
hydroponic growing systems, 268
hydroponics, 276-278, 285
    defined, 276, 415
hyperlinks, 113
    defined, 103, 415
hypermedia, *defined*, 126, 415
hypertext, 126-127
    defined, 103, 415

I

igloo, 159
illustrations, 73, 75
image editing program(s), 69, 71,
    75
images
    in animation, 85
    in holography, 360
immune system, 250, 263
immunization, *defined*, 252-253,
    415
impact(s), 24
    of applied physics, 407
    of bio-technology, 279. 281
    of CAD, 61
    of computer control systems,
        320
    of desktop publishing, 79
    of electronics and electricity,
        303

of engineering, 385
of fluid power systems, 240-241
of health technologies, 258-259
of Internet, 109
of lasers, 365
of manufacturing, 152-153
of multimedia, 131
of robot technology, 343
of structures, 174
of technology, 24-25
of television, 124
of transportation, 222-223
implants, artificial, 249
incandescent lamp, 406
inclined plane, 401 illus.
incoherent light, 352
Industrial Age, 16
industrial designer(s), 318, 375
industrial materials, 150
    defined, 141, 415
industrial products, 146
Industrial Revolution, 372
industries, use of fluid power
    systems in, 236, 238-240
inertia, 187, 398
    defined, 185, 396, 415
information, 16, 249, 252, 320
    gathering of, 34-35, 377
Information Age, 16, 17 illus., 320
infrastructure, 372
innovation, defined, 31, 415
input, 167, 207, 227, 356
    CNC data, 314-315, 318
    defined, 23, 415
input devices, computer, 50, 309
inputs, 303
    as part of a system, 159
inspection(s), 145 illus., 174
instrument drawing, 379
insulation, in a structure, 172
insulators, 24, 293, 301
intake stroke, 216, 218
interactive technology (IT), 342
interchangeable parts, 321
interface, robotic, 336
interior finishing, 172-173
intermodal transportation system,
    206
internal-combustion engine, 216
internal forces, 161, 179
Internet, 98-113, 131
    building the, 99-100
    getting on the, 100-101
    use of, 100, 101, 103, 113

Internet computers, 109
Internet relay chat (IRC), 106
Internet safety, and appropriate
    use, 101, 103
Internet service providers (ISPs),
    101
invention, defined, 31, 415
ion, positive or negative, 291
ionosphere, 117, 118
isotopes, 290

J

Jefferson, Thomas, 49
jet engine, 197 illus.
jet planes, 194, 196-197
job-lot production, defined, 149,
    415
job market, impacts on, 28, 320
job requirements, 244
jobs, 226, 320
    done by robots, 338-339 illus.
joints, 249, 250, 327, 330
joists, 171, 175
joule (J), 398
junction diode(s), 301, 302 illus.
just-in-time manufacturing, 320

K

keyboard, 50, 57, 260, 309
key frame, defined, 93
keyword, 105, 106
kinetic energy, 189
Knight, Phil, 31

L

labor-saving machines, 266
land transportation, 207-218
    and water transportation, 204-
    227
language
    changed by computers, 105
    on Internet, 101
languages, computer, 314
laser, 358, 369
    defined, 352, 415
laser applications, 360-361
laser beam, 356, 407
laser light, 359, 369
    and ordinary light, 352, 353
    illus.
laser printers, 50, 365
laser process, 359 illus.
lasers, 352-353, 358-361, 363-
    365, 407
    classes of, 353, 369

and fiber optics, 350-369
how they work, 358
types of, 358-359
laser safety, 353
laser system, 358 illus.
latex, 140
lathe, 313
law of charges, 291
law(s)
    for design of public places, 257
    for hiring, 178
    for Internet, 109
    zoning, 174
laws of motion. See Newton's laws
    of motion
leading, defined, 72, 415
lead-through programming, 337
LEDs, 292
lever, 22, 401 illus.
library of symbols, CAD, 309
lift, 187, 190 illus., 192, 194, 197,
    203, 222, 311
lifts, hydraulic, 236
light, 350, 369
    laser, 352, 353 illus.
    nature of, 351-352
    ordinary, 352, 353 illus.
    properties of, 406
    sources of, 406-407
    speed of, 358
    splitting of, 352
    use in plants, 266, 274
    visible, 406, 407
lightbulb, 31, 351, 406, 407 illus.
light emitting diodes (LEDs), 292
light energy, 404-406, 407
lighting, in animation, 93
light sensors, on CDs, 357
light wave(s), 352, 404-407
    characteristics of, 406 illus.
    sound waves and, 402-407
    and technology, 406-407
limbs, artificial, 19, 249, 258, 259
linear motion, 214, 232
line art, 73
lines
    in document design, 72
    drawn in CAD, 57
links, on the Web, 103
liquid(s)
    molecules of, 230
    as state of matter, 229, 289
    in transferring force, 234
live loads, 162
living areas, design of, 256-258

load-bearing walls, 171
load(s)
    in circuit, 297
    defined, 162, 415
    on structures, 179
locomotive, 206, 207, 208-210
logic, fuzzy, 287
lumber, 141

M

machine program code, 318
machine(s), 253-254, 266, 372,
    401, 403, 411
    computer numerical control,
        318, 325, 311, 313-314
    defined, 400, 415
    in manufacturing, 152, 313, 320
    mechanical advantage of, 401
    and quality of life, 259
    simple, 401 illus.
    and work, and power, 398-401
machine tool(s), 311, 313, 314,
    329
machinist, 311, 313, 318
maglev, 212-213
    defined, 212, 415
maglev train(s), 212 illus., 213,
    227
magnetic field(s), 212 illus., 227,
    295, 296
magnetic levitation, 212
magnetic resonance imaging
    (MRI), 254
magnetism, and electricity, 295,
    296
mailbox, for e-mail, 104
mailing lists, 107
manipulators, 330
manual data input (MDI), 315
manufacturing, 138-157
    automation in, 343
    computer use in, 308-325
    defined, 139, 415
    fluid power systems in, 236
    jobs in, 320
    just-in-time, 320
    lasers in, 363
    steps in, 147-151, 148 illus.
manufacturing engineers, 373
manufacturing system(s), 23, 146-
    151, 157, 319, 320, 325
manufacturing technologies, 18
marine transportation, 205, 219,
    227

mass production, 149, 213
mass transit, 211
material resources, in health
    technologies, 250
materials, 248
    boat-building, 220
    conductivity of, 293, 304
    development of, 374
    in electronics, 301
    genetic, 259
    for growing plants in, 276
    industrial, 141
    in manufacturing, 139-140,
        153, 156, 279
    for models, 36
    moving of, 236
    natural, 139, 140, 157
    and parts, in machining, 313
    in physical enhancements, 250
    processing of, 140-146, 363
    raw, 140
    semiconductor, 302
    smart, 175
    structural, 162
    synthetic, 139, 140, 157
    toxic, 250
    transport of, 211, 213
materials engineers, 374
mathematical models, 36
Mathematics, Linking to, 25, 39,
    52, 75, 90, 101, 124, 142,
    169, 188, 210, 241, 253, 269,
    297, 314, 333, 363, 382, 402
matter, 229, 289-290
measurement, use of lasers for,
    363
measurement system, for type, 71-
    72
measuring pressure, 236
    mathematics
mechanical advantage, 235
    defined, 234, 401, 416
mechanical conditioning, 146
mechanical energy, 206, 216, 218,
    295
mechanical engineers, 371-372,
    373
mechanical power, 229
mechanical processes, 141
mechanization, 265-266
medical problems, solving of, 249,
    253, 263
medicine, uses of lasers in, 364

medium, 402, 404
memory, of computers, 50
messages, on Internet, 104
metal(s), 139-140, 141, 143, 145-
    146
metric units of measure, 398
micrograph, 259 illus.
microorganisms, 279, 280
microphone, 117 illus.
microprocessors, 241, 259
microwaves, 407
microwave television system, 124
milling machine, 313
mining engineers, 372
mixing, 135
    as combining process, 144
    in radio broadcasting, 117
mixtures, 290
mode, 205
model animation, 89, 97
modeling, computer, 91, 310-311
modeling techniques, 36, 39
model(s), 36, 39, 45, 54
    in CAD, 36, 54, 55 illus.
    engineering, 379
    mathematical, 36
    working, 36, 379
modem, *defined*, 100, 416
    cost of using, 101 *mathematics*
modes, 205-206
mold, used in forming, 143
molecules, 229-230, 273, 290
monitor(s), 50, 71
monochromatic light, *defined*, 352,
    416
motion, 391, 393-395
    directions of, 214, 217, 232,
        234
    and forces, 188-189 illus.
    forces that oppose, 186-187
    joints and, 330
    laws of. *See* Newton's laws of
        motion
    simulating, 93
motor(s), 207, 209, 232, 332, 347
motor vehicles, 213, 216
mouse, computer, 50, 57, 309
movement, illusion of, 85
multimedia, 135
    and audio and video, 114-135
    defined, 115, 416
multimedia software, 126
multimedia video, 130
myoelectric prosthetics, 19, 258

N

nanotechnology, 330
NASA, 381
need, identifying, 32-34
needs, 15-16, 29, 161, 258
nematodes, 264, 269
netiquette, 101
netspeak, 103
network(s), 100, 113, 124, 125
neutrons, 290
newsgroups, on Internet, 107
newsletter(s), 76-77 illus., 80
    planning of, 78-79
Newton, Isaac, 185, 354, 395
newton-meter (N-m), 398
newton (N), 394 illus., 398
Newton's laws of motion, 185-
    186, 196, 395-398, 408, 411
nuclear engines, in rockets, 198
nucleus, as part of an atom, 290
number system, binary, 303
numerical control, 329. *See also*
    computer numerical control
nutrients, in hydroponics, 266,
    276, 278

O

object, in CAD, 57
ohm, 296
Ohm's law, 307
    defined, 296, 416
oil, 218, 232
on-line services, 101
on-site land transportation, 218
optical fibers, 360 illus.
optical scanners, 341
organs, 249, 259
origin, in Cartesian coordinate
    system, 52
orthographic projection, 62
output force, 235
output(s), 167, 207, 303, 357
    defined, 24, 416
    figuring, 25 *mathematics*
    as part of a system, 159
oxidation, 145
oxygen, 198, 216, 266, 278

P

packaging, product, 151, 375
page layout programs, *defined*, 68,
    69, 416
page layout software, 83
painters, in animation, 87
paint program(s), 90, 312

parallel circuit(s), 298, 299 illus.,
    307
part design, 318
partitions, 171
Pascal's Principle, *defined*, 230,
    234, 416
passenger service, 210-211
patent, 371 illus.
people movers, 218
periodic table of elements, 290
perlite, 276, 277
persistence of vision, *defined*, 85,
    416
personal computer (PC), 68, 69
pesticides, 270, 271
pH level, 273, 278
photocopier, 67
photographs, 68, 69 illus., 71, 73
photon(s), 405-406
    defined, 351, 416
photosynthesis, 267 illus., 274
    defined, 266, 416
photovoltaic cell, 297
physical enhancements, 250, 251
    defined, 249, 416
physics, 391
    applied, 390-411
pica, as type measure, 71-72
pilots, 192, 238
piston(s), 208, 216, 234
pitch, 194, 195 illus., 203
planes (surfaces), adjacent, 161
planes (vehicles). *See* airplanes;
    jet planes
plan preparation, for a structure,
    166-169
plans, for building a structure, 56,
    167-168, 169
    defined 167, 416
plant, genetically altered, 270,
    276
plant growth, 266
plastics, 140, 250
plot plan, 167
plotter(s), computer, 50, 309
plow, 266
plumbing systems, 171
plywood, 140, 171
pneumatic circuit, 237 illus.
pneumatic system(s), 235-236,
    245
    defined, 232, 416
    and hydraulic, in robotics, 336,
        333
    safety rules for using, 230, 232
points, as type measure, 71-72

polarity, 295
poles, of magnets, 212 illus., 213
pollination, 269, 272
pollutants, 207, 223, 279
pollution, 213, 223, 374
potential, of a charge, 293
potential difference, 294
    defined, 293, 416
potential energy, 189, 293
potentiometers, 299
power, 209-210, 216, 399-400,
    411
    calculating, 183, 399, 400
    defined, 399, 416
    fluid, 228-245
    hydroelectric, 183
    and machines, and work, 398-
        401
    for robotic movements, 332-
        333, 336
powering
    of maglev trains, 212 illus.
    of rail systems, 207-210
    of vessels, 221-222
power source, in circuit, 297
power stroke, 216, 218
power systems, 229
    fluid, 230, 233 illus., 232-236,
        238, 240, 245
pressure, 190, 230, 267
    calculating of, 230, 235
    defined, 230, 416
    measuring of, 236 *mathematics*
primary energy sources, 294
primary processes, 140-141, 141
    illus., 149, 157
    defined, 141, 416
primitives, *defined*, 91, 416
printers, 50, 71, 365
printing, 79
problem, statement of, 34, 45,
    377
problem solving, 45, 308
    design and, 30-45
    group method of, 35
problem-solving process, 30, 78,
    308, 377
process
    defined, 23, 416
    design, 31-41, 34-35 illus., 387
    engineering, 377-379, 381, 389
    feedback control, 340 illus.
    laser, 359 illus.
    manufacturing, 23
    in preparing plans, 167
    problem-solving, 377

publishing, 67
technological, 279
transportation, 207, 227
wastewater treatment, 280
   illus.
process colors, 78 illus.
processes
   biological, 280
   electronic communication, 115
   in electronic systems, 303
   manufacturing, 18, 311
   as part of a system, 159
   primary, 140-141, 149, 157
   secondary, 142-146, 149, 157
processing materials, 140-146
product design, 308, 309-310
product designers, 375
production
   in manufacturing, 149-150
   television, 120-122
production systems, 149, 157
productivity, 278, 373
products, 40, 146, 211, 213, 279,
   375
products, interactive, 126
programmer(s), 317, 318, 333,
   336, 337
programming, lead-through, 337
program(s)
   animation, 90, 97
   CAD, 52, 57, 58, 59, 60, 309,
      310
   CD-ROM, 126-127
   CNC, 315, 317, 318
   in desktop publishing, 68, 69,
      71, 75, 83
   engineering, 310
program signal, 117
propellers, 192, 222
prostheses, 249, 263
prosthesis, 251
   defined, 249, 416
prosthetics, myoelectric, 258
protons, 290
prototype(s), 36, 45, 54, 147,
   310, 379
   defined, 36, 148, 416
publication, planning of, 83
publishing
   defined, 67, 416
   desktop, 66-83
publishing process, 67
pulley, 401 illus.
pump, 232
punched tapes/cards, 313, 314,
   328, 329

punch list, in building, 174
puppets, 89
purchasing, in manufacturing, 149
purchasing agents, 109, 149

Q

quadrants, 52
qualifications, for hiring, 178
quality
   control systems, 173-174
   in desktop publishing, 71
quality control, 145 illus., 150
quality of life, 258, 259
questionable developments, in
   health technologies, 259

R

raceway aquaculture, 278
radiation, electromagnetic, 351
radio, 115, 116-119, 135
rafters, 171
railroad cars, 211 illus.
railroad, electric, 206
rail transportation, 207-213
   of today, 210-211
railways, 205, 206
rainbow, 352
random-access memory (RAM),
   defined, 50, 416
raw materials, defined, 140, 416
reaction
   action and, 397, 398
   chemical, in generating
electrical energy, 294, 295
reaction engines, 194, 196-197
read-only memory (ROM),
   defined, 50, 416
receiver, 115, 118, 232
reception, radio, 118-119
reciprocating motion, 232
recorders, videotape, 125
recycling, 47, 149, 152-153, 236
remedies, folk, 252
Remote Manipulator System
   (RMS), 341-342
rendering, in animation, 93
renderings, 36
research, 99, 109, 248, 252, 264,
   279, 281
reservoir, in hydraulic systems,
   232
residential structure, building of,
   166-173
resistance, 186, 188, 298
   defined, 294, 417
   finding of, 297 mathematics

and voltage and current, 293-
   294, 296
resistors, 298, 299
resolution, defined, 71, 417
resources, 16, 29, 253
   human technology, 249-250
   of technology, 16, 18, 23 illus.,
      139
rheostats, 299
robot generations, 342
robotic arm(s), 18 illus., 327, 330,
   331 illus., 341
robotic automation, 343
robotic hands, 330
robotics, 326-347
   defined, 327, 417
   development of, 327-329
robotic systems, 327, 330
   controlling, 336-337, 340-342
   modern, 329
robotic vehicle, solar powered,
   339 illus.
robot(s), 328, 329, 338-339 illus.,
   347
   defined, 327, 417
   interactive, 342
   modern, 327-328
   programming of, 337
   sensors in, 337, 340-342
   in work force, 330
rockets, 198-199, 203
rolling, 142, 143 illus.
rolling friction, 393, 394
rolling stock, 208
roof, framing of, 171
root pressure, 267
rotary motion, 214, 217, 232, 234
rotors, of helicopters, 194, 195
rough work, in construction, 171
rubber, 140, 146, 147
rudder, airplane, 192, 193, 203
rust, 145

S

safety
   fluid power system, 230, 232
   on Internet, 101
   with lasers, 353
   in manufacturing, 152
safety colors, 153
sailboats, 220, 221 illus.
sales, and distribution, 150-151
satellite networks, 124, 125
scale, drawing to, 169
   mathematics
scale models, 36
scanner(s), 341, 365

defined, 71, 417

schematic circuit diagrams, 236, 237 illus.

schematic symbols, 237 illus., 302 illus.

science, 391

Science, Linking to, 24, 41, 54, 79, 85, 107, 117, 145, 166, 190, 217, 236, 258, 267, 294, 311, 330, 352, 382, 398

science web sites, 107

scientific method, 41, 391 illus.

scientists, 109, 253, 259, 272, 301, 330, 359, 391

screw, as simple machine, 401 illus.

search engines, for using Internet, 105, 106, 113

seat belts, 396

secondary processes, 142-146, 149, 157
   defined, 142, 417

section drawings, 168, 169 illus.

seeds, 265, 266, 269

selective breeding, 272, 417

semiconductors, 301, 302

senses, enhanced, 259

sensor(s), 122, 175, 328, 357
   in robots, 337, 340-342

separating, 143

separation processes, 311

series circuit(s), 297-298 illus., 307

shading, lighting, and rendering, in animation, 93

shear, as force, 161 illus., 179

shearing, 143

sheathing, 171

shells, valence, 291, 301

ship(s), 205 illus., 206, 219

signal(s), electronic, 122, 135, 301

silicon, 140

silicone, 250

simple circuit, 297

simple machines, 401 illus.

simulation software, *defined*, 310

site plan, 167

site preparation, 169

site selection, 166

sketching, 35, 378

skills
   in animation, 89-90
   required for jobs, 244

smart buildings, 175

smart materials, *defined*, 175

society, impacts on, 24-25, 152, 222-223

software, 52, 130, 131
   animation, 89, 90, 93, 311
   CAD, 50, 52, 65, 309-311, 319, 325
   CAM, 319
   communication, 101
   for desktop publishing, 68-69, 83
   and hardware, 68-69, 71, 100-101
   for Internet, 104, 106
   multimedia, 126
   robotic, 337
   simulation, 310

softwood, 139

solar cell, 297

soldering, 144

solid models, in CAD, 54, 55 illus.

solid(s)
   molecules of, 229-230
   as state of matter, 229, 289

solution (liquid), nutrient, 276, 277, 278

solution(s) (answers), 389
   developing alternative, 35-36, 378
   evaluation of, 40-41, 379, 381
   implementing of, 39-40, 379
   selecting, 36, 39, 378-379

sonar, 403

sound, 117, 253, 402, 403
   defined, 115, 417

sound energy, 259, 402, 403

sound waves, 402-403
   and light waves, 402-407

space flight, as engineering case study, 381-385

Space Shuttle orbiter, 198, 199 illus., 341-342

spark plug, 217

speaker, radio, 119 illus.

special effects, 85

specifications, *defined*, 168, 417

spectrum, electromagnetic, 351 illus.

speed, 187
   of electromagnetic waves, 124
   of light, 116, 406
   of modem, 100
   of sound, 403
   of vehicles, 210, 213, 220

sphere, volume of, 333
   mathematics

spreadsheet, 42

stamping, as forming process, 143

standard, *defined*, 379, 417

standardization, 321

static load, 162

steam engines, on ships, 221-222

steel, 220

stepper motor(s), 332, 347

stock, as workpiece, 311, 313

storage, 206

storyboard(s), 91, 97, 121
   defined, 86 illus., 417

strokes, in engine operation, 216, 217, 218

structural materials, 162

structural member(s), 163, 171, 179
   defined, 163, 417

structural shapes, 166

structure(s), 158-179, 160 illus., 161, 163, 176
   building a residential, 166-173
   defined, 159, 417
   finishing of, 172-173
   forces on, 161-162

studio, 116

studs, 171

substructure, 171

subsystems, 206
   of lasers, 358, 369
   systems and, 24
   of transportation, 206-207

superproductivity, in farming, 281

superstructure, 169, 170 illus., 179
   defined, 171, 417

supports, 163, 164-165, 171

surface friction, 393

surface models, in CAD, 54, 55 illus.

surgery, 255, 327, 342, 364

survey, of a site, 169

surveying, 363, 364

switches, 298, 299, 302

symbols
   in drawings, 236, 237 illus.
   in electrical measures, 296
   in electronics, 302 illus.
   library of, in CAD, 309

synthetic materials, 139, 140, 157

system, 29, 206
   binary number, 303
   Cartesian coordinate, 52-53, 54, 55 illus.
   in construction, 176
   defined, 23, 417

feedback-control, 24
governmental, 206
holographic, 361 illus.
home-heating, 23-24
immune, 250, 263
laser, 358 illus.
manufacturing, 23, 320, 325
mass transit, 211
measurement, for type, 71-72
parts of, 23-24, 29, 159, 179
requirements of, 166
technological, 206, 265
system diagrams, 236, 237 illus.
system drawings, 168
systems, 241, 360
aeroponic, 276, 277 illus.
aggregate, 276, 277 illus.
basic power, 229
CAD, 36, 50-52, 57, 60, 61, 65
CAD/CAM, 319-320, 325
communication, 99, 126
computer-aided manufacturing, 319-320
computer control, 308-325
control, in construction, 173-174, 175
electronic, 303
elements of, 207
fluid power, 229-230, 232-236, 238, 240, 245
hydroponics, 268, 276
numerical control, 314-315
power, 183
robotic, 329-332,336-337, 340-342
and subsystems, 24
in technology, 23-24
telephone, 109
in transportation, 205, 206-210, 218, 223
utility, in a building, 171
water culture, 276, 277 illus.
systems engineers, 372

T

tankers, 219, 220
team(s)
of engineers, 383, 384
publishing, 80, 81
of scientists, 328
working as, 33, 94
teamwork, 153
technician, 122
technological process, 279
technological system, 206, 265

technologies, 20-21
bio-related, 246-285
communication, 19, 46-135
construction, 18-19
control, 286-347
development of, 15-16
environmental, 264-285
health, 248-263
integrated, 348-411
manufacturing, 18
new, 373, 374
power, 180-245
production, 136-179
resources of, 18, 29
transportation, 19
technology, 29
activities of, 18-21, 29
biorelated, 19, 249, 265, 273, 279, 281, 285
in cereal making, 247
computer, 100
control system, 327
defined, 15, 417
friction and, 393-394
gravitational force and, 395
health care, 252-255
how it works, 14-29
interactive (IT), 342
light waves and, 406-407
resources of, 16, 18, 23 illus., 139
and society, 13, 47, 137, 181, 247, 287, 349
systems in, 23-24
tools of, 15-16
in transportation, 204, 222
technology resources, human, 249-250
telecast, 115
teleconferencing, video, 119
telephone system(s), 109, 360
television, interactive, 125
television cameras, 122-124
television (HDTV), high-definition, 125
television picture tube, 125 illus.
television program, producing of, 120-122
television reception, 124
television(s), 109, 115, 119-125
televisions, Web, 109
television transmission, 123 illus., 124, 135
tempering, 145
tension, as force, 161 illus., 179

terminals
battery, 293
in airports, 218
testing, 40, 54, 310-311
text, in word processing, 69
text area, figuring, 75 mathematics
textiles, 140
thaumatrope, 92
thermal conditioning, 145
thermal process, 141
thermoplastics, 140
thermoset plastics, 140
three-dimensional (3D)
animation, 90, 91
three-dimensional drawings, in CAD, 65, 309
three-dimensional images, in holography, 360
three-dimensional models, 36
thrust, 190, 194, 196-197, 198, 203
defined, 187
time, in animation, 89, 90, 97
Titanic, and robotic divers, 328
titanium, 250, 252
tool bar, 104 illus., 105 illus.
tool(s), 15, 49, 144 illus., 236, 309
machine, 311, 313, 314, 329
of medical technology, 254
of technology, 15-16
used in diagnosis, 254
for using Internet, 105
top-level domains (TLDs), 103
torsion, 161 illus., 179
tractor-trailer, 206
trade, international, 321
trade-off, 378
train, high-speed, 204 illus., 210 illus.
train(s), 204 illus., 206, 210, 211, 227
trait, 267, 272
transistor(s), 307, 328
defined, 301, 417
uses and types of, 301-303
transmission lines, 234
transmitter, 117, 118, 122
transportation
defined, 205, 417
fluid power systems in, 238
land and water, 204-227
modes of, 205-206
on-site, 218
rail, 207-213

systems in, 206-207
transportation system(s), 16, 205, 206-207, 223
transportation technologies, 19
treatment, 252, 255
triangle, as structural shape, 166
ttrusses, 165, 166, 175
turbines, 295
two-dimensional (2D) animation, 90
two-dimensional drawings, in CAD, 53-55
two-dimensional models, 36
type, 71-72
typeface(s), 79
    defined, 72, 417

U

UHF, defined, 124, 417
ultrasound, 254 illus., 403
Uniform Product Code (UPC), 365
Uniform Resource Locator (URL), 110
    defined, 103, 417
units of measure, in electricity, 296
UPC symbol, 365
upthrust, 220, 221 illus.
utility commands, in CAD, 60
utility systems, 171

V

vaccines, 240, 252, 253
vacuum, 404
vacuum tube, 301
valence electrons, 291, 293, 301
valence shell(s), 291, 301
valve(s), 216, 232, 250 illus.
vegetative propagation, 272, 273 illus.
vehicles, 205, 206, 213, 216, 222, 223
velocity, escape, 395
Venture Star, 349
vermiculite, 276, 277
vessel(s), 205, 219-222
    defined, 219, 417
VHF, defined, 124, 417
vibrations, and sound, 117
video, 135
    defined, 115, 417

on Internet, 109
and multimedia and audio, 114-135
video cameras, 125
video compression, 131
video editing, digital, 130-131
video recorders, 125, 126 illus.
video recorders (VCRs), 125
videotape, 130
video teleconferencing, 119
virtual reality, 40 illus., 61
visible light, 406
vision, persistence of, 85
volt, 296
voltage, 295, 297, 298, 299, 300
    and current and resistance, 293-294, 296
    in transistors, 302
voltage source, 294
volume, 230
vulcanization, 146

W

wall(s), 171, 172
    section drawing of, 169 illus.
warehouse, 206, 320
wastewater, treating of, 279, 280 illus.
water culture systems, 276, 277 illus.
water environment, 278
waterjets, in hydrofoils, 222
water pressure, as a force, 162
water transportation, 219-222
Watt, James, 402
watt (W), 400
wave diagram, 115 illus.
wavelength, 115, 351 illus., 406
wave(s), 402
    electromagnetic, 351, 369, 406, 407
    light, 352, 404-407
    radio, 118
    sound, 402-403
ways, in transportation, 205
Web, the, 103
web browsers, 104, 105, 107
web site(s), 103, 107, 110
Web telephone service, 109
wedge, 401 illus.
weight, 394-395

wet cells, 294
wheel, invention of, 207
wheel and axle, 401 illus.
wind power, 182, 221
wind pressure, as a force, 162
wind turbine, 182
wing(s), 189-190, 203, 222
wireframe drawings, 54, 55 illus.
wood, 139, 220
    figuring usable, 142 mathematics
    processing of, 141 illus.
wood frame, 171
word processing, 69
work, 398-399, 411
    calculating, 398, 399
    defined, 398, 417
    of engine, 216
    performed by fluid power, 229, 245
    potential to do, 293
    and power, and machines, 398-401
    result of electricity, 289
work envelope 330, 347
    defined, 332, 417
    volume of, 333 mathematics
workers
    construction, 174
    in health technologies, 249, 262
    in manufacturing, 152
working areas, design of, 256-258
working drawings, 167, 309, 325
working model, 36, 379
workpiece, in CNC, 311, 315
Workplace, Linking to, 28, 44, 64, 82, 96, 112, 134, 156, 178, 202, 226, 244, 262, 284, 306, 324, 346, 368, 388, 410
workstation(s), 149, 255-256, 320
World Wide Web, 103-107, 113
    searching the, 105-106

X Y Z

X, Y, Z axes, 65, 313, 315-317
X-ray, 250 illus., 254 illus., 351
yeast, 279
zoetrope, 85, 94
zone, for e-mail, 104
zoning laws, 174

Photo Credits

INTERIOR DESIGN: DesignNet

PHOTO CREDITS
Cover: The Stock Market/Ron Lowery
Cover Inset: Mark Romine

Adventure Photo & Film/Tom Sanders: 185l
Arnold & Brown: 37b, 185r
Autodesk, Inc.: 19
Vince Backeberg, Permission by Will Vinton Studios: 89r
Donna Coveney/MIT: 338l, 338-339
Whitney Cranshaw: 270tl
DAVID R. FRAZIER Photolibrary: 54, 114, 153, 158, 218, 219l, 300tr; Aaron Haupt 116, 370
Dinamation Dinosaurs ©1997 Dinamation International Corp.: 11, 228
Flexfoot®, Inc.: 248
FPG International LLC: Ron Chapple 14; Jim Cummins 78; Jeff Kaufman 41; Alan E. McGee 17tr; Jeffrey Sylvester 13; Arthur Tilley 107; Tom Wilson 17cr
IS Robotics: 342
Merle H. Jensen: 274b
©The Walt Disney Company, Courtesy Kobal: 84, 93
Laser Alignment, Inc.: 364t
Laser Images Inc. "Laserium"¨: 365
Nick Merrick © Hedrich Blessing, Courtesy of Herman Miller, Inc.: 255r
NASA: 341, 339br,349
Olympus America Inc.: 71
Tom Pantages: 51t, (Courtesy WMUR-TV) 130, 175, 300tl, (Courtesy Aavld) 315l
Peter Arnold, Inc.: Manfred Kage 259; Dagmar Schilling 363; Christopher Swann 327

Photo Researchers:
Bill Bachmann/Science Source 48; Scott Camazine/Science Source 254br; Robert Chase 254t; Tim Davis/Science Source 55bl; José Dupont/Explorer 223; Simon Fraser/SPL 17br; George Haling 213; Adam Hart-Davis/Science Source 300b; Richard Hensen 139; James Holmes/Rover/SPL 319; James King-Holmes/SPL 61; Mehau Kulyk/SPL 254bl; Dick Luria/Science Source 51b; Rafael Macia 162, 303; Will & Deni McIntyre 364b; Lawrence Migdale 15l; Hank Morgan/Science Source 339tr; NASA/Science Source 383; NCSA/University of Illinois/SPL 99; David Parker/600-Group/SPL 18; Alfred Pasieka/SPL 55tr; Princess Margaret Rose Hospital/SPL 249; Rosenfeld Images Ltd./SPL 55br, 377; Kaj R. Svensson/SPL 372; U.S. Department of Energy/Science Source 326; Peter Yates/SPL 40
Images PhotoDisc, Inc.: 198-199, 329
PHOTOTAKE: CNRI 258; GJLP/CNRI 250; Ansel Horn 252; Yoav Levy 350
Picturesque/Steve Murray: 68, 69
Railway Technical Research Institute: 204
Society of Nematologists: 264
SuperStock, Inc.: 137, 138, 145, 149, 159t, 240, 241, 247, 253, 278, 309, 373, 393; Bill Barley 308, 311; Phil Cantor 5; Chigmaroff/Davison 313; Francisco Cruz 30; Ron Dahlquist 186t; Franz Edson 15r; Malcolm Fife 219r; Roger Allyn Lee 375; Robert Llewellyn 39, 159b; D.C. Lowe 206; Mia & Klaus Matthes 205; Jack Novak 310; Charles Orrico 379; Herve Pelletier 374; Tom

Rosenthal 25; William Strode 321, 340b; J.L. Woody Wooden 288; Yagi Studio 402

The Stock Market: Jon Feingersh 131; John Henley 66; Michael Keller 100; Mug Shots 47; Jeff Zaruba 122

Uniphoto/Carl Yarbrough: 184

United States Patent Office: 371

Copyright 1997. USA TODAY. Reprinted with Permission.: 67

Westlight: Craig Aurness 186b, 207, 390; Steve Chenn 34; Kenneth Garrett 270-271t; Dallas & John Heaton 17cl; Walter Hodges 31; Ian Lloyd 181; Warren Morgan 17bl; Charles O'Rear 196, 281; P.L.I. 343; Jim Pickerell 49; Roda/Premium 210

Whirlpool Corporation: 17tl

Will Vinton Studios: 89l

Illustration Credits

Ken Clubb: 20, 21, 86, 87, 118, 123, 141, 148, 160, 182-183, 188-189, 211, 221b, 222, 229, 267, 269, 270b, 271b, 273t, 315r, 356, 357, 384, 394, 395, 396, 397, 399, 400, 401, 403

Design Associates: 98, 104, 105, 110, 111

Thomas Gagliano: 217l

Gorman Typesetting: 55tl, 74, 80, 237, 302b

Courtesy of Herman Miller, Inc.: 255

Jody James: 56, 57, 59, 60, 63

Steven Karp: 22, 27, 72, 85, 92, 95, 117, 119, 125, 126, 127, 132, 142, 143, 144, 154, 161, 163, 164, 165, 172, 176, 187, 190, 191, 193, 194, 195, 197, 198, 200, 201, 208, 209, 212t, 214, 215, 216, 217r, 220, 221t, 221c, 224l, 224, 225, 230, 231, 233, 234, 235, 238, 242, 256, 260, 265, 273b, 280, 282, 289, 290, 291, 293, 294, 295, 297, 298, 299, 302t, 305, 331, 332, 333, 334, 336, 340t, 344, 345, 351, 353, 354, 358, 359, 360, 361, 362, 366, 367, 376, 382, 387, 405, 407l, 409

McGraw-Hill Home Interactive: 91

Mirto Art Studio: 23, 106

William P. Spence: 167, 168, 169

Jeffrey J. Stoecker: 115, 128, 129, 170, 212 b, 277, 287, 391, 406, 407r